(CLASS)
"A" DISK

(F8 ?)

QB
C:\SIGPRO

DISCRETE-TIME

SIGNALS

AND SYSTEMS

DISCRETE-TIME SIGNALS AND SYSTEMS

Nasir Ahmed

Department of Electrical and Computer Engineering
University of New Mexico
Albuquerque, New Mexico

T. Natarajan

ARCO Oil and Gas Company
Dallas, Texas

A RESTON BOOK
PRENTICE-HALL, INC., Englewood Cliffs, New Jersey 07632

Library of Congress Cataloging in Publication Data

Ahmed, Nasir.
    Discrete-time signals and systems.

    Includes bibliographical references and
index.
    1. Discrete-time systems.   I. Natarajan, T.
II. Title.
QA402.A355   1983   621.38   82–9146
ISBN 0–8359–1375–9   AACR2

Editorial/production supervision and
interior design by Norma M. Karlin

© 1983 by Prentice-Hall, Inc.
A Division of Simon & Schuster
Englewood Cliffs, New Jersey 07632

10   9   8   7   6   5   4   3

Printed in the United States of America

Contents

6 INFINITE IMPULSE RESPONSE DISCRETE-TIME FILTERS 226

7 FINITE IMPULSE RESPONSE DISCRETE-TIME FILTERS 279

Preface

Rapid advances in computer technology in the past decade have had a tremendous impact on a variety of disciplines. The need for one to be familiar with discrete-time signals and systems is thus ever increasing. To this end, the main objective of this book is to introduce the reader to the fundamentals of discrete-time signals and systems and to provide a working knowledge of the same.

The material included in this book evolved from a senior level elective course offered by the Department of Electrical Engineering at Kansas State University during the past several years. Typically, most of the material in Chapters 2 through 7 and selected topics in Chapters 1 and 8 were covered. With the background so acquired and a basic course in random signals, students will be well prepared to undertake graduate-level work in different areas involving discrete-time signals and systems (e.g., signal processing, communications, and control systems).

This book consists of eight chapters, the first of which introduces terminology related to signals and systems and then presents a review of the fundamentals of continuous-time signals and systems. The reader who is already familiar with the review material in Chapter 1 may prefer to skip it and proceed to Chapter 2, which concerns the elements of difference equations. A transform method for discrete-time systems is introduced in Chapter 3, and Chapter 4 deals with the spectral analysis of discrete-time signals. The notion of associating transfer functions with discrete-time systems is developed in Chapter 5. Some useful design methods related to a special class of discrete-time systems, called filters, are introduced in Chapters 6 and 7. The last chapter addresses implementation aspects of discrete-time systems using fixed-point arithmetic. An in-depth study of this subject matter would require a background in random signal analysis, which is not a prerequisite for this book. However, our main goal in Chapter 8 is to familiarize the reader with potential implementation problems and then to present some useful techniques to account for them. To this end, a variety of illustrative examples are presented in this chapter.

The authors believe that this book can be used for the following purposes: (1) as a text for a formal senior level elective course; (2) as a source of material for the practicing engineer who wishes to acquire a working knowledge of discrete-time signals and systems on a self-study basis. Thus, a large number of example problems are included.

The leading author wishes to thank a number of students (present and former) at Kansas State University for their help and valuable suggestions for improvements. Some of these students are: W. Blasi, J. Devore, M. Flickner, J. Fogler, D. Haran, B. Harms, D. Hein, D. Hush, M. Junod, N. Magotra, D. Martz, K. Scarbrough, J. Schmalzel, S. Steps, S. Vijayandra and D. Youn. Thanks are also due to G. C. Carter of the Naval Underwater Systems Center, New London, Connecticut, and to G. R. Elliott of Sandia Laboratories, Albuquerque, New Mexico, for their suggestions.

The authors also are very grateful to their wives Esther and Santhi for their patience and encouragement. And to Ms. Brenda Merryman we pay a special tribute for typing the manuscript rapidly and accurately.

Nasir Ahmed

T. Natarajan

List of Acronyms

CT	Continuous-time
DT	Discrete-time
FS	Fourier series
FT	Fourier transform
IFT	Inverse Fourier transform
LT	Laplace transform
ILT	Inverse Laplace transform
ZT	Z-transform
IZT	Inverse Z-transform
DFT	Discrete Fourier transform
IDFT	Inverse discrete Fourier transform
FFT	Fast Fourier transform
IFFT	Inverse fast Fourier transform
PDS	Power density spectrum
FIR	Finite impulse response
IIR	Infinite impulse response
BLT	Bilinear transform
BFP	Block floating-point

DISCRETE-TIME

SIGNALS

AND SYSTEMS

1

Terminology
and
Review

This chapter serves two purposes. The first is to introduce terminology related to signals and systems, and the second is to review some fundamentals of continuous-time signals and systems.

1.1 TERMINOLOGY

The signals we encounter can be divided into four categories, as follows:

1. Analog signals

2. Continuous-time signals

3. Discrete-time signals or sequences

4. Digital signals

These signals are defined with respect to an independent variable, which we will assume is *time*.

An *analog signal* is one that is defined over a continuum of values of time, and whose amplitude can assume a continuous range of values, as illustrated in Fig. 1.1-1a. The term "analog" was perhaps introduced in connection with analog computation, where voltages and currents are used to represent physical quantities (e.g., velocity, displacement).

Signals that are defined at a continuum of times and whose amplitudes either range over a continuous range of values or a finite num-

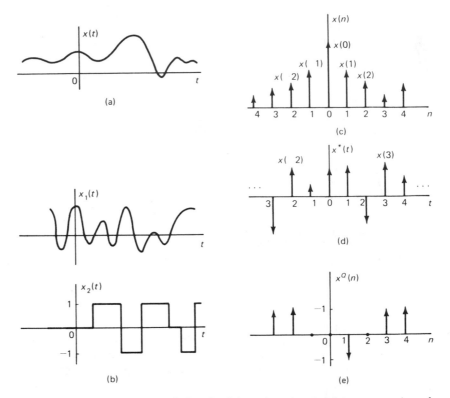

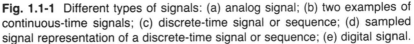

Fig. 1.1-1 Different types of signals: (a) analog signal; (b) two examples of continuous-time signals; (c) discrete-time signal or sequence; (d) sampled signal representation of a discrete-time signal or sequence; (e) digital signal.

ber of possible values are called *continuous-time* signals (see Fig. 1.1-1b). Thus analog signals can be viewed as being a special case of continuous-time signals. In practice, however, one tends to use the terms "analog" and "continuous-time" interchangeably.

Discrete-time signals are defined at only a particular set of values of time, which means that such signals can be represented as *sequences* of numbers. A sequence of numbers x, in which the nth member is denoted by $x(n)$, may be denoted as

$$x = \{x(n)\}, \qquad |n| < \infty \qquad (1.1\text{-}1)$$

Although (1.1-1) implies that $x(n)$ is actually the nth member of the sequence x, it is convenient to denote the sequence itself by $x(n)$. Thus we can represent a discrete-time signal or sequence $x(n)$ in graphical form as shown in Fig. 1.1-1c. We note that $x(n)$ is defined for integer values only; that is, it is undefined for noninteger values of n. Again, it is sometimes convenient to represent the information in a given sequence $x(n)$ in the form of a *sampled signal* $x^*(t)$, which is defined as

$$x^*(t) = \sum_{n=-\infty}^{\infty} x(n)\, \delta(t - n) \qquad (1.1\text{-}2)$$

where $\delta(t - n)$ denotes an impulse function of strength $x(n)$ at $t = n$; see Fig. 1.1-1d. The presence of impulse functions in $x^*(t)$ implies that it is zero for all values of t except for $t = \pm n$, where the impulses are located.

The last category of signals is *digital signals*, for which both time and amplitude are discrete, as illustrated in Fig. 1.1-1e. Here the digital signal $x^Q(n)$ takes one of three values, $-1, 0$, or 1, at only integer values of n.

Let us now turn our attention to the different types of *systems* one can identify. They can be classified along the same lines as signals. Therefore, *analog systems* are systems for which both the input and output are analog signals. Systems whose input and output signals are discrete-time signals (or sequences) are called *discrete-time systems*. Again, *digital systems* are those systems whose inputs and outputs are digital signals.

In the interest of being concise, the acronyms CT and DT will be used to denote "continuous time" and "discrete time," respectively. Also, it will be assumed that the term "system" refers to the class of systems that are *linear* and *time invariant*.

The remainder of this chapter is devoted to a review of some fundamentals of continuous-time signals and systems. In particular, the following topics will be discussed:

- Fourier series
- Fourier transform
- Relation between the Fourier series and the Fourier transform
- Convolution and windowing
- Energy, power, amplitude, and phase spectra
- Laplace transform
- Time-domain considerations of continuous-time systems
- Laplace transform considerations of continuous-time systems
- Poles, zeros, and stability
- Sinusoidal steady-state response

Depending upon their backgrounds, some readers may wish to skip some or all of this review material and proceed to Chapter 2.

1.2 FOURIER SERIES REPRESENTATION

Given a periodic function $x_p(t)$ with period L (seconds), its Fourier series (FS) expansion is given by

$$x_p(t) = \frac{1}{L} \sum_{n=-\infty}^{\infty} c_n e^{jn\omega_0 t} \qquad (1.2\text{-}1)$$

where c_n is the nth FS coefficient and $\omega_0 = 2\pi/L$ is the fundamental radian frequency. Also, $\omega_0 = 2\pi f_0$, where f_0 is the fundamental frequency in hertz (Hz). The frequencies nf_0 and $n\omega_0$ are called *harmonics*, since they are integral multiples of the fundamental frequencies f_0 and ω_0, respectively.

For the FS expansion in (1.2-1) to exist, we assume that $x_p(t)$ satisfies the following properties:

1. $\displaystyle\int_L |x_p(t)|\,dt < \infty$ $\qquad\qquad\qquad\qquad\qquad\qquad$ (1.2-2)

where the notation $\displaystyle\int_L$ denotes integration over one period of $x_p(t)$.

2. $x_p(t)$ has at most a finite number of discontinuities in one period.

3. $x_p(t)$ has at most a finite number of maxima and minima in one period.

With these assumptions, it can be shown that the right-hand side of (1.2-1) converges uniformly to $x_p(t)$, in each closed interval containing no discontinuity. At points of discontinuity (e.g., at $t = t_0$), $x_p(t)$ converges to the mean value $[x_p(t_0^-) + x_p(t_0^+)]/2$. The FS coefficients c_n in (1.2-1) can be computed by using the fact that the set of exponential functions $\{e^{jn\omega_0 t}\}$ are *orthogonal;* that is, they satisfy the relation

$$\int_L e^{j(m-n)\omega_0 t}\,dt = \begin{cases} L, & m = n \\ 0, & m \neq n \end{cases} \qquad (1.2\text{-}3)$$

Using (1.2-1) and (1.2-3), it is straightforward to show that the nth FS coefficient is given by

$$c_n = \int_L x_p(t) e^{-jn\omega_0 t}\,dt, \qquad |n| < \infty \qquad (1.2\text{-}4)$$

Equations (1.2-1) and (1.2-4) are referred to as a FS pair.

Example 1.2-1: Let $x_p(t)$ be the pulse train shown in Fig. 1.2-1, where L is the period. Find the FS expansion of $x_p(t)$.

Solution: From (1.2-4) we have

$$c_n = \int_L x_p(t)e^{-jn\omega_0 t}\, dt, \qquad |n| < \infty$$

where $\omega_0 = 2\pi/L$ radians/second.

It is convenient to integrate over the interval $(-L/2, L/2)$ which equals one period. This results in

$$c_n = \int_{-L/2}^{L/2} x_p(t)e^{-jn\omega_0 t}\, dt$$

$$= \int_{-\tau/2}^{\tau/2} (1)e^{-jn\omega_0 t}\, dt \qquad (1.2\text{-}5)$$

Carrying out the integration in (1.2-5), we obtain

$$c_n = \frac{2}{n\omega_0}\left[\frac{e^{jn\omega_0\tau/2} - e^{-jn\omega_0\tau/2}}{2j}\right] \qquad (1.2\text{-}6)$$

Applying the identity

$$\sin\theta = \frac{e^{j\theta} - e^{-j\theta}}{2j}$$

to (1.2-6), there results

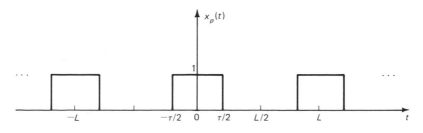

Fig. 1.2-1 Given periodic function, $x_p(t)$.

$$c_n = \tau \left[\frac{\sin(n\omega_0 \tau/2)}{(n\omega_0 \tau/2)} \right], \quad |n| < \infty \qquad (1.2\text{-}7)$$

Thus the desired FS expansion is given by (1.2-1) to be

$$x_p(t) = \frac{1}{L} \sum_{n=-\infty}^{\infty} c_n e^{jn\omega_0 t}$$

where (1.2-7) yields the c_n and $\omega_0 = 2\pi/L$ radians/second.

We note the FS coefficients in (1.2-7) are in the form of a $\sin(x)/x$ function, as illustrated in Fig. 1.2-2.

1.3 FOURIER TRANSFORM REPRESENTATION

Example 1.2-1 provides a convenient transition from the FS to the Fourier transform representation of a given function. Suppose L becomes large and approaches infinity. Then from Fig. 1.2-1 it is apparent that the pulse train reduces to a *single pulse*, which is no longer periodic (i.e., aperiodic). Let us denote this aperiodic function by $x(t)$. Also, we observe that as L increases the individual lines that represent the c_n's in Fig. 1.2-2 begin to crowd in and ultimately approach the continuous $\sin(x)/x$ function.

Next we examine the behavior of the FS coefficient c_n in (1.2-5); that is,

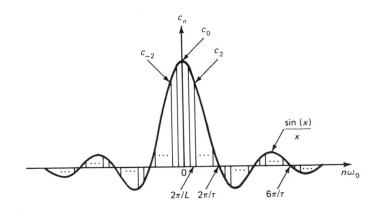

Fig. 1.2-2 Plot of the Fourier series coefficients of the pulse train in Fig. 1.2-1.

$$c_n = \int_{-L/2}^{L/2} x_p(t)e^{-jn\omega_0 t}\, dt$$

It follows that as L tends to infinity the fundamental angular frequency $\omega_0(= 2\pi/L)$ becomes a differential angular frequency $d\omega$, while $n\omega_0$ becomes a continuous angular frequency ω. Thus we have

$$\lim_{L\to\infty} \{c_n\} = \int_{-\infty}^{\infty} x(t)e^{-j\omega t}\, dt \qquad (1.3\text{-}1)$$

The right-hand side of (1.3-1) is defined as the *Fourier transform* (FT) of $x(t)$, which we denote by $X(\omega)$; that is,

$$X(\omega) = \int_{-\infty}^{\infty} x(t)e^{-j\omega t}\, dt \qquad (1.3\text{-}2)$$

Let us now study the effect of letting L tend to infinity on (1.2-1), which can be written as

$$x_p(t) - \frac{1}{2\pi} \sum_{n=-\infty}^{\infty} c_n e^{jn\omega_0 t}\, (\omega_0)$$

since $L = 2\pi/\omega_0$. As L tends to infinity, the summation over the harmonics becomes an integration over the entire continuous range $(-\infty, \infty)$. Also, ω_0 becomes $d\omega_0$; $n\omega_0$ becomes ω; c_n becomes $X(\omega)$, and $x_p(t)$ becomes $x(t)$. Thus the limiting form of (1.2-1) is given by

$$x(t) = \frac{1}{2\pi} \int_{-\infty}^{\infty} X(\omega)e^{j\omega t}\, d\omega \qquad (1.3\text{-}3)$$

which is defined as the *inverse Fourier transform* (IFT) of $x(t)$.

Equations (1.3-2) and (1.3-3) are called a FT pair. A *sufficient condition* for the existence of the FT of an aperiodic function $x(t)$ is that

$$\int_{-\infty}^{\infty} |x(t)|\, dt < \infty \qquad (1.3\text{-}4)$$

That is, $x(t)$ is *absolutely integrable*.

Example 1.3-1: If $x(t) = e^{-at}u(t)$, where $u(t)$ is the unit step function and $a > 0$, find $X(\omega)$.

Solution: Substituting $x(t)$ in (1.3-2), we get

$$X(\omega) = \int_0^\infty e^{-(a+j\omega)t}\, dt$$

which yields $\quad\quad\quad\quad X(\omega) = \dfrac{1}{a + j\omega}$ $\quad\quad\quad\quad$ (1.3-5)

Similarly, one can find the Fourier transforms (FTs) of other functions. Some properties of the FT that are useful to do so are listed in Table 1.3-1.

Example 1.3-2: Given $y(t) = e^{-a|t|}$, $a > 0$. Find $Y(\omega)$.

Solution: The sketch of $y(t)$ in Fig. 1.3-1a shows that it can be written as

$$y(t) = x(-t) + x(t)$$ $\quad\quad\quad\quad$ (1.3-6)

Table 1.3-1   Some Properties of the Fourier Transform

Time Function	Fourier Transform
1. $z(t) = x(t) \pm y(t)$	$Z(\omega) = X(\omega) \pm Y(\omega)$
2. $cx(t)$, where c is a constant	$cX(\omega)$
3. $\dfrac{dx(t)}{dt}$	$j\omega X(\omega)$
4. $\displaystyle\int_{-\infty}^{t} x(\tau)\, d\tau$	$\dfrac{X(\omega)}{j\omega}$, if $\displaystyle\int_{-\infty}^{\infty} x(t)\, dt = 0$
5. $tx(t)$	$j\dfrac{dX(\omega)}{d\omega}$
6. $x(-t)$	$X(-\omega)$
7. $x(at)$	$\dfrac{1}{|a|} X\left(\dfrac{\omega}{a}\right)$
8. $x(t - a)$	$e^{-j\omega a} X(\omega)$
9. $x(t) \cos \omega_0 t$	$\dfrac{1}{2}[X(\omega + \omega_0) + X(\omega - \omega_0)]$
10. $x(t) \sin \omega_0 t$	$\dfrac{1}{2j}[X(\omega - \omega_0) - X(\omega + \omega_0)]$
11. $e^{-at}x(t)$	$X(\omega + a)$

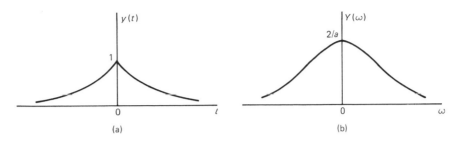

Fig. 1.3-1 Plots pertaining to Example 1.3-2.

where $$x(t) = e^{-a|t|}u(t)$$

From lines 1 and 6 of Table 1.3-1, it follows that the FT of $y(t)$ can be expressed as

$$Y(\omega) = X(-\omega) + X(\omega) \tag{1.3-7}$$

We note that the *FT* of $x(t)$ in (1.3-6) was evaluated in Example 1.3-1. Hence substitution of (1.3-5) in (1.3-7) leads to

$$Y(\omega) = \frac{1}{a - j\omega} + \frac{1}{a + j\omega}$$

Thus $$Y(\omega) = \frac{2a}{\omega^2 + a^2} \tag{1.3-8}$$

which is sketched in Fig. 1.3-1b.

Other Issues

It is important to note that the condition in (1.3-4) need only be *sufficient*. That is, functions that do not satisfy (1.3-4) can still have FTs. Some of these which are of interest to us can be expressed in terms of the unit impulse function $\delta(t)$, which may be defined as follows:

$$\int_{-\infty}^{\infty} \delta(t) \, dt = 1 \tag{1.3-9a}$$

where $\delta(t) = 0$ for $t \neq 0$, and

$$\int_{-\infty}^{\infty} \delta(t - t_0)x(t) \, dt = x(t_0) \tag{1.3-9b}$$

where t_0 is a real number.

Equation (1.3-9a) means that $\delta(t)$ is infinite at $t = 0$, but the area enclosed by it equals 1. Hence the name *unit* impulse function. The area enclosed is also referred to as the *strength* of the impulse function. Thus, $k\delta(t)$ represents an impulse of strength k, since

$$\int_{-\infty}^{\infty} k\delta(t) \, dt = k \tag{1.3-10}$$

Let us consider the case when the FT of a certain function $x_1(t)$ is given by

$$X_1(\omega) = 2\pi\delta(\omega)$$

Then, from the definition of the IFT in (1.3-3), we have

$$x_1(t) = \frac{1}{2\pi} \int_{-\infty}^{\infty} 2\pi\delta(\omega) \, e^{j\omega t} d\omega$$

$$= 1, \qquad |t| < \infty$$

Thus we have the result

$$FT\{1\} = 2\pi\delta(\omega) \tag{1.3-11}$$

where $FT\{\cdot\}$ denotes the FT of the quantity in the parentheses. Similarly, if $X_2(\omega) = 2\pi\delta(\omega-\omega_0)$, one obtains the result

$$x_2(t) = \frac{1}{2\pi} \int_{\infty}^{\infty} 2\pi\delta(\omega-\omega_0) \, e^{j\omega t} d\omega$$

$$= e^{j\omega_0 t}$$

That is

$$FT\{e^{j\omega_0 t}\} = 2\pi\delta(\omega-\omega_0) \tag{1.3-12}$$

Equations (1.3-11) and (1.3-12) are commonly used for evaluating the FTs of some useful functions that are not absolutely integrable—that is, they do not satisfy (1.3-4). To illustrate, let

$$x_3(t) = \cos\omega_0 t \tag{1.3-13}$$

Then we have

$$X_3(\omega) = \text{FT}\{\cos\omega_0 t\} \tag{1.3-14}$$

Substitution of

$$\cos\omega_0 t = \frac{1}{2}(e^{j\omega_0 t} + e^{-j\omega_0 t})$$

in (1.3-14), there results

$$X_3(\omega) = \frac{1}{2}\text{FT}\{e^{j\omega_0 t}\} + \frac{1}{2}\text{FT}\{e^{-j\omega_0 t}\} \tag{1.3-15}$$

From (1.3-12) it follows that the right-hand side of (1.3-15) can be expressed as a sum of two impulse functions to obtain

$$X_3(\omega) = \pi[\delta(\omega-\omega_0) + \delta(\omega+\omega_0)] \tag{1.3-16}$$

Equation (1.3-16) implies that the FT of the cosine function in (1.3-13) consists of two impulses at $\omega = \omega_0$ and $\omega = -\omega_0$, respectively. The strength of each of these impulses is π. Similarly we can find the FTs of additional useful functions that are not absolutely integrable. A rigorous treatment of this approach is beyond the scope of this book. An excellent discussion of the same is available in a book by Papoulis [2], to which the interested reader may refer.

We conclude this section by presenting two additional tables. The FTs of some functions that are commonly encountered are listed in Table 1.3-2, while Table 1.3-3 includes a set of useful identities pertaining to the impulse function.

1.4 RELATION BETWEEN THE FOURIER SERIES AND THE FOURIER TRANSFORM

We now enquire into the relation between the FS and the FT by considering a function $x(t)$ that is defined over a finite interval L, as illustrated in Fig. 1.4-1. This function is zero for $|t| > L/2$.

From (1.3-2) it follows that the FT of $x(t)$ in Fig. 1.4-1 is given by

$$X(\omega) = \int_{-L/2}^{L/2} x(t)e^{-j\omega t}\,dt \tag{1.4-1}$$

Table 1.3-2 Some Functions and Their Fourier Transforms

$x(t)$	$X(\omega)$
1. $\delta(t)$	1
2. 1	$2\pi\delta(\omega)$
3. $u(t)$	$\pi\delta(\omega) + \dfrac{1}{j\omega}$
4. $\cos \omega_0 t$	$\pi[\delta(\omega + \omega_0) + \delta(\omega - \omega_0)]$
5. $\sin \omega_0 t$	$j\pi[\delta(\omega + \omega_0) - \delta(\omega - \omega_0)]$
6. $\cos \omega_0 t u(t)$	$\dfrac{\pi}{2}[\delta(\omega + \omega_0) + \delta(\omega - \omega_0)] - \dfrac{j\omega}{\omega^2 - \omega_0^2}$
7. $\sin \omega_0 t u(t)$	$\dfrac{\pi}{2j}[\delta(\omega - \omega_0) - \delta(\omega + \omega_0)] - \dfrac{\omega_0}{\omega^2 - \omega_0^2}$
8. $e^{-at}u(t),\ a > 0$	$\dfrac{1}{a + j\omega}$
9. $e^{-a|t|},\ a > 0$	$\dfrac{2a}{\omega^2 + a^2}$
10a. $e^{j\omega_0 t}$	$2\pi\delta(\omega - \omega_0)$
10b. $e^{j\omega_0 t}u(t)$	$\pi\delta(\omega - \omega_0) + \dfrac{1}{j(\omega - \omega_0)}$
11. $\mathrm{sgn}\ (t) = \begin{cases} 1, & t \geq 0 \\ -1, & t < 0 \end{cases}$	$\dfrac{2}{j\omega}$
12. $|t|$	$\dfrac{-2}{\omega^2}$
13. $\begin{cases} 1, & |t| < T \\ 0, & \text{elsewhere} \end{cases}$	$\dfrac{2 \sin \omega T}{\omega}$
14. $\displaystyle\int_{-\infty}^{t} x(\tau)\ d\tau$	$\dfrac{X(\omega)}{j\omega} + \pi X(0)\delta(\omega)$

Table 1.3-3 Some Useful Identities Related to Impulse Functions

1. $\delta(t) = \lim\limits_{\epsilon \to 0} \dfrac{1}{\epsilon}[u(t) - u(t - \epsilon)]$

2. $\delta(t) = \lim\limits_{\omega \to \infty} \dfrac{\sin \omega t}{\pi t}$

3. $\delta(t) = \lim\limits_{\omega \to \infty} \dfrac{2 \sin^2 (\omega t/2)}{\pi \omega t^2}$

4. $\delta(t) = \lim\limits_{\epsilon \to 0} \dfrac{e^{-t^2/\epsilon}}{\sqrt{\pi\epsilon}}$

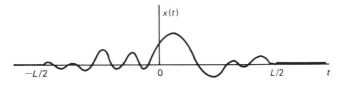

Fig. 1.4-1 Function $x(t)$ defined over an interval L.

Substituting $n\omega_0$ for ω in (1.4-1), there results

$$X(n\omega_0) = \int_{-L/2}^{L/2} x(t)e^{-jn\omega_0 t}\, dt$$

or
$$X(n\omega_0) = \int_L x(t)e^{-jn\omega_0 t}\, dt, \qquad |n| < \infty \qquad (1.4\text{-}2)$$

where we define $\omega_0 = 2\pi/L$.

Next, if $x_p(t)$ denotes the L-periodic extension of $x(t)$, then its FS representation is given by (1.2-1) and (1.2-4) to be

$$x_p(t) = \frac{1}{L} \sum_{n=-\infty}^{\infty} c_n e^{jn\omega_0 t}$$

where
$$c_n = \int_L x_p(t)e^{-jn\omega_0 t}\, dt, \qquad |n| < \infty \qquad (1.4\text{-}3)$$

A comparison of (1.4-2) with (1.4-3) leads to the following fundamental relation:

$$c_n = X(n\omega_0), \qquad |n| < \infty \qquad (1.4\text{-}4)$$

Thus, if a function is defined over a finite interval L, its FT $X(\omega)$ at a set of *equally spaced* points on the ω-axis is exactly specified by the FS. The distance between these points of specification is $2\pi/L$ radians.

To illustrate this basic concept, consider the function $x(t)$ shown in Fig. 1.4-2a. Then its FT is given by (1.3-2) to be

$$X(\omega) = \int_{-\epsilon}^{\epsilon} x(t)e^{-j\omega t}\, dt$$

or
$$X(\omega) = \int_{-\epsilon}^{\epsilon} x(t)\cos \omega t\, dt - j \int_{-\epsilon}^{\epsilon} x(t)\sin \omega t\, dt \qquad (1.4\text{-}5)$$

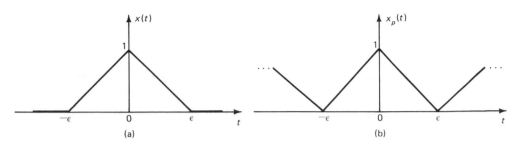

Fig. 1.4-2 Function $x(t)$ and its periodic extension $x_p(t)$.

Inspection of Fig. 1.4-2a reveals that $x(t)$ is an even function in the region $|t| \leq \epsilon$. Thus (1.4-5) simplifies to yield

$$X(\omega) = 2 \int_0^\epsilon x(t) \cos \omega t \, dt$$

where $x(t) = 1 - \dfrac{t}{\epsilon}, \qquad 0 \leq t \leq \epsilon$

Evaluating the preceding expression, we obtain

$$X(\omega) = \frac{2[1 - \cos(\omega\epsilon)]}{\omega^2 \epsilon} \tag{1.4-6}$$

Now suppose we wish to evaluate the FS of the periodic extension of $x(t)$, as depicted in Fig. 1.4-2b. Then from (1.4-4) we have

$$c_n = X(n\omega_0) = \frac{2[1 - \cos(n\omega_0\epsilon)]}{n^2\omega_0^2\epsilon}, \qquad n \neq 0 \tag{1.4-7}$$

From Fig. 1.4-2b it is apparent that the period of $x_p(t)$ is 2ϵ; that is, $L = 2\epsilon$. Hence

$$\omega_0 = \frac{2\pi}{L} = \frac{\pi}{\epsilon} \tag{1.4-8}$$

Substitution of (1.4-8) in (1.4-7) leads to

$$c_n = \frac{2\epsilon[1 - \cos(n\pi)]}{n^2\pi^2} \tag{1.4-9}$$

which can be written as

$$c_n = \frac{2\epsilon}{n^2 \pi^2} [1 - (-1)^n] \qquad (1.4\text{-}10)$$

From (1.4-10) it is apparent that

$$c_n = \begin{cases} 0, & n \text{ even} \\ \dfrac{4\epsilon}{n^2 \pi^2}, & n \text{ odd} \end{cases} \qquad (1.4\text{-}11a)$$

For $n = 0$, we make the simple observation in (1.4-5) that

$$X(0) = \int_{-\epsilon}^{\epsilon} x(t)\, dt = \epsilon$$

which is the area enclosed by $x(t)$ in Fig. 1.4-2a.
    Hence (1.4-4) yields

$$c_0 = X(0) = \epsilon \qquad (1.4\text{-}11b)$$

Thus the FS expansion of $x_p(t)$ in Fig. 1.4-2b is [see (1.2-1)]

$$x_p(t) = \frac{1}{2\epsilon} \sum_{n=-\infty}^{\infty} c_n e^{j(n\pi/\epsilon)t}$$

where the c_n's are given by (1.4-11) to be as follows:

$$c_n = \begin{cases} \epsilon, & n = 0 \\ 0, & n \text{ even} \\ \dfrac{4\epsilon}{n^2 \pi^2}, & n \text{ odd} \end{cases}$$

In conclusion, we remark that there are other forms of the FS and FT pairs, which can be obtained by simple modifications of those defined by (1.2-1), (1.2-4) and (1.3-2), (1.3-3), respectively. It is convenient to summarize them in tabular form, as shown in Table 1.4-1.

Table 1.4-1   Different Forms of Fourier Series and Fourier Transform Pairs

Form Number[†]	Fourier Series Pair	Fourier Transform Pair
I	$$x_p(t) = \frac{1}{L} \sum_{n=-\infty}^{\infty} c_n e^{jn\omega_0 t}$$ $$c_n = \int_L x_p(t) e^{-jn\omega_0 t} \, dt$$	$$x(t) = \frac{1}{2\pi} \int_{-\infty}^{\infty} X(\omega) e^{j\omega t} \, d\omega$$ $$X(\omega) = \int_{-\infty}^{\infty} x(t) e^{-j\omega t} \, dt$$
II	$$x_p(t) = \frac{1}{\sqrt{L}} \sum_{n=-\infty}^{\infty} c_n e^{jn\omega_0 t}$$ $$c_n = \frac{1}{\sqrt{L}} \int_L x_p(t) e^{-jn\omega_0 t} \, dt$$	$$x(t) = \frac{1}{\sqrt{2\pi}} \int_{-\infty}^{\infty} X(\omega) e^{j\omega t} \, d\omega$$ $$X(\omega) = \frac{1}{\sqrt{2\pi}} \int_{-\infty}^{\infty} x(t) e^{-j\omega t} \, dt$$
III	$$x_p(t) = \sum_{n=-\infty}^{\infty} c_n e^{jn\omega_0 t}$$ $$c_n = \frac{1}{L} \int_L x_p(t) e^{-jn\omega_0 t} \, dt$$	$$x(t) = \int_{-\infty}^{\infty} X(\omega) e^{j\omega t} \, d\omega$$ $$X(\omega) = \frac{1}{2\pi} \int_{-\infty}^{\infty} x(t) e^{-j\omega t} \, dt$$

[†]Three more forms can be obtained by replacing j by $-j$ in each of the series and transform pairs.

1.5 CONVOLUTION AND WINDOWING

Convolution

We often encounter the product of two FTs in the form

$$Z(\omega) = X(\omega)Y(\omega) \tag{1.5-1}$$

In such cases we usually wish to acquire $z(t)$. To this end we use the IFT in (1.3-3) and write

$$z(t) = \frac{1}{2\pi} \int_{-\infty}^{\infty} Z(\omega) e^{j\omega t} \, d\omega \tag{1.5-2}$$

Substituting (1.5-1) in (1.5-2), there results

$$z(t) = \frac{1}{2\pi} \int_{-\infty}^{\infty} X(\omega)Y(\omega) e^{j\omega t} \, d\omega \tag{1.5-3}$$

Next, by definition of the FT, we have

$$Y(\omega) = \int_{-\infty}^{\infty} y(\tau)e^{-j\omega\tau}\, d\tau \qquad (1.5\text{-}4)$$

Substitution of (1.5-4) in (1.5-3) leads to

$$z(t) = \frac{1}{2\pi} \int_{-\infty}^{\infty} X(\omega)e^{j\omega t}\left[\int_{-\infty}^{\infty} y(\tau)e^{-j\omega\tau}\, d\tau\right] d\omega$$

Interchanging the order of integration, we get

$$z(t) = \int_{-\infty}^{\infty} y(\tau)\left[\frac{1}{2\pi}\int_{-\infty}^{\infty} X(\omega)e^{j\omega t}e^{-j\omega\tau}\, d\omega\right] d\tau$$

which becomes

$$z(t) = \int_{-\infty}^{\infty} y(\tau)\left[\frac{1}{2\pi}\int_{-\infty}^{\infty} X(\omega)e^{j(t-\tau)\omega}\, d\omega\right] d\tau \qquad (1.5\text{-}5)$$

From the definition of the IFT in (1.3-3), it is apparent that the quantity in brackets in (1.5-5) is merely $x(t - \tau)$. Thus we have

$$z(t) = \int_{-\infty}^{\infty} x(t - \tau)y(\tau)\, d\tau \qquad (1.5\text{-}6a)$$

It is readily seen that (1.5-6a) can also equivalently be written as

$$z(t) = \int_{-\infty}^{\infty} x(\tau)y(t - \tau)\, d\tau \qquad (1.5\text{-}6b)$$

The right-hand side of (1.5-6) is known as the *convolution integral*. It states that $z(t)$ can be evaluated by *convolving* $x(t)$ and $y(t)$. The graphical implication of convolution is illustrated in Fig. 1.5-1, with respect to (1.5-6b), for the case when

$$x(t) = \begin{cases} e^{-2t}, & t \geq 0 \\ 0, & \text{elsewhere} \end{cases} \qquad (1.5\text{-}7a)$$

and

$$y(t) = \begin{cases} 1, & -1 < t \leq 0 \\ e^{-t}, & t \geq 0 \\ 0, & \text{elsewhere} \end{cases} \qquad (1.5\text{-}7b)$$

First we note that $y(t - \tau)$ is essentially a "mirror image" of $y(t)$; see Fig. 1.5-1c. Next, from Figs. 1.5-1d through 1.5-1f, it is apparent that there are three regions of integration associated with the convolution process in (1.5-6b), as far as this example is concerned. These regions are identified next.

REGION 1. See Fig. 1.5-1d; $t < -1$.

$$z(t) = 0, \quad t < -1 \tag{1.5-8a}$$

since there is no overlap of the functions $x(\tau)$ and $y(t - \tau)$.

REGION 2. See Fig. 1.5-1e; $-1 < t \leq 0$.

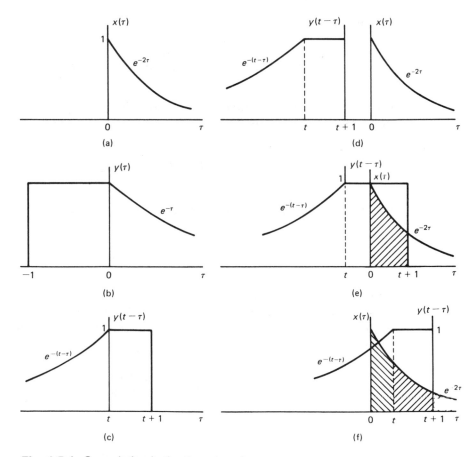

Fig. 1.5-1 Convolution in the time domain.

$$z(t) = \int_0^{t+1} e^{-2\tau}(1)\, d\tau = \frac{1}{2}\left[1 - e^{-2(t+1)}\right], \qquad -1 < t \le 0 \quad (1.5\text{-}8b)$$

REGION 3. See Fig. 1.5-1f; $t \ge 0$.

$$z(t) = \left[\int_0^t e^{-2\tau}e^{-(t-\tau)}\, d\tau\right] + \left[\int_t^{t+1} e^{-2\tau}(1)\, d\tau\right]$$

$$= \left[e^{-t}\int_0^t e^{-\tau}\, d\tau\right] + \left[\int_t^{t+1} e^{-2\tau}\, d\tau\right]$$

which simplifies to become

$$z(t) = -\frac{1}{2}e^{-2t} - \frac{1}{2}e^{-2(t+1)} + e^{-t}, \qquad t \ge 0 \quad (1.5\text{-}8c)$$

Equation (1.5-8) thus yields the convolution function $z(t)$ for any value of t. From the preceding discussion it is apparent that the convolution process involves successive shifts and integration of one of the given functions with the mirror image of the other. The reader is encouraged to rework the example on convolution using (1.5-6a); that is, by working with $y(t)$ and the mirror image of $x(t)$, and hence verifying that their convolution agrees with $z(t)$ in (1.5-8); see Problem 1-3.

Equation (1.5-6) represents *time-domain convolution* in that $x(t)$ and $y(t)$ are both functions of time, t. Thus (1.5-1) and (1.5-6) imply that *convolution* in the *time domain* is equivalent to *multiplication* in the *FT (frequency) domain*.

Using a method similar to that used to derive (1.5-6), we can show that if

$$z(t) = x(t)y(t) \qquad\qquad (1.5\text{-}9)$$

then

$$Z(\omega) = \frac{1}{2\pi}\int_{-\infty}^{\infty} X(\alpha)Y(\omega - \alpha)\, d\alpha \qquad\qquad (1.5\text{-}10a)$$

or equivalently

$$Z(\omega) = \frac{1}{2\pi}\int_{-\infty}^{\infty} X(\omega - \alpha)Y(\alpha)\, d\alpha \qquad\qquad (1.5\text{-}10b)$$

The derivation of (1.5-10) starting with (1.5-9) is left as an exercise (Problem 1-4).

Equation (1.5-10) is again a convolution integral as was the case in (1.5-6). There is a basic difference, however, in that (1.5-10) represents the convolution of two FTs. The important conclusion related to (1.5-9) and (1.5-10) is that *multiplication* in the *time domain* is equivalent to *convolution* in the *FT (frequency) domain*.

We conclude this discussion by working an example related to convolution in the frequency domain. To this end, let $x(t)$ and $r(t)$ be as shown in Figs. 1.5-2a and 1.5-2b, respectively. Then their product, $z(t)$, is as depicted in Fig. 1.5-2c; that is,

$$z(t) = \begin{cases} \cos \omega_0 t, & |t| < T \\ 0, & \text{elsewhere} \end{cases} \tag{1.5-11}$$

We wish to evaluate $Z(\omega)$ via the convolutional integral in (1.5-10a); that is,

$$Z(\omega) = \frac{1}{2\pi} \int_{-\infty}^{\infty} X(\alpha) R(\omega - \alpha) \, d\alpha$$

where $X(\omega)$ and $R(\omega)$ are the FTs of $x(t)$ and $r(t)$ shown in Figs. 1.5-2a and 1.5-2b, respectively.

From the FT definition in (1.3-2) we have

$$R(\omega) = \int_{-T}^{T} (1) e^{-j\omega t} \, dt$$

which yields (see line 13 of Table 1.3-2)

$$R(\omega) = 2 \frac{\sin \omega T}{\omega} \tag{1.5-12}$$

Again, the *FT* of $x(t) = \cos \omega_0 t$ is given by line 4 of Table 1.3-2 to be

$$X(\omega) = \pi[\delta(\omega + \omega_0) + \delta(\omega - \omega_0)] \tag{1.5-13}$$

Substitution of (1.5-12) and (1.5-13) in (1.5-10a) leads to

$$Z(\omega) = \int_{-\infty}^{\infty} [\delta(\alpha + \omega_0) + \delta(\alpha - \omega_0)] \frac{\sin (\omega - \alpha)T}{\omega - \alpha} \, d\alpha \tag{1.5-14}$$

That is,

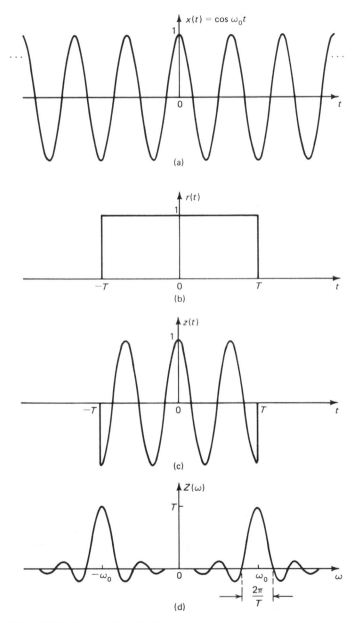

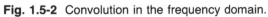

Fig. 1.5-2 Convolution in the frequency domain.

$$Z(\omega) = \int_{-\infty}^{\infty} \delta(\alpha + \omega_0) \frac{\sin(\omega - \alpha)T}{\omega - \alpha} \, d\alpha$$

$$+ \int_{-\infty}^{\infty} \delta(\alpha - \omega_0) \frac{\sin(\omega - \alpha)T}{\omega - \alpha} \, d\alpha$$

which yields

$$Z(\omega) = \left. \frac{\sin(\omega - \alpha)T}{\omega - \alpha} \right|_{\alpha = -\omega_0} + \left. \frac{\sin(\omega - \alpha)T}{\omega - \alpha} \right|_{\alpha = \omega_0}$$

by virtue of the property of impulse functions given in (1.3-9b).
Thus we have the desired result

$$Z(\omega) = \frac{\sin(\omega + \omega_0)T}{\omega + \omega_0} + \frac{\sin(\omega - \omega_0)T}{\omega - \omega_0} \tag{1.5-15}$$

a sketch of which is shown in Fig. 1.5-2d.

Windowing

The notion of convolution is relevant to a fundamental problem in Fourier analysis. It is the problem of determining the FT $X(\omega)$ of a function $x(t)$ using a given segment

$$x_T(t) = \begin{cases} x(t), & |t| \le T \\ 0, & \text{elsewhere} \end{cases} \tag{1.5-16}$$

of $x(t)$; see Fig. 1.5-3. It is apparent that $X(\omega)$ cannot be determined *exactly* since we have no knowledge of $x(t)$ for $|t| > T$. Hence $X(\omega)$ can only be *estimated*. The problem is to estimate $X(\omega)$ using $X_T(\omega)$, which is the FT of the segment $x_T(t)$ in (1.5-16). Windowing is a technique that enables us to do so, such that the estimation error

$$E_T(\omega) = X_T(\omega) - X(\omega) \tag{1.5-17}$$

is reduced.

The basic idea is to first multiply the segment $x_T(t)$ by a window function $w(t)$ that is *even* about $t = 0$ and is *zero* for $|t| > T$. Then we have (see Fig. 1.5-3c)

$$\hat{x}(t) = x_T(t)w(t) = x(t)w(t) \tag{1.5-18}$$

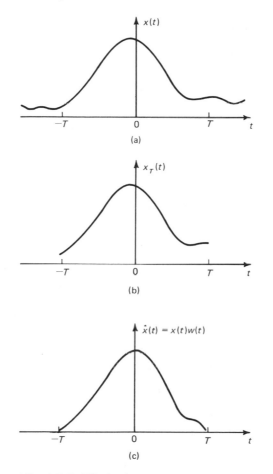

Fig. 1.5-3 Windowing.

The next step is to use $\hat{X}(\omega)$ to estimate $X(\omega)$. To this end we note that (1.5-18) represents a product in the time domain, which is equivalent to convolving $X(\omega)$ and $W(\omega)$ using (1.5-10a); that is,

$$\hat{X}(\omega) = \frac{1}{2\pi} \int_{-\infty}^{\infty} X(\alpha)W(\omega - \alpha) \, d\alpha \qquad (1.5\text{-}19)$$

A natural question that arises at this point is whether there is an *optimum* window function. This question is easily answered by examining (1.5-19) and noting that if

$$W(\omega - \alpha) = 2\pi\delta(\omega - \alpha) \qquad (1.5\text{-}20)$$

then we have the desired result

$$\hat{X}(\omega) \equiv X(\omega) \qquad (1.5\text{-}21)$$

Thus, if $w_0(t)$ denotes the optimum window function, then (1.5-20) implies that

$$W_0(\omega) = 2\pi\delta(\omega) \qquad (1.5\text{-}22)$$

where $W_0(\omega)$ is the FT of $w_0(t)$.

The optimum window is now obtained by taking the IFT of $W_0(\omega)$ using line 2 in Table 1.3-2. This results in

$$w_0(t) = 1, \quad |t| < \infty \qquad (1.5\text{-}23)$$

as the optimum solution. However, from (1.5-23) it is clear that this optimum window is an impractical solution since it is nonzero for *all* *t*. As such it requires a knowledge of $x(t)$ in the range $|t| > T$. Thus we are motivated to seek *suboptimum* windows that have the following properties:

1. They are even functions about $t = 0$.

2. They are zero in the range $|t| > T$.

3. Their transforms $W(\omega)$ have narrow main lobes and small side lobes, as illustrated in Fig. 1.5-4, and are such that

$$\lim_{T \to \infty} W(\omega) = W_0(\omega) = 2\pi\delta(\omega) \qquad (1.5\text{-}24)$$

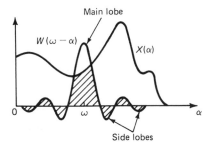

Fig. 1.5-4 Convolution of Fourier transform of window function and Fourier transfrom of $x(t)$.

Then, for all finite values of T, we obtain the result

$$\hat{X}(\omega) \approx X(\omega) \tag{1.5-25}$$

which is an approximation to (1.5-21).

Examples of some commonly used window functions are listed in Table 1.5-1 along with their FTs. Using the list of identities in Table 1.3-3, it is straightforward to verify that these FTs satisfy the requirement in (1.5-24); see Problem 1-6. For example, interchanging the variables ω and t in line 3 of Table 1.3-3, we obtain

$$\lim_{t \to \infty} \frac{2 \sin^2 (\omega t/2)}{\omega^2 t} = \pi\delta(\omega) \tag{1.5-26}$$

Table 1.5-1   A Set of Window Functions

Window Function	Fourier Transform
1. Rectangular $$w_r(t) = \begin{cases} 1, & |t| < T \\ 0, & \text{elsewhere} \end{cases}$$	$$W_r(\omega) = 2 \frac{\sin \omega T}{\omega}$$
2. Triangular (Bartlett) $$w_{tr}(t) = \begin{cases} 1 - \dfrac{|t|}{T}, & |t| \le T \\ 0, & \text{elsewhere} \end{cases}$$	$$W_{tr}(\omega) = T\left[\frac{\sin(\omega T/2)}{\omega T/2}\right]^2$$
3. Hanning (Tukey or raised cosine) $$w_{ha}(t) = \begin{cases} \dfrac{1}{2}(1 + \cos \dfrac{\pi t}{T}), & |t| \le T \\ 0, & \text{elsewhere} \end{cases}$$	$$W_{ha}(\omega) = \left(\frac{\sin \omega T}{\omega}\right)\left[\frac{\pi^2}{\pi^2 - \omega^2 T^2}\right]$$
4. Hamming $$w_H(t) = \begin{cases} 0.54 + 0.46 \cos \dfrac{\pi t}{T}, & |t| \le T \\ 0, & \text{elsewhere} \end{cases}$$	$$W_H(\omega) = \frac{2 \sin \omega T}{\omega}\left[\frac{0.54\pi^2 - 0.08\omega^2 T^2}{\pi^2 - \omega^2 T^2}\right]$$
5. Kaiser $$w_K(t) = \frac{I_0[\theta\sqrt{1 - (t/T)^2}]}{I_0(\theta)}, \quad |t| \le T$$	$$W_K(\omega) = \frac{T \sin [\sqrt{\omega^2 T^2 - \theta^2}]}{2I_0(\theta)\sqrt{\omega^2 T^2 - \theta^2}}$$

where I_0 is the modified Bessel function of the first kind and zero order; θ is a parameter that can be adjusted to obtain a narrow main lobe and small side lobes for $W_K(\omega)$.

Again, replacing t by T in (1.5-26), one has

$$\lim_{T\to\infty} \frac{2\sin^2(\omega T/2)}{\omega^2 T} = \pi\delta(\omega)$$

which is equivalent to

$$\lim_{T\to\infty} T\left[\frac{\sin(\omega T/2)}{\omega T/2}\right]^2 = 2\pi\delta(\omega) \tag{1.5-27}$$

Equation (1.5-27) and line 2 of Table 1.5-1 yield

$$\lim_{T\to\infty} W_{tr}(\omega) = 2\pi\delta(\omega)$$

which means that the triangular or Bartlett window satisfies (1.5-24).

We conclude the section by presenting plots of the rectangular, triangular, hanning, and Hamming windows that are listed in Table 1.5-1. These plots are displayed in Fig. 1.5-5, while illustrative decibel plots of their FTs are plotted in Fig. 1.5-6 for the case $T = 1$. The decibel plots were obtained by computing

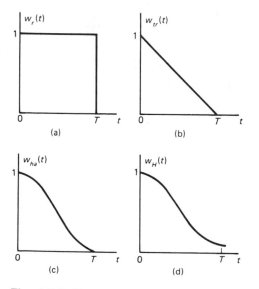

Fig. 1.5-5 Plots of four window functions: (a) rectangular; (b) triangular; (c) raised cosine, Tukey, or hanning; (d) Hamming.

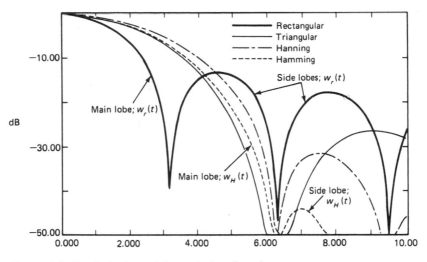

Fig. 1.5-6 Decibel plots of four window functions.

$$dB = 20 \log_{10} \left| \frac{W(\omega)}{W(0)} \right| \tag{1.5-28}$$

From Fig. 1.5-6 it is apparent that the Hamming window offers a good compromise between having a narrow main lobe and very small side lobes. We also note that the rectangular window results in the largest side lobes.

More recently a family of window functions was developed by Kaiser [1, 3, 5, 8], some information related to which is given in Table 1.5-1. Kaiser windows are extremely flexible in that the main lobe width and size of the side lobes of their FTs can be varied by adjusting a parameter θ; see Table 1.5-1. Thus θ can be adjusted to obtain the desired narrow main lobe and very small side lobes.

1.6 ENERGY, POWER, AMPLITUDE, AND PHASE SPECTRA

Let us consider the case when $x(t)$ represents the voltage across a 1-Ω resistor. Then the *total energy* E_T^x, and the *average power* P_{av}^x, dissipated in the resistor are defined as follows:

$$E_T^x = \int_{-\infty}^{\infty} x^2(t)\, dt \tag{1.6-1a}$$

and
$$P_{av}^x = \lim_{J \to \infty} \frac{1}{J} \int_{-J/2}^{J/2} x^2(t) \, dt \qquad (1.6\text{-}1b)$$

where J is the length of a segment of $x(t)$ in seconds. We note that the units of E_T^x and P_{av}^x are joules and watts, respectively.

From (1.6-1) it is apparent that if $x(t)$ has finite energy its average power will be zero. Conversely, if the power is not zero, then the energy is infinite. As such, in a given problem we are interested in either energy or power, but not both.

For the special case of *periodic signals*, the average power is defined as

$$P_{av}^x = \frac{1}{L} \int_L x^2(t) \, dt \qquad (1.6\text{-}2)$$

where L is the period of $x(t)$. For example, if

$$x(t) = A \sin (\omega_0 t + \theta)$$

where A, ω_0, and θ are constants, then (1.6-2) yields

$$P_{av}^x = \frac{1}{L} \int_0^L A^2 \sin^2 (\omega_0 t + \theta) \, dt$$

with $L = 2\pi/\omega_0$ which is easily simplified to obtain the familiar result

$$P_{av}^x = \frac{A^2}{2}$$

Energy Density Spectrum

Suppose we define

$$z(t) = x^2(t)$$

Then, from the definition of the FT, we have

$$Z(\omega) = \int_{-\infty}^{\infty} x^2(t) e^{-j\omega t} \, dt \qquad (1.6\text{-}3)$$

Again the convolution integral in (1.5-10a) yields

$$Z(\omega) = \frac{1}{2\pi} \int_{-\infty}^{\infty} X(\alpha)X(\omega - \alpha) \, d\alpha \tag{1.6-4}$$

Equating the right-hand sides of (1.6-3) and (1.6-4), and letting $\omega = 0$, we obtain

$$\int_{-\infty}^{\infty} x^2(t) \, dt = \frac{1}{2\pi} \int_{-\infty}^{\infty} X(\alpha)X(-\alpha) \, d\alpha$$

which is equivalent to

$$\int_{-\infty}^{\infty} x^2(t) \, dt = \frac{1}{2\pi} \int_{-\infty}^{\infty} |X(\omega)|^2 \, d\omega \tag{1.6-5}$$

$$\equiv E_T^x, \quad \text{the total energy in } x(t);$$

see (1.6-1a). Equation (1.6-5) is known as *Parseval's theorem*. It is a fundamental result since it leads to the following definition of the *energy density spectrum*, $\epsilon^x(\omega)$:

$$\epsilon^x(\omega) = \frac{1}{2\pi} |X(\omega)|^2 \tag{1.6-6}$$

It is important to note that $\epsilon^x(\omega)$ does *not* represent the amount of energy in $x(t)$ at a specific frequency, say f_0. Rather, it is the *area* under the curve $\epsilon^x(\omega)$ that represents energy. Thus, for example, the energy in $x(t)$ contained in the frequency band $\omega_1 \leq \omega \leq \omega_2$ is given by

$$\epsilon^x(\omega_1, \omega_2) = \int_{-\omega_2}^{-\omega_1} \epsilon^x(\omega) \, d\omega + \int_{\omega_1}^{\omega_2} \epsilon^x(\omega) \, d\omega \tag{1.6-7}$$

where the first term accounts for negative frequencies. For the case when $x(t)$ is real, it is easy to show that (Problem 1-9) $|X(\omega)|^2$ is an even function of ω, and hence (1.6-7) simplifies to yield

$$\epsilon^x(\omega_1, \omega_2) = 2 \int_{\omega_1}^{\omega_2} \epsilon^x(\omega) \, d\omega \tag{1.6-8}$$

As an illustrative example, let $x(t)$ be the rectangular pulse in Fig. 1.6-1a. Then (1.6-1a) yields the total pulse energy to be

$$E_T^x = \int_{-1}^{1} (1)^2 \, dt = 2 \tag{1.6-9}$$

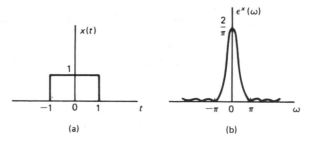

Fig. 1.6-1 Rectangular pulse and its energy density spectrum.

Next (1.5-12) implies that

$$X(\omega) = 2\,\frac{\sin \omega}{\omega}$$

since $T = 1$.

Thus the energy density spectrum of the rectangular pulse is given by (1.6-6) as

$$\epsilon^x(\omega) = \frac{2}{\pi}\left[\frac{\sin \omega}{\omega}\right]^2, \qquad |\omega| < \infty \qquad (1.6\text{-}10)$$

A plot of $\epsilon^x(\omega)$ is shown in Fig. 1.6-1b, from which it is apparent that most of the pulse energy is confined to the $|\omega| \leq \pi$ frequency region. It can also be shown that $\epsilon^x(\omega)$ in (1.6-10) satisfies Parseval's relation in (1.6-5); see Problem 1-10.

Fourier Series Power Spectrum

We now seek to define a power spectrum for the FS representation of a real periodic signal $x_p(t)$ whose period is L. To this end, let us use form III of the FS pair in Table 1.4-1; that is,

$$x_p(t) = \sum_{n=-\infty}^{\infty} c_n e^{jn\omega_0 t} \qquad (1.6\text{-}11)$$

where $\quad c_n = \dfrac{1}{L}\displaystyle\int_L x_p(t)e^{-jn\omega_0 t}\,dt, \qquad |n| < \infty$

Equation (1.6-11) implies that

$$x_p(t) = \sum_{n=-\infty}^{\infty} c_{-n}e^{-jn\omega_0 t} \qquad (1.6\text{-}12)$$

Multiplying both sides of (1.6-11) and (1.6-12), we obtain

$$x_p^2(t) = \sum_{n=-\infty}^{\infty} |c_n|^2 + \sum_{s=-\infty}^{\infty}\sum_{q=-\infty}^{\infty} c_s c_{-q}e^{j(s-q)\omega_0 t}, \qquad s \ne q \quad (1.6\text{-}13)$$

where $|c_n|^2 = c_n c_{-n}$. Integration of both sides of (1.6-13) over one period leads to

$$\int_L x_p^2(t)\,dt$$

$$= L\sum_{n=-\infty}^{\infty}|c_n|^2 + \sum_{s=-\infty}^{\infty}\sum_{q=-\infty}^{\infty} c_s c_{-q}\int_L e^{j(s-q)\omega_0 t}\,dt, \qquad s \ne q \quad (1.6\text{-}14)$$

From the orthogonality relation in (1.2-3), it is apparent that the double summation term in (1.6-14) vanishes, and hence

$$\int_L x_p^2(t)\,dt = L\sum_{n=-\infty}^{\infty}|c_n|^2 \qquad (1.6\text{-}15)$$

or

$$\frac{1}{L}\int_L x_p^2(t)\,dt = \sum_{n=-\infty}^{\infty}|c_n|^2$$

Since $x(t)$ is real, and

$$c_n = \frac{1}{L}\int_L x_p(t)e^{-jn\omega_0 t}\,dt$$

it readily follows that c_0 is real, and $|c_{-n}| = |c_n|$. Hence (1.6-15) can be written as

$$\frac{1}{L}\int_L x_p^2(t)\,dt = c_0^2 + 2\sum_{n=1}^{\infty}|c_n|^2 \qquad (1.6\text{-}16)$$

which is *Parseval's theorem* for a FS representation of $x_p(t)$, where $x_p(t)$ is periodic with period L. The left-hand side of (1.6-16) is recognized to be the average power P_{av}^x in $x_p(t)$, as defined in (1.6-2); that is, it is the power dissipated in a 1-Ω resistor if $x_p(t)$ is the voltage across it. Hence it is natural to define c_0^2 as the dc power component, and $2|c_n|^2$ as the power associated with the nth harmonic; that is,

$$P_0^x = c_0^2$$

and
$$P_n^x = 2|c_n|^2, \qquad n \geq 1$$

(1.6-17)

is the desired *FS power spectrum*.

COMMENT. If $x_p(t)$ is *complex valued*, then Parseval's relation corresponding to (1.6-16) is given by (1.6-15) to be

$$\frac{1}{L} \int_L |x_p(t)|^2 \, dt = |c_0|^2 + \sum_{n=1}^{\infty} \left[|c_n|^2 + |c_{-n}|^2 \right]$$

(1.6-18)

since $|c_{-n}|^2 \neq |c_n|^2$, in general. In this case the FS power spectrum is defined as

$$P_0^x = |c_0|^2$$

and
$$P_n^x = |c_n|^2 + |c_{-n}|^2, \qquad n \geq 1$$

(1.6-19)

Amplitude and Phase Spectra

It is also useful to define amplitude and phase spectra for the FS and FT. The basic idea is to realize that c_n and $X(\omega)$ are complex numbers and hence can be expressed in polar form. In the case of the FS we have

$$c_n = \text{Re } [c_n] + j \text{ Im } [c_n]$$

(1.6-20)

where Re $[c_n]$ and Im $[c_n]$ denote the real and imaginary parts of c_n. Using (1.6-20) we define

$$p_n^x = |c_n|$$

(1.6-21a)

as the *FS amplitude spectrum*, and

$$\psi_n^x = \tan^{-1} \{ \text{Im } [c_n]/\text{Re } [c_n] \}, \ |n| < \infty$$

(1.6-21b)

as the *FS phase spectrum*.

Since the c_n in (1.6-21) are given by

$$c_n = \frac{1}{L} \int_L x_p(t) e^{-jn\omega_0 t} \, dt$$

$$= \frac{1}{L} \left[\int_L x_p(t) \cos n\omega_0 t \, dt - j \int_L x_p(t) \sin n\omega_0 t \, dt \right]$$

(1.6-22)

it can be shown that (Problem 1-11) when $x_p(t)$ is real, p_n^x and ψ_n^x are even and odd functions, respectively, about $n = 0$; see Fig. 1.6-2.

The notion of amplitude and phase spectra for the FT follows naturally from the preceding discussion since the FT is a limiting case of the FS, and can be written as

$$X(\omega) = \text{Re}\,[X(\omega)] + j\,\text{Im}\,[X(\omega)]$$

(1.6-23)

Thus (1.6-23) yields the definitions

$$p^x(\omega) = |X(\omega)|$$

(1.6-24a)

and

$$\psi^x(\omega) = \tan^{-1} \left\{ \frac{\text{Im}\,[X(\omega)]}{\text{Re}\,[X(\omega)]} \right\}$$

(1.6-24b)

as the *FT amplitude* and *phase spectra,* respectively. It is apparent FT spectra are *continuous* functions of ω, while the FS spectra in (1.6-21) are *discrete* (noncontinuous) functions of ω, and hence are sometimes referred to as *line spectra.*

Two examples of FT spectra are depicted in Fig. 1.6-3 for $\omega \geq 0$; details are left as an exercise (Problem 1-12). As was the case with FS spectra, the FT amplitude and phase spectra are even and odd functions, respectively, about $\omega = 0$ when $x(t)$ is *real;* see Problem 1-13.

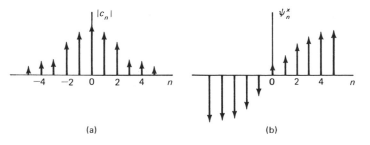

(a)   (b)

Fig. 1.6-2 Even and odd structure of Fourier series amplitude and phase spectra.

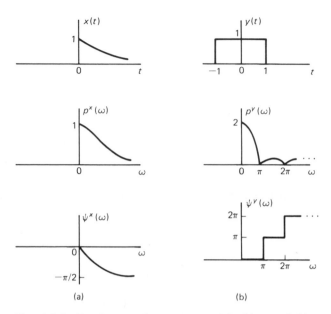

Fig. 1.6-3 Fourier transform spectra: (a) $x(t) = e^{-t}u(t)$;
(b) $y(t) = 1$, $|t| < 1$, and 0 elsewhere

Power Density Spectrum

We now consider a real-valued $x(t)$ that does not die down and exists over the infinite interval. As such, the total energy is infinite, and the average power is nonzero. In such cases we are usually interested in the manner in which the power is spread over the frequency range. To this end, we define the *power density spectrum* to be

$$G_x(\omega) = \frac{1}{2\pi} \lim_{J \to \infty} \frac{1}{J} |X(\omega)|^2 \qquad (1.6\text{-}25)$$

where $X(\omega)$ denotes the FT of a segment of $x(t)$, J seconds long.
   Again, the average power in $x(t)$ is given by (1.6-1b) to be

$$P_{av}^x = \lim_{J \to \infty} \frac{1}{J} \int_{-J/2}^{J/2} x^2(t)\, dt \qquad (1.6\text{-}26)$$

The power density spectrum we have defined has the property that

$$P_{av}^x = \int_{-\infty}^{\infty} G_x(\omega)\, d\omega \qquad (1.6\text{-}27)$$

Equation (1.6-27) implies that the total *area* under $G_x(\omega)$ equals the average power P_{av}. Hence the name power "density" spectrum.

It follows that the power in a specified bandwidth (ω_1, ω_2) is given by

$$P^x_{\omega_1, \omega_2} = \int_{-\omega_2}^{-\omega_1} G_x(\omega) \, d\omega + \int_{\omega_1}^{\omega_2} G_x(\omega) \, d\omega$$

Now, since $|X(\omega)|^2$ is an even function about $\omega = 0$ for real $x(t)$ (see Problem 1-13), (1.6-25) implies that $P^x_{\omega_1, \omega_2}$ simplifies to yield

$$P^x_{\omega_1, \omega_2} = 2 \int_{\omega_1}^{\omega_2} G_x(\omega) \, d\omega \tag{1.6-28}$$

Power density spectra are particularly useful when dealing with *random* signals, since they provide a meaningful measure for the distribution of average power in such signals.

1.7 LAPLACE TRANSFORM

This section will be devoted to a discussion of the *Laplace transform*, which also plays a significant role in connection with CT signals and systems. In particular, it is a powerful tool for the representation and analysis of CT systems. This aspect of the Laplace transform will be addressed in Section 1.9.

Given an $x(t)$, we define its Laplace transform (LT) $X(s)$ as

$$X(s) = \int_0^\infty x(t)e^{-st} \, dt \tag{1.7-1}$$

where $s = \sigma + j\omega$ is a complex variable, and $x(t)$ is a *causal* signal; that is, $x(t) = 0$, for $t < 0$. The quantity σ serves as a convergence factor.

Equation (1.7-1) can be written as

$$X(s) = \int_0^\infty x(t)e^{-\sigma t}e^{-j\omega t} \, dt \tag{1.7-2}$$

From (1.7-2) it is apparent that the LT is actually the FT of the function $g(t) = e^{-\sigma t}x(t)$ for those $x(t)$ that are casual. In addition, if $x(t)$ is also absolutely integrable, then $X(\omega)$ equals $X(s)$ with $s = j\omega$. To illustrate, consider the following $x(t)$, which is causal and is absolutely integrable:

$$x(t) = e^{-at} u(t) \qquad (1.7\text{-}3)$$

where $a > 0$.

Then, substituting (1.7-3) into (1.7-1), there results

$$X(s) = \int_0^\infty e^{-(a+s)t} \, dt$$

which yields

$$X(s) = \frac{1}{s + a} \qquad (1.7\text{-}4)$$

With $s = \sigma + j\omega$ in (1.7-4), we have

$$X(s) = \frac{1}{\sigma + j\omega + a}$$

Hence

$$\lim_{\sigma \to 0} X(s) = \frac{1}{a + j\omega}$$

which is exactly equal to the FT of $x(t)$ in (1.7-3); see line 8 of Table 1.3-2.

Convergence Considerations

For the LT to exist, the right-hand side of (1.7-1) must converge. This convergence occurs when the quantity $|x(t)e^{-st}|$ in (1.7-1) is absolutely integrable; that is,

$$I = \int_0^\infty |x(t)e^{-st}| \, dt < \infty \qquad (1.7\text{-}5a)$$

Since $|e^{j\omega t}| = 1$, (1.7-5a) simplifies to yield

$$I = \int_0^\infty |x(t)|e^{-\sigma t} \, dt < \infty \qquad (1.7\text{-}5b)$$

To illustrate, let $x(t)$ be the exponential in (1.7-3). Then (1.7-5b) leads to

$$I = \int_0^\infty e^{-(a+\sigma)t}\, dt = -\frac{e^{-(a+\sigma)t}}{(a+\sigma)}\Big|_0^\infty$$

which is finite only if $\sigma > -a$. Hence $X(s) = 1/(s + a)$ in (1.7-4) exists for all $\sigma > -a$. These values of σ for which $X(s)$ exists are said to define a *region of convergence*, as illustrated in Fig. 1.7-1. Similarly, it can be shown that the LT of $x(t) = e^{t^2}$ does not exist; that is, it is not possible to find σ that is large enough to satisfy the condition in (1.7-5b); see Problem 1-14.

Corresponding to the LT in (1.7-1), the *inverse Laplace transform* (ILT) is given by the complex inversion integral

$$x(t) = \frac{1}{2\pi j} \int_C X(s)e^{st}\, ds \qquad (1.7\text{-}6)$$

where C is a contour that is chosen to include all singularities of $X(s)$.

Evaluation of $x(t)$ via the contour integral in (1.7-6) is considerably involved, except for rather simple functions. However, many transforms can be readily inverted with the aid of transform pairs, such as those listed in Table 1.7-1. This approach for evaluating the ILT involves the partial fraction method and will be considered in Section 1.9.

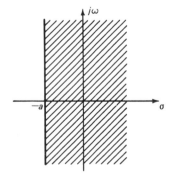

Fig. 1.7-1 Hatched region denotes the region of convergence for the Laplace transform of $x(t) = e^{-at}u(t)$, $a > 0$.

Table 1.7-1   Some Laplace Transform Pairs

Time Function	Laplace Transform
1. $cx(t)$, where c is a constant	$cX(s)$
2. $x(t) + y(t)$	$X(s) + Y(s)$
3. $\dfrac{dx(t)}{dt}$	$sX(s) - x(0)$
4. $\dfrac{d^n x(t)}{dt^n}$	$s^n X(s) - \displaystyle\sum_{i=0}^{n-1} s^{n-1-i} \dfrac{d^i x(0)}{dt^i}$
5. $\displaystyle\int_0^t x(\tau)\, d\tau$	$\dfrac{X(s)}{s}$
6. $tx(t)$	$-\dfrac{dX(s)}{ds}$
7. e^{-at}, $a > 0$	$\dfrac{1}{s+a}$
8. $e^{-at}x(t)$, $a > 0$	$X(s+a)$
9. $x(at)$	$\dfrac{1}{a} X\left(\dfrac{s}{a}\right)$
10. $\delta(t)$	1
11. $u(t)$	$\dfrac{1}{s}$
12. $\cos \omega_0 t$	$\dfrac{s}{s^2 + \omega_0^2}$
13. $\sin \omega_0 t$	$\dfrac{\omega_0}{s^2 + \omega_0^2}$
14. t^n	$\dfrac{n!}{s^{n+1}}$
15. $x(t - a)u(t - a)$, $a \geq 0$	$e^{-as}X(s)$

Convolution Properties

As was the case with the FT, there are convolution integrals associated with the LT. For example, if

$$Z(s) = X(s)Y(s) \tag{1.7-7}$$

then

$$z(t) = \int_0^\infty x(\tau)\, y(t - \tau)\, d\tau \qquad (1.7\text{-}8a)$$

or, equivalently,

$$z(t) = \int_0^\infty x(t - \tau) y(\tau)\, d\tau \qquad (1.7\text{-}8b)$$

The right-hand side of (1.7-8) represents the convolution integral in the time domain. Equations (1.7-8) and (1.7-7) imply that *convolution* in the *time domain* is equivalent to *multiplication* in the *LT (frequency)* domain.

For illustration purposes, let us derive (1.7-8b) starting from (1.7-7). To this end, we start with the definition of the ILT in (1.7-6) and write

$$z(t) = \frac{1}{2\pi j} \int_C Z(s) e^{st}\, ds \qquad (1.7\text{-}9)$$

Substitution of (1.7-7) in (1.7-9) results in

$$z(t) = \frac{1}{2\pi j} \int_C X(s) Y(s) e^{st}\, ds$$
$$= \frac{1}{2\pi j} \int_C X(s) e^{st} \left[\int_0^\infty y(\tau) e^{-s\tau}\, d\tau \right] ds \qquad (1.7\text{-}10)$$

Interchanging the order of integration in (1.7-10), there results

$$z(t) = \int_0^\infty y(\tau) \left[\frac{1}{2\pi j} \int_C X(s) e^{st} e^{-s\tau}\, ds \right] d\tau$$
$$= \int_0^\infty y(\tau) \left[\frac{1}{2\pi j} \int_C X(s) e^{(t-\tau)s}\, ds \right] d\tau \qquad (1.7\text{-}11)$$

We recognize that the term in brackets is $x(t - \tau)$. This follows from the ILT definition

$$x(t) = \frac{1}{2\pi j} \int_C X(s) e^{st}\, ds$$

Thus (1.7-11) becomes

$$z(t) = \int_0^\infty x(t - \tau)y(\tau) \, d\tau$$

which is the desired convolution integral in (1.7-8b).

Corresponding to (1.7-7) and (1.7-8), we also have the result that if

$$z(t) = x(t) \, y(t) \qquad (1.7\text{-}12)$$

then

$$Z(s) = \frac{1}{2\pi j} \int_C X(\alpha)Y(s - \alpha) \, d\alpha \qquad (1.7\text{-}13\text{a})$$

or, equivalently,

$$Z(s) = \frac{1}{2\pi j} \int_C X(s - \alpha)Y(\alpha) \, d\alpha \qquad (1.7\text{-}13\text{b})$$

The pertinent derivation is left as an exercise for the reader; see Problem 1-15.

The implication of (1.7-12) and (1.7-13) is that *multiplication* in the *time domain* is equivalent to *convolution* in the *LT (frequency) domain*.

We conclude our discussion pertaining to the LT with two additional formulas:

$$\lim_{s \to \infty} sX(s) = x(0) \qquad (1.7\text{-}14)$$

which is known as the *initial value theorem*.

$$\lim_{s \to 0} sX(s) = x(\infty) \qquad (1.7\text{-}15)$$

which is known as the *final value theorem*, where $sX(s)$ is assumed to have no poles in the right-half s-plane or on the $j\omega$-axis.

1.8 CONTINUOUS-TIME SYSTEMS: TIME-DOMAIN CONSIDERATIONS

We consider a linear, time-invariant CT system that is subjected to a single input $x(t)$, as illustrated in Fig. 1.8-1a. The corresponding output (response) is $y(t)$. The input-output relationship of this class of systems may be described in terms of linear differential equations with *constant* coefficients, as follows:

$$b_k \frac{d^k y(t)}{dt^k} + b_{k-1} \frac{d^{k-1}y(t)}{dt^{k-1}} + \cdots + b_0 y(t)$$

$$= a_\ell \frac{d^\ell x(t)}{dt^\ell} + a_{\ell-1} \frac{d^{\ell-1}x(t)}{dt^{\ell-1}} + \cdots + a_0 x(t) \quad (1.8\text{-}1)$$

where k and ℓ are positive integers. Without loss of generality, we restrict our attention to the case $k = \ell$, where k specifies the order of the system.

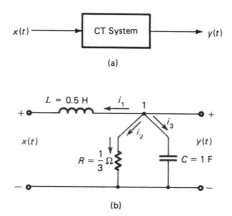

Fig. 1.8-1 (a) Block diagram representation for continuous-time system; (b) a
simple RLC circuit.

As an example, let us consider the simple circuit shown in Fig. 1.8-1b. Applying Kirchhoff's current law to node 1, we obtain

$$i_1 + i_2 + i_3 = 0$$

or $\quad \dfrac{1}{L} \displaystyle\int_{-\infty}^{t} [y(\tau) - x(\tau)]\, d\tau + \dfrac{y(t)}{R} + C \dfrac{dy(t)}{dt} = 0 \qquad (1.8\text{-}2)$

Substitution for L, R, and C in (1.8-2) leads to

$$2 \int_{-\infty}^{t} [y(\tau) - x(\tau)]\, d\tau + 3y(t) + \frac{dy(t)}{dt} = 0 \qquad (1.8\text{-}3)$$

Term-by-term differentiation of (1.8-3) results in

$$2[y(t) - x(t)] + 3\frac{dy(t)}{dt} + \frac{d^2y(t)}{dt^2} = 0$$

$$(1.8\text{-}4)$$

or

$$\frac{d^2y(t)}{dt^2} + 3\frac{dy(t)}{dt} + 2y(t) = 2x(t)$$

Comparing (1.8-4) with (1.8-1), we note that $k = 2$, and hence the RLC circuit of Fig. 1.8-1b represents a second-order system. The response of this system can be found by solving the differential equation for a specified input $x(t)$. Classical methods for obtaining such solutions are well known and are discussed in introductory texts related to circuit theory and differential equations.

Another way of representing the CT system in Fig. 1.8-1a is via the *convolution integral*

$$y(t) = \int_0^\infty x(t - \tau)h(\tau)\, d\tau, \qquad t \geq 0 \qquad (1.8\text{-}5a)$$

or

$$y(t) = \int_0^\infty x(\tau)h(t - \tau)\, d\tau, \qquad t \geq 0 \qquad (1.8\text{-}5b)$$

where $h(t)$ is called the *impulse response* of the system. That is, $h(t)$ is the system response when the input is the unit impulse function $\delta(t)$.

It is clear that the convolution integral in (1.8-5) is very similar to that in (1.5-6), which is associated with the FT. Thus $y(t)$ in (1.8-5) can also be evaluated via successive shifts and integration of $h(t)$ with the mirror image of $x(t)$, or vice versa. This aspect is illustrated in Fig. 1.8-2 for a system whose impulse response and input are as follows:

$$h(t) = \begin{cases} e^{-t}, & t \geq 0 \\ 0, & t < 0 \end{cases}$$

$$(1.8\text{-}6)$$

and

$$x(t) = \begin{cases} 1, & 0 \leq t \leq 1 \\ 0, & \text{elsewhere} \end{cases}$$

To summarize, there are at least two time-domain methods for evaluating the system response $y(t)$ in Fig. 1.8-2. These are (1) by solving differential equations, and (2) by evaluating the convolution integral. In the next section we shall discuss some aspects of finding the system response via a transform-domain method that employs the LT.

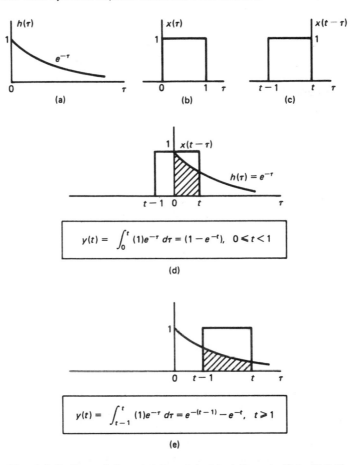

Fig. 1.8-2 Convolution details related to $h(t)$ and $x(t)$ in (1.8-6).

1.9 CONTINUOUS-TIME SYSTEMS: LAPLACE-TRANSFORM CONSIDERATIONS

For all practical purposes, it suffices to restrict our attention to CT systems whose impulse responses are such that

$$h(t) = 0, \qquad t < 0 \qquad\qquad (1.9\text{-}1a)$$

Such systems are said to be *causal;* that is, they cannot produce an output before being subjected to an input, which in this case is $\delta(t)$. We also assume that the input signals are *causal;* that is,

$$x(t) = 0, \quad t < 0 \tag{1.9-1b}$$

Let us now use the LT to solve for the output $y(t)$ when a given system is represented by a differential equation as in (1.8-1). The key property that enables us to do so is given in line 4 of Table 1.7-1; that is,

$$L\left\{\frac{d^n x(t)}{dt^n}\right\} = s^n X(s) - \sum_{i=0}^{n-1} s^{n-1-i} \frac{d^i x(0)}{dt^i} \tag{1.9-2}$$

where L denotes the Laplace transform. For example, with $n = 1$ and 2, (1.9-2) yields the following equations:

$$n = 1: \quad L\left\{\frac{dx(t)}{dt}\right\} = sX(s) - x(0) \tag{1.9-3a}$$

$$n = 2: \quad L\left\{\frac{d^2 x(t)}{dt^2}\right\} = s^2 X(s) - sx(0) - x'(0) \tag{1.9-3b}$$

where $\quad x'(0) = \dfrac{dx(t)}{dt}\bigg|_{t=0}$

The effect of applying (1.9-2) to the differential equation in (1.8-1) is to get an equivalent *algebraic equation* that can be used to solve for $Y(s)$. The desired system response $y(t)$ can then be obtained by finding the ILT of $Y(s)$ via the partial fraction expansion method, where $Y(s)$ is assumed to be in the form of a rational function. These concepts are best explained by means of a simple example.

Example 1.9-1: Find the response $y(t)$ of the *RLC* circuit in Fig. 1.8-1b when $x(t) = 5u(t)$, and the initial conditions are $y(0) = 1$ and $y'(0) = 0$.

Solution: The differential equation that describes the *RLC* circuit in Fig. 1.8-1b is given by (1.8-4) to be

$$\frac{d^2 y(t)}{dt^2} + 3\frac{dy(t)}{dt} + 2y(t) = 2x(t)$$

From this differential equation and (1.9-3) it follows that

$$[s^2 Y(s) - sy(0) - y'(0)] + 3[sY(s) - y(0)] + 2Y(s) = 2\,L\{x(t)\} \tag{1.9-4}$$

Since $x(t) = 5u(t)$, from line 11 of Table 1.7-1 we have

$$X(s) = \frac{5}{s} \qquad\qquad (1.9\text{-}5)$$

Also, the initial conditions are given to be

$$y(0) = 1 \quad \text{and} \quad y'(0) = 0 \qquad\qquad (1.9\text{-}6)$$

Substitution of (1.9-5) and (1.9-6) in (1.9-4) leads to

$$s^2Y(s) - s + 3[sY(s) - 1] + 2Y(s) = \frac{10}{s}$$

which simplifies to yield

$$(s^2 + 3s + 2)Y(s) = \frac{10}{s} + s + 3$$

Hence $\qquad\qquad Y(s) = \dfrac{s^2 + 3s + 10}{s(s^2 + 3s + 2)} = \dfrac{s^2 + 3s + 10}{s(s + 1)(s + 2)} \qquad (1.9\text{-}7)$

Next we obtain a partial fraction expansion of (1.9-7) by writing

$$Y(s) = \frac{A}{s} + \frac{B}{s + 1} + \frac{C}{s + 2} \qquad\qquad (1.9\text{-}8)$$

Evaluating the expansion coefficients, we have

$$A = sY(s)\,|_{s=0} = 5$$
$$B = (s + 1)Y(s)\,|_{s=-1} = -8 \qquad\qquad (1.9\text{-}9)$$
and $\qquad\qquad C = (s + 2)Y(s)\,|_{s=-2} = 4$

Substitution of (1.9-9) into (1.9-8) leads to

$$Y(s) = \frac{5}{s} - \frac{8}{s + 1} + \frac{4}{s + 2}$$

We now find the ILT of $Y(s)$ by referring to Table 1.7-1. From lines 7 and 11 it is apparent that

$$y(t) = 5 - 8e^{-t} + 4e^{-2t}, \qquad t \geq 0 \qquad (1.9\text{-}10)$$

which is the desired response of the circuit in Fig. 1.8-1b to the step input $5u(t)$.

The reader is encouraged to verify that $y(t)$ in (1.9-10) does indeed satisfy the initial conditions $y(0) = 1$ and $y'(0) = 0$.

Transfer Function

We recall that the system in Fig. 1.8-1a can be represented by the convolution integral in (1.8-5). Thus we have

$$y(t) = \int_0^\infty x(t - \tau)h(\tau)\, d\tau, \qquad t \geq 0$$

where $h(t)$ is the impulse response of the system. However, from (1.7-7) and (1.7-8) it follows that this convolution integral is equivalent to the following product relationship:

$$Y(s) = H(s)X(s) \qquad (1.9\text{-}11)$$

where $X(s)$ and $Y(s)$ are the LTs of the input and output, respectively, and

$$H(s) = L\{h(t)\} \qquad (1.9\text{-}12a)$$

is the LT of the system impulse response.

The quantity $H(s)$ is defined as the *transfer function* of the system. From (1.9-12a) it is apparent that the impulse response can be obtained from the transfer function via the relation

$$h(t) = L^{-1}\{H(s)\} \qquad (1.9\text{-}12b)$$

where L^{-1} denotes the ILT.

It is instructive to introduce the transfer function concept via the differential equation in (1.8-1). We assume that $k = \ell$ in (1.8-1) and that *all initial conditions are zero;* that is, the system is *initially relaxed.* Then line 4 of Table 1.7-1 implies that

$$L\left\{\frac{d^n x(t)}{dt^n}\right\} = s^n X(s)$$

and

$$L\left\{\frac{d^n y(t)}{dt^n}\right\} = s^n Y(s)$$

As such, taking the LT of each term in (1.8-1), we obtain

$$(b_k s^k + b_{k-1} s^{k-1} + \cdots + b_0)X(s) = (a_k s^k + a_{k-1} s^{k-1} + \cdots + a_0)Y(s)$$

That is,

$$Y(s) = \left[\frac{b_k s^k + b_{k-1} s^{k-1} + \cdots + b_0}{a_k s^k + a_{k-1} s^{k-1} + \cdots + a_0} \right] X(s) \qquad (1.9\text{-}13)$$

Comparing (1.9-13) with (1.9-11), we have

$$H(s) = \frac{b_k s^k + b_{k-1} s^{k-1} + \cdots + b_0}{a_k s^k + a_{k-1} s^{k-1} + \cdots + a_0} \qquad (1.9\text{-}14)$$

which is again the transfer function of the system.

Example 1.9-2: For the *RLC* circuit in Fig. 1.8-1b, find the transfer function

$$H(s) = \frac{Y(s)}{X(s)} \qquad (1.9\text{-}15)$$

where $Y(s)$ and $X(s)$ are the LTs of $y(t)$ and $x(t)$, respectively. Also find the impulse response $h(t)$.

Solution: The transfer function $H(s)$ can be found via two methods. First, we can start with the differential equation in (1.8-4), which describes the *RLC* circuit in Fig. 1.8-1b. The LT of each term in (1.8-4) can be expressed as

$$L\left\{ \frac{d^2 y(t)}{dt^2} \right\} + 3L\left\{ \frac{dy(t)}{dt} \right\} + 2L\{y(t)\} = 2L\{x(t)\} \qquad (1.9\text{-}16)$$

Since the transfer function implies zero initial conditions, line 4 of Table 1.7-1 yields

$$L\left\{ \frac{d^n y(t)}{dt^n} \right\} = s^n Y(s), \qquad \text{for } n = 1, 2$$

Thus (1.9-16) leads to

$$(s^2 + 3s + 2)Y(s) = 2X(s)$$

which means that the transfer function is

$$H(s) = \frac{Y(s)}{X(s)} = \frac{2}{s^2 + 3s + 2} \qquad (1.9\text{-}17)$$

The second method we discuss avoids the necessity of first deriving the differential equation in (1.8-4) to find $H(s)$. In this approach we replace the R, L, and C elements by their transform impedances, which are R, Ls, and $1/Cs$, respectively. This results in the following circuit.

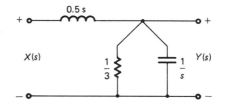

Now the voltage-divider principle implies that

$$\frac{Y(s)}{X(s)} = H(s) = \frac{Z(s)}{Z(s) + 0.5s} \qquad (1.9\text{-}18)$$

where $Z(s)$ is the impedence of the resistance in parallel with the capacitance; that is,

$$Z(s) = \frac{(1/3)(1/s)}{(1/3) + (1/s)} = \frac{1}{s + 3} \qquad (1.9\text{-}19)$$

Substituting (1.9-19) in (1.9-18), there results

$$H(s) = \frac{2}{s^2 + 3s + 2}$$

which agrees with $H(s)$ in (1.9-17).

Next the impulse response $h(t)$ is obtained by finding the ILT of $H(s)$:

$$h(t) = L^{-1}\left\{\frac{2}{s^2 + 3s + 2}\right\}$$

$$= L^{-1}\left\{\frac{2}{(s + 1)(s + 2)}\right\} \qquad (1.9\text{-}20)$$

A partial fraction expansion of $2/(s + 1)(s + 2)$ results in

$$\frac{2}{(s + 1)(s + 2)} = \frac{2}{s + 1} - \frac{2}{s + 2} \qquad (1.9\text{-}21)$$

Equations (1.9-20) and (1.9-21) imply that

$$h(t) = 2\left[L^{-1}\left\{ \frac{1}{s + 1} \right\} - L^{-1}\left\{ \frac{1}{s + 2} \right\} \right] \qquad (1.9\text{-}22)$$

From line 7 of Table 1.7-1 and (1.9-22), it is apparent that

$$h(t) = 2[e^{-t} - e^{-2t}], \qquad t \geq 0 \qquad (1.9\text{-}23)$$

is the desired impulse response.

From the discussion in this and the previous section, it is clear that the concepts of impulse response and transfer function play a fundamental role in connection with CT systems. This is because one can find the system response to *any* input if its impulse response or transfer function is known. In addition, they also enable us to determine whether a given system is stable or not. This aspect will be addressed in the next section.

1.10 POLES, ZEROS, AND STABILITY

Let us consider the general form of a transfer function. It is given by

$$H(s) = \frac{b_k s^k + b_{k-1} s^{k-1} + \cdots + b_0}{a_\ell s^\ell + a_{\ell-1} s^{\ell-1} + \cdots + a_0} = \frac{N(s)}{D(s)} \qquad (1.10\text{-}1)$$

It is apparent that $N(s)$ and $D(s)$ can be factored into k and ℓ roots, respectively. These roots may be real or complex numbers.

The roots of $N(s)$ are called the *zeros* of the transfer function $H(s)$, while the roots of $D(s)$ are called its *poles*. Again, all the poles and zeros of $H(s)$ are said to be its *critical frequencies*. For example, the transfer function

$$H(s) = \frac{s(s - 1)}{(s + 1)(s^2 + s + 1)} \qquad (1.10\text{-}2)$$

has zeros at $s = 0$ and $s = 1$, a pole at $s = -1$, and a complex-conjugate pole pair at $s = -0.5 \pm j\,(\sqrt{3}/2)$. It is also considered to have a zero at infinity, in the sense that $H(s)$ goes to zero as s tends to infinity. It is convenient to plot the finite-valued poles and zeros of $H(s)$ as depicted in Fig. 1.10-1, where the symbols "$\times$" and "0" denotes poles and zeros, respectively.

The relation

$$Y(s) = H(s)X(s)$$

implies that

$$y(t) = L^{-1}\{H(s)X(s)\} \tag{1.10-3}$$

where $y(t)$ is the output when the system with transfer function $H(s)$ is subjected to the input $x(t)$.

From (1.10-3) it is apparent that poles in $Y(s)$ can arise from two sources: (1) poles of the transfer function $H(s)$, and (2) poles of the input function $X(s)$. The portion of the output $y(t)$ due to the poles of $H(s)$ is defined as the *natural response*. On the other hand, the *forced response* is the portion of $y(t)$ that is due to the poles of $X(s)$. If the system has all its poles in the left-half s-plane, then the natural response goes to zero as $t \to \infty$. In such cases, the natural response is called the *transient response*. Also, if the input to such systems is periodic, then the corresponding forced response is referred to as the *steady-state response*.

Stability Considerations

By a *stable* system we mean a system that yields a finite output for every finite input. Stability conditions for a system can be represented in terms of its impulse response $h(t)$ or its transfer function $H(s)$, as follows:

1. If

$$\lim_{t \to \infty} h(t) = 0 \tag{1.10-4}$$

then the system is *stable*. This condition is equivalent to requiring all the poles of $H(s)$ to be in the *left-half* s-plane. For example,

$$H(s) = \frac{1}{s + a}, \qquad a > 0$$

represents a stable system; see Fig. 1.10-2a.

2. If

$$\lim_{t \to \infty} h(t) \to \infty \tag{1.10-5}$$

then the system is *unstable*. This condition is equivalent to saying that systems with one or more poles in the *right-half* s-plane, or with multiple-order pole(s) on the $j\omega$-axis, are unstable. For example,

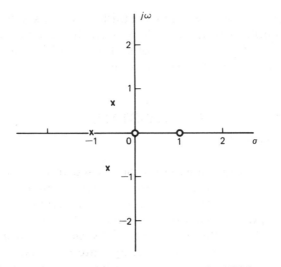

Fig. 1.10-1 Pole-zero s-plane plot for $H(s)$ in (1.10-2).

$$H(s) = \frac{1}{s - a}, \qquad a > 0$$

represents an *unstable* system; see Fig. 1.10-2b.

3. As t tends to infinity, if $h(t)$ approaches a nonzero value or a bounded oscillation, then the system is *marginally stable*. For example,

$$H(s) = \frac{1}{s^2 + \omega_0^2}$$

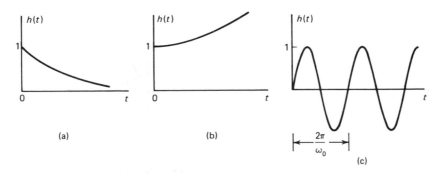

Fig. 1.10-2 Impulse responses of systems that are stable, unstable, and marginally stable.

represents a marginally stable system; see Fig. 1.10-2c. The first-order complex conjugate pole pair *on* the $j\omega$ axis causes $h(t)$ to be oscillatory.

These concepts of stability will be discussed again in Chapter 5 in the context of DT systems.

1.11 SINUSOIDAL STEADY-STATE RESPONSE

Perhaps the most commonly used input to a system is the sinusoidal signal, which may be in the form of sin ωt, cos ωt, or $e^{j\omega t}$. Thus a great deal of attention is paid to the corresponding system output. In particular, we are usually interested in the steady-state portion of this output, which is called the *sinusoidal steady-state response.*

The sinusoidal steady-state response is very widely used in analytical and experimental studies pertaining to CT as well as DT systems. Its relevance to DT systems will be discussed in Chapter 5.

The input is considered to be of the form

$$x(t) = Xe^{j(\omega t + \psi_x)} \tag{1.11-1}$$

where ω is the radian frequency and X and ψ_x are constants. This input is conveniently represented using *phasor* notation as depicted in Fig. 1.11-1, where X and ψ_x represent the *magnitude* and *phase* of the input phasor $X^{\dagger}$. Depending upon whether a sine or cosine input is desired, we can specifiy either the imaginary or the real part of $x(t)$ in (1.11-1).

It is important to realize that we are interested only in the *steady-state* portion of the output $y(t)$. Thus, if

$$y(t) = y_{ss}(t) + y_{tr}(t) \tag{1.11-2}$$

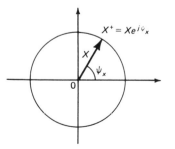

Fig. 1.11-1 Input using phasor notation.

where $y_{ss}(t)$ and $y_{tr}(t)$ denote the steady-state and transient components, we seek only $y_{ss}(t)$. Since the system we consider is *linear* and *time invariant*, $y_{ss}(t)$ can also be expressed as a phasor whose frequency is the *same* as that of the input phasor, but whose magnitude and phase may be different. Thus we have

$$y_{ss}(t) = Ye^{j(\omega t + \psi_y)} \tag{1.11-3}$$

The steady-state output also has a phasor interpretation as depicted in Fig. 1.11-2. From the above discussion it is apparent that we need only determine Y and ψ_y in terms of X and ψ_x to obtain $y_{ss}(t)$.

We assume that $x(t)$ in (1.11-1) is the input to a stable system whose transfer function is $H(s)$; that is, the poles of $H(s)$ are all in the left-half s-plane. Next the LT of $x(t)$ in (1.11-1) results in

$$X(s) = Xe^{j\psi_x}L\{e^{j\omega t}\} \tag{1.11-4}$$

$$= \frac{Xe^{j\psi_x}}{s - j\omega}$$

Again, since $\qquad\qquad Y(s) = H(s)X(s) \tag{1.11-5}$

(1.11-4) and (1.11-5) yield

$$Y(s) = \frac{Xe^{j\psi_x}}{s - j\omega} H(s) \tag{1.11-6}$$

The ILT of (1.11-6) yields $y(t)$ as the sum of the transient and steady-state components as in (1.11-2). The steady-state portion is due to the

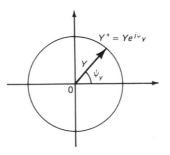

Fig. 1.11-2 Sinusoidal steady-state using phasor notation.

pole of the input $X(s)$. Hence a partial fraction expansion of $Y(s)$ in (1.11-6) can be expressed as

$$Y(s) = \frac{A}{s - j\omega} + \{\text{other terms}\}$$

where $A = (s - j\omega)Y(s)|_{s=j\omega} = H(\omega)Xe^{j\psi_x}$

and $H(\omega)$ denotes $H(j\omega)$. That is,

$$y(t) = Ae^{j\omega t} + \{\text{transient components}\}$$

which means that

$$y_{ss}(t) = H(\omega)Xe^{j(\omega t + \psi_x)} \tag{1.11-7}$$

Thus from (1.11-3) and (1.11-7) we obtain

$$Ye^{j\psi_y} = H(\omega)Xe^{j\psi_x} \tag{1.11-8}$$

Since $H(\omega)$ is a complex number, it can be expressed as

$$H(\omega) = |H(\omega)|\, e^{j\beta(\omega)} \tag{1.11-9}$$

Substitution of (1.11-9) in (1.11-8) leads to

$$Ye^{j\psi_y} = |H(\omega)|\, Xe^{j[\beta(\omega) + \psi_x]} \tag{1.11-10}$$

In phasor notation (1.11-10) becomes

$$Y^\dagger = Ye^{j\psi_y} \tag{1.11-11}$$

where $Y = |H(\omega)|\, X$

$\psi_y = \beta(\omega) + \psi_x$

are the magnitude and phase of the output, as illustrated in Fig. 1.11-2.

Equation (1.11-11) yields the desired sinusoidal steady-state response and also leads to the following useful definitions pertaining to CT systems:

1. $H(\omega)$: *steady-state transfer function.*
2. $|H(\omega)|$: *magnitude (amplitude) response.*

3. $|H(\omega)|^2$: *squared-magnitude response.*

4. $\beta(\omega)$: *phase response.*

These amplitude and phase responses can be evaluated from a given steady-state transfer function using the relations

$$|H(\omega)| = [R^2(\omega) + I^2(\omega)]^{1/2} \qquad (1.11\text{-}12a)$$

and

$$\beta(\omega) = \tan^{-1}\left[\frac{I(\omega)}{R(\omega)}\right] \qquad (1.11\text{-}12b)$$

where $R(\omega)$ and $I(\omega)$ are the real and imaginary parts of $H(\omega)$. The values of $|H(\omega)|$ and $\beta(\omega)$ for a *particular* ω are called the system *gain* and *phase shift*, respectively. Given an $H(s)$, we can evaluate $|H(\omega)|$ and $\beta(\omega)$ using the computer program in Appendix 1.1.

Example 1.11-1: Find $y_{ss}(t)$ if

$$x(t) = 20 \sin (100t + 30°) \qquad (1.11\text{-}13)$$

in the following circuit for the case $R = 10 \text{ k}\Omega$ and $C = 2 \text{ }\mu\text{F}$.

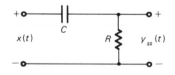

Solution: The transfer function of this circuit is given by

$$H(s) = \frac{Y(s)}{X(s)} = \frac{R}{R + (1/Cs)} = \frac{RCs}{RCs + 1}$$

The corresponding steady-state transfer function is

$$H(\omega) = \frac{j\omega RC}{1 + j\omega RC} \qquad (1.11\text{-}14)$$

From (1.11-13) it is apparent that the input frequency is $\omega = 100$ radians/second. Substituting $\omega = 100$, $R = 10,000$, and $C = 2 \times 10^{-6}$ in (1.11-14), we obtain

$$H(100) = \frac{j2}{1 + j2}$$

which is equivalent to

$$H(100) = 0.8 + j0.4 \qquad (1.11\text{-}15)$$

Next, from (1.11-15) and (1.11-12), we obtain

$$|H(100)| = 0.89$$

and

$$\beta(100) = 26.57°$$

The steady-state output is thus

$$y_{ss}(t) = Y \sin (100t + \psi_y)$$

where Y and ψ_y are obtained via (1.11-11); that is,

$$Y = |H(100)| X = (0.89)(20) = 17.8$$

and

$$\psi_y = 26.57° + 30° = 56.57°$$

since $X = 20$ and $\psi_x = 30°$; see (1.11-13). That is,

$$y_{ss}(t) = 17.8 \sin (100t + 56.57°) \qquad (1.11\text{-}16)$$

This output and the corresponding input in (1.11-13) are displayed in phasor form in Fig. 1.11-3.

Time-Delay Function

Given a CT system whose phase function is $\beta(\omega)$, its *time-delay function* $T_d(\omega)$ is defined as

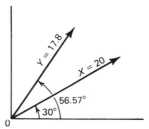

Fig. 1.11-3 Phasor representation of input and steady-state output related to Example 1.11-1.

$$T_d(\omega) = -\frac{d\beta(\omega)}{d\omega} \qquad (1.11\text{-}17)$$

The time-delay function represents the delay required by a sinusoid of frequency ω to pass through the system. An important class of systems is one for which the phase function $\beta(\omega)$ is a *linear* function of ω; that is,

$$\beta(\omega) = -c\omega \qquad (1.11\text{-}18)$$

where c is a positive constant; see Fig. 1.11-4. From (1.11-17) and (1.11-18) it follows that the time-delay function of this class of systems is given by

$$T_d(\omega) = c \qquad (1.11\text{-}19)$$

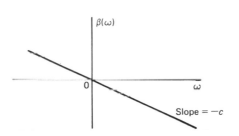

Fig. 1.11-4 Linear phase function.

which is a *constant* for *all* frequencies; see Fig. 1.11-5. This is a very important attribute in that it causes *all* the frequency components in the input to the system to be delayed the *same* amount. As such, signals that pass through linear phase or constant time-delay systems do *not* experience *phase distortion*. We shall discuss more about such systems in Chapter 7 with reference to the design of DT filters.

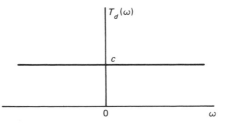

Fig. 1.11-5 Constant time-delay function.

Input–Output Spectra

We conclude this section by deriving expressions for the output amplitude and phase spectra in terms of the corresponding input spectra and the system magnitude and phase responses. To this end we substitute $s = j\omega$ in (1.11-5) to obtain

$$Y(\omega) = H(\omega)X(\omega) \qquad (1.11\text{-}20)$$

Next we can write

$$Y(\omega) = |Y(\omega)|\ e^{j\psi_y(\omega)}$$

$$X(\omega) = |X(\omega)|\ e^{j\psi_x(\omega)} \qquad (1.11\text{-}21)$$

and

$$H(\omega) = |H(\omega)|\ e^{j\beta(\omega)}$$

where $|X(\omega)|, \psi_x(\omega)$ = input amplitude and phase spectra, respectively

$|Y(\omega)|, \psi_y(\omega)$ = output amplitude and phase spectra, respectively

$|H(\omega)|, \beta(\omega)$ = system magnitude and phase responses, respectively

Substitution of (1.11-21) in (1.11-20) leads to

$$|Y(\omega)| = |H(\omega)||X(\omega)| \qquad (1.11\text{-}22a)$$

and

$$\psi_y(\omega) = \psi_x(\omega) + \beta(\omega) \qquad (1.11\text{-}22b)$$

From (1.11-22) it is clear that we can easily determine the output spectra, given the input spectra and the system magnitude and phase responses.

1.12 SUMMARY

We have introduced some basic terminology in this chapter as it relates to signals and systems. Many of the concepts that were discussed in the context of CT signals and systems will be extended to their DT counterparts in forthcoming chapters.

1–1 Use (1.2-1) and (1.2-4) to find the FS expansion of the following periodic functions.

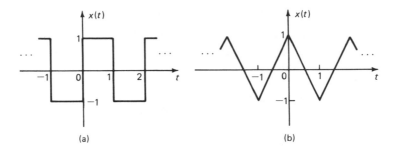

(a)   (b)

1–2 If $x(t) = \sin \omega_0 t$, $|t| < \infty$, show that its FT is as given in line 5 of Table 1.3-2.

1–3 Given:

$$x(t) = \begin{cases} e^{-2t}, & t \geq 0 \\ 0, & \text{elsewhere} \end{cases}$$

and

$$y(t) = \begin{cases} 1, & -1 < t \leq 0 \\ e^{-t}, & t \geq 0 \\ 0, & \text{elsewhere} \end{cases}$$

Evaluate the convolution integral

$$z(t) = \int_{-\infty}^{\infty} x(t - \tau)y(\tau) \, d\tau$$

using the step-by-step graphical approach summarized in Fig. 1.5-1. Verify that $z(t)$ so obtained agrees with that given by (1.5-8).

1–4 Starting with (1.5-9), derive the convolution integral in (1.5-10a) or (1.5-10b).

1–5 Let $x(t)$ be the triangular pulse shown in Fig. 1.4-2a, and

$$y(t) = \cos \omega_0 t, \qquad |t| < \infty$$

Evaluate the FT of

$$z(t) = x(t)y(t)$$

using the convolution integral

$$Z(\omega) = \frac{1}{2\pi} \int_{-\infty}^{\infty} Y(\alpha)X(\omega - \alpha) \, d\alpha$$

Hint: The FT of $x(t)$ is given by (1.4-6).

1–6   Using Table 1.3-3, show that the FTs of the rectangular, hanning, and Hamming windows defined in Table 1.5-1 satisfy the requirement in (1.5-24); that is,

(a) $\lim\limits_{T\to\infty} W_r(\omega) = 2\pi\delta(\omega)$.

(b) $\lim\limits_{T\to\infty} W_{ha}(\omega) = 2\pi\delta(\omega)$.

Hint: Starting with $w_{ha}(t)$, first show that its FT can be expressed as

$$W_{ha}(\omega) = \frac{\sin \omega T}{\omega} + 0.5 \left[\frac{\sin (\omega - \pi/T)T}{\omega - \pi/T} + \frac{\sin (\omega + \pi/T)T}{\omega + \pi/T} \right]$$

(c) $\lim\limits_{T\to\infty} W_H(\omega) = 2\pi\delta(\omega)$

Hint: Starting with $w_H(t)$, first show that its FT can be expressed as

$$W_H(\omega) = 1.08 \frac{\sin \omega T}{\omega} + 0.46 \left[\frac{\sin (\omega - \pi/T)T}{\omega - \pi/T} + \frac{\sin (\omega + \pi/T)T}{\omega + \pi/T} \right]$$

1–7   Using the FTs of $x(t) = 1$ and $x(t) = \mathrm{sgn}\,(t)$, which are given in Table 1.3-2, derive the FT of $u(t)$ and verify that it agrees with that given in line 3 of Table 1.3-2.

1–8   This problem concerns $W_K(\omega)$, which is the FT of the Kaiser window defined in Table 1.5-1; that is,

$$W_K(\omega) = \frac{T \sin \left[\sqrt{\omega^2 T^2 - \theta^2}\right]}{2I_0(\theta)\sqrt{\omega^2 T^2 - \theta^2}} \tag{P1-8-1}$$

where I_0 is the modified Bessel function of the first kind and zero order. Show that when $(\omega T)^2 < \theta^2$, (P1-8-1) can be expressed as

$$W_K(\omega) = \frac{T \sinh \left[\sqrt{\theta^2 - \omega^2 T^2}\right]}{2I_0(\theta)\sqrt{\theta^2 - \omega^2 T^2}}$$

1–9 If $x(t)$ is real, show that $|X(\omega)|^2$ is an even function about $\omega = 0$. Hence show that $\epsilon^x(\omega_1, \omega_2)$ in (1.6-7) simplifies to that in (1.6-8).

1–10 The total energy in the pulse $x(t)$ shown in Fig. 1.6-1a is given by (1.6-9) to be 2 joules. Also, its energy density spectrum is given by (1.6-10) to be

$$\epsilon^x(\omega) = \frac{2}{\pi}\left[\frac{\sin \omega}{\omega}\right]^2$$

Verify Parseval's theorem in (1.6-5) by showing that

$$\int_{-\infty}^{\infty} \epsilon^x(\omega) \, d\omega = 2$$

1–11 Show that p_n^x in (1.6-21a) is an even function about $n = 0$ by establishing that $c_{-n} = \tilde{c}_n$ via (1.6-22), where the symbol ~ denotes complex conjugate. Also use (1.6-22) to show that ψ_n^x given by (1.6-21b) is an odd function about $n = 0$.

1–12 With respect to $x(t)$ and $y(t)$ in Fig. 1.6-3, show that

$$p^x(\omega) = \frac{1}{\sqrt{1 + \omega^2}}$$

$$\psi^x(\omega) = -\tan^{-1} \omega$$

and

$$p^y(\omega) = 2\left[\frac{\sin \omega}{\omega}\right]$$

$$\psi^y(\omega) = k\pi, \quad k\pi \le \omega \le (k + 1) \pi$$

for $k = 0, 1, 2, \ldots$.

1–13 From the definition of the FT, we have

$$X(\omega) = \int_{-\infty}^{\infty} x(t) \cos \omega t \, dt - j \int_{-\infty}^{\infty} x(t) \sin \omega t \, dt$$

Verify that $X(-\omega) = \tilde{X}(\omega)$ when $x(t)$ is real, where the symbol ~ denotes complex conjugate. Hence show that $p^x(\omega)$ in (1.6-24a) is an even function about $\omega = 0$. Also show that $\psi^x(\omega)$ in (1.6-24b) is an odd function about $\omega = 0$.

1–14 Show that the LT of $x(t) = e^{t^2}$, $t \ge 0$, does not exist.

1–15 Starting with (1.7-12), derive (1.7-13b).

1–16 Consider the following *RC* circuit.

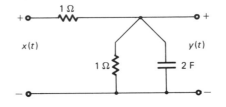

(a) Show that the differential equation that describes this circuit is

$$\frac{dy(t)}{dt} + y(t) = 0.5x(t)$$

(b) Use the LT to find $y(t)$ if

$$x(t) = \begin{cases} e^{-2t}, & t \geq 0 \\ 0, & t < 0 \end{cases}$$

and $y(0) = 0.5$.

(c) Identify the natural and forced responses obtained in part (b).

1–17 **(a)** Find the transfer function $H(s)$ and impulse response $h(t)$ for the *RC* circuit in Problem 1-16.

(b) Sketch the magnitude and phase responses, $|H(\omega)|$ and $\beta(\omega)$, respectively.

(c) If $x(t) = 10 \sin(100t + 60°)$, find $y_{ss}(t)$.

1–18 **(a)** Use the graphical procedure shown in Fig. 1.8-2 to evaluate the response $y(t)$ of the circuit in Problem 1-16 if the input is

$$x(t) = \begin{cases} t, & 0 \leq t \leq 1 \\ 0, & \text{elsewhere} \end{cases}$$

(b) Verify that the LT method yields the same $y(t)$.

1–19 **(a)** If $x(t) = y(t) \cos \omega_0 t$, show that

$$X(\omega) = \frac{Y(\omega - \omega_0) + Y(\omega + \omega_0)}{2}$$

(b) If $x(t) = y(t) \sin \omega_0 t$, show that

$$X(\omega) = \frac{Y(\omega - \omega_0) - Y(\omega + \omega_0)}{2j}$$

Comment: See lines 9 and 10 of Table 1.3-1.

1–20   The transfer function of a certain CT system is

$$H(s) = \frac{s(s - 5)(s^2 + s + 1)}{(s + 1)(s + 2)(s + 3)(s^2 + ks + 5)}$$

where $k \geq 0$.

(a) Show a plot of the poles and zeros of $H(s)$.

(b) Comment on the stability of this system for the case when $k = 0$ and $k > 0$.

1–21   The FS coefficients of $x_p(t)$ in Fig. 1.4-2b are given by (1.4-11) to be

$$c_n = \begin{cases} \epsilon, & n = 0 \\ 0, & n \text{ even.} \\ \dfrac{4\epsilon}{n^2\pi^2}, & n \text{ odd} \end{cases}$$

Verify that this information satisfies Parseval's theorem in (1.6-16).

1–22   Given $H(s) = (s - a)/(s + a)$, where $a > 0$. Show that

$$|H(\omega)| = 1$$

$$\beta(\omega) = \pi - 2\tan^{-1}\left(\frac{\omega}{a}\right), \qquad |\omega| < \infty$$

and

$$T_d(\omega) = 2/a\left[1 + \left(\frac{\omega}{a}\right)^2\right]$$

Comment: Since this magnitude response is constant for *all* ω, $H(s)$ is said to be an "all-pass" function. Also note that, if $(\omega/a)^2 << 1$, then the corresponding system has a constant time-delay function.

1–23   Consider a system whose impulse response is given by

$$h(t) = \begin{cases} e^{-at}\cos\omega_0 t, & t \geq 0 \\ 0, & \text{elsewhere} \end{cases}$$

(a) Show that its transfer function is

$$H(s) = \frac{s + a}{(s + a)^2 + \omega_0^2}$$

(b) Suppose $a = 0.1$ and $\omega_0 = 100$, and let the input to $H(s)$ be $x(t)$ as given by (1.11-13). Find the steady-state response $y_{ss}(t)$.

1–24   The input to the following *RC* circuit is

$$x(t) = \begin{cases} 1, & 0 \le t \le 1 \\ 0, & \text{elsewhere} \end{cases}$$

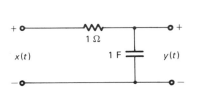

Show that $y(t) = (1 - e^{-t})u(t) - [1 - e^{-(t-1)}]u(t - 1)$

APPENDIX 1.1† COMPUTER PROGRAM FOR EVALUATING THE MAGNITUDE AND PHASE RESPONSES OF CONTINUOUS-TIME SYSTEMS

The program listed in this appendix yields the magnitude and phase responses of a given CT transfer function $H(s)$ in the form

$$H(s) = \frac{a_1 s^k + a_2 s^{k-1} + \cdots + a_k s + a_{k+1}}{b_1 s^m + b_2 s^{m-1} + \cdots + b_m s + b_{m+1}}$$

The number of numerator and denominator coefficients of $H(s)$ is $(k + 1)$ and $(m + 1)$, *respectively. As an illustration, let*

$$H(s) = \frac{7853.95s + 2{,}467{,}381}{s^2 + 628s + 2{,}467{,}381} \tag{A.1-1}$$

and suppose magnitude and phase responses are desired at 50 points between 0 and 500 Hz. Then the queries and related responses are as follows:

---

†Program developed by Myron Flickner and David Martz.

```
CONTINUOUS TRANSFER FUNCTION RESPONSE
NUMBER OF NUMERATOR COEFFICIENTS (1-128) ? 2
       A( 1)*S** 1 = 7853.95
       A( 2)*S** 0 = 2467381
NUMBER OF DENOMINATOR COEFFICIENTS (1-128) ? 3
       B( 1)*S** 2 = 1
       B( 2)*S** 1 = 628
       B( 3)*S** 0 = 2467381
HIGHEST FREQUENCY NEEDED (HERTZ) ? 500
NUMBER OF POINTS (MAXIMUM OF 1023) ? 50

POWER RATHER THAN AMPLITUDE (Y/N) ? N
```

The output that results is listed next, from which it follows that $H(s)$ in (A.1-1) is the transfer function of a CT system which is resonant at approximately 250 Hz.

FREQ	MAGNITUDE	PHASE
.00	1.0000	.0000
10.20	1.0222	.1850
20.41	1.0868	.3547
30.61	1.1889	.4997
40.82	1.3233	.6177
51.02	1.4854	.7106
61.22	1.6728	.7822
71.43	1.8843	.8363
81.63	2.1205	.8761
91.84	2.3834	.9041
102.04	2.6761	.9218
112.24	3.0030	.9306
122.45	3.3701	.9310
132.65	3.7848	.9232
142.86	4.2565	.9071
153.06	4.7968	.8819
163.27	5.4193	.8464
173.47	6.1397	.7989
183.67	6.9730	.7370
193.88	7.9297	.6573
204.08	9.0042	.5561
214.29	10.1544	.4296
224.49	11.2763	.2756
234.69	12.1916	.0964
244.90	12.6968	-.0986
255.10	12.6796	-.2943
265.31	12.1984	-.4754
275.51	11.4283	-.6326
285.71	10.5480	-.7635
295.92	9.6761	-.8704
306.12	8.8712	-.9572
316.33	8.1537	-1.0279
326.53	7.5242	-1.0858
336.73	6.9746	-1.1339
346.94	6.4950	-1.1741

FREQ	MAGNITUDE	PHASE
357.14	6.0752	-1.2080
367.34	5.7061	-1.2370
377.55	5.3797	-1.2619
387.75	5.0895	-1.2836
397.96	4.8304	-1.3025
408.16	4.5975	-1.3192
418.36	4.3873	-1.3340
428.57	4.1966	-1.3471
438.77	4.0229	-1.3590
448.98	3.8640	-1.3696
459.18	3.7182	-1.3792
469.38	3.5837	-1.3880
479.59	3.4594	-1.3960
489.79	3.3442	-1.4033
499.99	3.2370	-1.4100

```
C*****************************************************************************
C
      COMPILER FREE
      PARAMETER ITTO=10,ITTI=11
      DIMENSION A(128),B(128),AMAG(1024),PHASE(1024),IANS(1),ILABEL(5)
      COMPLEX OMEGA,CTEMPN,CTEMPD,CVALUE
100   CONTINUE
      PI=3.14159
      PIT2=2*PI
      ANGLE=0.0
      YLAST=0.0
      WRITE(ITTO,1)
1     FORMAT('<33><14> CONTINUOUS TRANSFER FUNCTION RESPONSE ')
C
C     READ IN NUMERATOR COEFFICIENTS
C
      WRITE(ITTO,2)
2     FORMAT(' NUMBER OF NUMERATOR COEFFICIENTS (1-128) ? ',Z)
      READ(ITTI) NN
      IF(NN.LT.1.OR.NN.GT.128) GO TO 100
C
          DO 1000 I=1,NN
          I1=NN-I
          WRITE(ITTO,3) I,I1
3         FORMAT(' ',10X,'A(',I3,')*S**',I3,' = ',Z)
          READ(ITTI) A(I)
1000      CONTINUE
C
C     READ IN DENOMINATOR COEFFICIENTS
C
110   CONTINUE
      WRITE(ITTO,4)
4     FORMAT(' NUMBER OF DENOMINATOR COEFFICIENTS (1-128) ? ',Z)
      READ(ITTI) ND
      IF(ND.LT.1.OR.ND.GT.128) GO TO 110
```

```
C
                DO 1010 I=1,ND
                I1=ND-I
                WRITE(ITTO,5) I,I1
5               FORMAT(' ',10X,'B(',I3,')*S**',I3,' = ',Z)
                READ(ITTI) B(I)
1010            CONTINUE
C
C       READ IN UPPER FREQUENCY LIMIT
C
        WRITE(ITTO,6)
6       FORMAT('HIGHEST FREQUENCY NEEDED (HERTZ) ? ',Z)
        READ(ITTI) HF
        HFF=HF
        HF=2*HF*PI
C
C       READ IN NUMBER OF POINTS
C
        WRITE(ITTO,7)
7       FORMAT(' NUMBER OF POINTS (MAXIMUM OF 1023) ? ',Z)
        READ(ITTI) NP
        NP=NP-1
C
C       CALCULATE STEP SIZE
C
        DELTA=HF/FLOAT(NP)
C
C       DETERMINE IF POWER OF AMPLITUDE RESPONSE IS WANTED
C
        TYPE'<12>POWER RATHER THAN AMPLITUDE (Y/N) ? '
        READ(11,8) IQ
8       FORMAT(S1)
C
C       DETERMINE POWER AND PHASE RESPONSE
C
        START=DELTA
        NP=NP+1
C
C       MAKE SURE FUNCTION CAN BE EVALUATED AT ZERO
C
        IF(B(ND).EQ.0.0) B(ND)=1.0E-20
        AMAG(1)=A(NN)*A(NN)/(B(ND)*B(ND))

        PHASE(1)=0.0
C
                DO 1020 I=2,NP
                OMEGA=CMPLX(0.0,START)
C
C               EVALUATE THE NUMERATOR
C
                CTEMPN=(0.0,0.0)
C
                        DO 1030 K=1,NN
                        CTEMPN=CTEMPN+A(K)*OMEGA**(NN-K+1)
1030                    CONTINUE
C
C               EVALUATE THE DENOMINATOR
```

```
C
              CTEMPD=(0.0,0.0)
                  DO 1040 K=1,ND
                  CTEMPD=CTEMPD+B(K)*OMEGA**(ND-K+1)
1040              CONTINUE
C
              CVALUE=CTEMPN/CTEMPD
              AMAG(I)=REAL(CVALUE*(CONJG(CVALUE)))
C
C     CALCULATE PHASE RESPONSE
C
              Y=AIMAG(CVALUE)
              X=REAL(CVALUE)
              IF(X.LT.0.0.AND.Y.GT.0.0.AND.YLAST.LT.0.0) ANGLE=ANGLE-PIT
              IF(X.LT.0.0.AND.Y.LT.0.0.AND.YLAST.GT.0.0) ANGLE=ANGLE+PIT
              PHASE(I)=ATAN2(Y,X)+ANGLE
              YLAST=Y
                  START=START+DELTA
1020              CONTINUE
C
C     CALCULATE AMPLITUDE RESPONSE IF NEEDED
C
              IF(IQ.EQ.'Y<40>') GO TO 130
                  DO 1050 I=1,NP
                  TEMP=AMAG(I)
                  AMAG(I)=SQRT(TEMP)
1050              CONTINUE
C
C     PRINT OUT VALUES
C
130           STEPF=HFF/(NP-1)
              FREQ=0.0
              WRITE(12,15)
              WRITE(12,16)
15            FORMAT(3X,'FREQ',9X,'MAGNITUDE',7X,'PHASE')
16            FORMAT('<12>')
17            FORMAT('<12>',F8.2,5X,F10.4,5X,F10.4)
              DO 2000 K=1,NP
              WRITE(12,17) FREQ,AMAG(K),PHASE(K)
              FREQ=FREQ+STEPF
2000          CONTINUE
              WRITE(12,18)
18            FORMAT('<33><14>')
              STOP***NORMAL TERMINATION***
              END
```

REFERENCES

1. A. V. OPPENHEIM and R. W. SCHAFER, *Digital Signal Processing*, Prentice-Hall, Englewood Cliffs, N.J., 1975.

2. A. PAPOULIS, *The Fourier Integral and Its Applications*, McGraw-Hill, New York, 1962.

3.   J. F. KAISER, "Digital Filters" in *System Analysis by Digital Computer*, F. F. Kuo and J. F. Kaiser, eds., Wiley, New York, 1966.

4.   A. H. KOSCHMANN, unpublished class notes, University of New Mexico, Albuquerque, N.M., 1962.

5.   W. D. STANLEY, *Digital Signal Processing*, Reston, Reston, Va., 1975.

6.   A. PAPOULIS, *Signal Analysis*, McGraw-Hill, New York, 1977.

7.   S. D. STEARNS, *Digital Signal Analysis*, Hayden, Rochelle Park, N.J., 1975.

8.   L. R. RABINER and B. GOLD, *Theory and Application of Digital Signal Processing*, Prentice-Hall, Englewood Cliffs, N.J., 1975.

9.   N. AHMED and K. R. RAO, *Orthogonal Transforms for Digital Signal Processing*, Springer-Verlag, New York, 1975.

10.   R. A. GABEL and R. A. ROBERTS, *Signals and Linear Systems*, 2nd ed., Wiley, New York, 1980.

2

Elements of Difference Equations

In this chapter we present an elementary discussion of linear difference equations with constant coefficients. Our motivation for doing so is that such difference equations will be used in Chapter 5 to describe and analyze discrete-time (DT) systems. Two methods for solving this class of difference equations will be included in this chapter, while a third method will be discussed in Chapter 3.

2.1 INTRODUCTORY REMARKS

The notion of linear difference equations with constant coefficients is best introduced by means of the simple resistive network that is shown in Fig. 2.1-1, where $V(n)$ denotes the voltage at the nth node, for $-2 \leq n \leq 3$. We wish to describe this network by means of a difference equation. To this end, we consider a typical section (below Fig. 2.1-1) of this network, where I_1, I_2, and I_3 denote currents leaving the node $n - 1$. Application of Kirchhoff's current law to node $n - 1$ leads to the equation

$$I_1 + I_2 + I_3 = 0$$

Substituting for I_1, I_2, and I_3 in the preceding equation, we obtain

$$\frac{V(n - 1) - V(n)}{1} + \frac{V(n - 1) - V(n - 2)}{1} + \frac{V(n - 1) - 0}{1} = 0$$

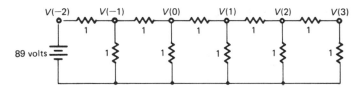

Fig. 2.1-1 Resistance network; each element value is 1 ohm.

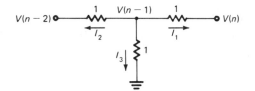

which simplifies to yield

$$V(n) - 3V(n - 1) + V(n - 2) = 0, \qquad 0 \le n \le 3 \qquad (2.1\text{-}1)$$

Equation (2.1-1) is the desired difference equation that describes the network in Fig. 2.1-1 in terms of its node voltages. We observe that it is a *second-order* difference equation since the voltage at node n [i.e., $V(n)$] is expressed as a linear combination of the voltages at *two* previous node voltages $V(n - 1)$ and $V(n - 2)$.

2.2 SOLUTION OF DIFFERENCE EQUATIONS

A logical question that arises at this point is how one can solve (2.1-1) to obtain $V(n)$. Since (2.1-1) represents a second-order difference equation, we would require *two* known voltages, say $V(-2)$ and $V(-1)$, to obtain the rest. To illustrate,

$$V(-2) = 89 \text{ volts}$$

and

$$V(-1) = 34 \text{ volts}$$

Then a simple procedure for obtaining the remaining $V(n)$ for $0 \le n \le 3$ would be a *recursive* method, since (2.1-1) implies that

$$V(n) = 3V(n - 1) - V(n - 2), \qquad 0 \le n \le 3 \qquad (2.2\text{-}1)$$

With $n = 0, 1, 2,$ and 3, (2.2-1) yields the desired voltages to be as follows:

$$V(0) = 3V(-1) - V(-2) = 13 \text{ volts}$$

$$V(1) = 3V(0) - V(-1) = 5 \text{ volts}$$

$$V(2) = 3V(1) - V(0) = 2 \text{ volts}$$

and $$V(3) = 3V(2) - V(1) = 1 \text{ volt}$$

We shall refer to the preceding scheme as the *recursive method* for solving difference equations. It is observed that, although this method yields each $V(n)$ in a simple recursive manner, it does not provide a *closed-form* solution, that is, a solution which yields $V(n)$ without having to first compute $V(0), V(1), \ldots, V(n - 1)$. If a closed-form solution is desired, one can solve difference equations using the *method of undetermined coefficients*, which parallels the classical method of solving linear differential equations with constant coefficients.

Method of Undetermined Coefficients

We illustrate this method via examples. Suppose we seek the general solution of the second-order difference equation

$$y(n) - \frac{5}{6} y(n - 1) + \frac{1}{6} y(n - 2) = 5^{-n}, \qquad n \geq 0 \qquad (2.2\text{-}2)$$

with initial conditions $y(-2) = 25$ and $y(-1) = 6$.

In (2.2-2), $y(n)$ may be interpreted as the response (output) of a DT system to the input (forcing) function 5^{-n} for $n \geq 0$, where n is a time index. It is apparent that (2.2-2) is a second-order difference equation since it expresses the output $y(n)$ at time n as a linear combination of *two* previous outputs $y(n - 1)$ and $y(n - 2)$.

The general (or closed-form) solution $y(n)$ of (2.2-2) is obtained in three steps that are similar to those used for solving second-order differential equations. They are as follows:

1. Obtain the *complementary solution* $y_c(n)$ in terms of two arbitrary constants c_1 and c_2.

2. Obtain the *particular solution* $y_p(n)$, and write

$$y(n) = y_c(n) + y_p(n) = f(c_1, c_2) + y_p(n) \qquad (2.2\text{-}3)$$

where $y_c(n) = f(c_1, c_2)$ implies that $y_c(n)$ is a function of c_1 and c_2.

3. Solve for c_1 and c_2 in (2.2-3) using two given initial conditions.

In what follows, we elaborate on the preceding steps.

STEP 1. We assume that the complementary solution $y_c(n)$ has the form

$$y_c(n) = c_1 a_1^n + c_2 a_2^n \tag{2.2-4}$$

where the a_i are real constants.
Next substitute $y(n) = a^n$ in the homogeneous equation to get

$$a^n - \frac{5}{6} a^{n-1} + \frac{1}{6} a^{n-2} = 0 \tag{2.2-5}$$

Dividing both sides of (2.2-5) by a^{n-2}, we obtain

$$a^2 - \frac{5}{6} a + \frac{1}{6} = 0$$

or

$$\left(a - \frac{1}{2} \right)\left(a - \frac{1}{3} \right) = 0$$

which yields the *characteristic roots*

$$a_1 = \frac{1}{2} \quad \text{and} \quad a_2 = \frac{1}{3}$$

Thus the complementary solution is

$$y_c(n) = c_1 2^{-n} + c_2 3^{-n}$$

where c_1 and c_2 are arbitrary constants.

STEP 2. The particular solution $y_p(n)$ is assumed to be

$$y_p(n) = c_3 5^{-n}$$

since the forcing function is 5^{-n}; see (2.2-2).
Substitution of $y(n) = y_p(n) = c_3 5^{-n}$ in (2.2-2) leads to

$$c_3[5^{-n} - \left(\frac{5}{6} \right) 5^{-(n-1)} + \left(\frac{1}{6} \right) 5^{-(n-2)}] = 5^{-n}$$

Dividing both sides of this equation by 5^{-n}, we obtain

$$c_3[1 - \left(\frac{5}{6}\right)5 + \left(\frac{1}{6}\right)5^2] = 1$$

which implies that $c_3 = 1$. Thus

$$y(n) = y_c(n) + y_p(n) \tag{2.2-6}$$

$$= c_1 2^{-n} + c_2 3^{-n} + 5^{-n}$$

STEP 3. Since the initial conditions are

$$y(-2) = 25 \quad \text{and} \quad y(-1) = 6$$

(2.2-6) yields the simultaneous equations

$$4c_1 + 9c_2 = 0 \tag{2.2-7}$$

and

$$2c_1 + 3c_2 = 1$$

Solving (2.2-7) for c_1 and c_2, we obtain

$$c_1 = \frac{3}{2} \quad \text{and} \quad c_2 = -\frac{2}{3}$$

Thus the desired general solution is given by (2.2-6) to be

$$y(n) = \frac{3}{2}(2^{-n}) - \frac{2}{3}(3^{-n}) + 5^{-n}, \quad n \geq 0 \tag{2.2-8}$$

As mentioned earlier, $y(n)$ can be interpreted as the output of a DT system when it is subjected to the exponential input (forcing function) 5^{-n}, which is the right-hand side of the given difference equation in (2.2-2).

RULES FOR CHOOSING PARTICULAR SOLUTIONS. As is the case with the solution of differential equations, there are a set of rules one must follow to form appropriate particular solutions while solving difference equations, as summarized in Table 2.2-1. For example, the form of the particular solution related to the difference equation in (2.2-2) was $c_3 5^{-n}$, which agrees with line 3 of Table 2.2-1. We will illustrate the use of this table by means of more examples.

Table 2.2-1   Rules for Choosing Particular Solutions

Terms in forcing function	Choice of particular solution[†]
1. A constant	c; c is a constant
2. $b_1 n^k$; b_1 is a constant	$c_0 + c_1 n + c_2 n^2 + \cdots + c_k n^k$; the c_i are constants
3. $b_2 d^{\pm n}$; b_2 and d are constants	Proportional to $d^{\pm n}$
4. $b_3 \cos (n\omega)$ b_3 and b_4 are	$c_1 \sin (n\omega) + c_2 \cos (n\omega)$
5. $b_4 \sin (n\omega)$ constants	

[†]If a term in any of the particular solutions in this column is a part of the complementary solution, it is necessary to modify the corresponding choice by multiplying it by n before using it. If such a term appears r times in the complementary solution, the corresponding choice must be multiplied by n^r.

Example 2.2-1: Solve the second-order difference equation

$$y(n) - \frac{3}{2} y(n-1) + \frac{1}{2} y(n-2) = 1 + 3^{-n}, \qquad n \geq 0 \quad (2.2\text{-}9)$$

with the initial conditions $y(-2) = 0$ and $y(-1) = 2$.

Solution: The solution consists of three steps.

STEP 1. Assume the complementary solution as $y_c(n) = c_1 a_1^n + c_2 a_2^n$. Substituting $y(n) = a^n$ in the homogeneous counterpart of (2.2-9), we obtain the characteristic equation

$$a^2 - \frac{3}{2} a + \frac{1}{2} = 0$$

the roots of which are $a_1 = \frac{1}{2}$ and $a_2 = 1$.
Thus

$$y_c(n) = c_1 2^{-n} + c_2 1^n = c_1 2^{-n} + c_2 \quad (2.2\text{-}10)$$

STEP 2. To choose an appropriate particular solution, we refer to Table 2.2-1. From the given forcing function and lines 1 and 3 of Table 2.2-1, it follows that a choice for the particular solution is $c_3 + c_4 3^{-n}$. However, we observe that this choice for the particular solution and $y_c(n)$ in (2.2-10) have common terms, each of which is a constant; that is, c_3 and c_2, respectively. Thus in accordance with the footnote of Table 2.2-1, we modify the choice $c_3 + c_4 3^{-n}$ to obtain

$$y_p(n) = c_3 n + c_4 3^{-n} \qquad (2.2\text{-}11)$$

Next, substitution of $y_p(n)$ in (2.2-11) into (2.2-9) leads to

$$c_3 n + c_4 3^{-n} - \frac{3}{2} c_3 n + \frac{3}{2} c_3 - \frac{9}{2} c_4 3^{-n}$$
$$+ \frac{1}{2} c_3 n - c_3 + \frac{9}{2} c_4 3^{-n} = 3^{-n} + 1 \qquad (2.2\text{-}12)$$

From (2.2-12) it follows that

$$\frac{1}{2} c_3 = 1$$

and
$$c_4 \left[1 - \frac{9}{2} + \frac{9}{2} \right] 3^{-n} = 3^{-n}$$

which results in

$$c_3 = 2 \quad \text{and} \quad c_4 = 1$$

Thus, combining (2.2-10) and (2.2-11), we get

$$y(n) = c_1 2^{-n} + c_2 + 2n + 3^{-n} \qquad (2.2\text{-}13)$$

STEP 3. To evaluate c_1 and c_2 in (2.2-13), the given initial conditions are used; that is, $y(-2) = 0$ and $y(-1) = 2$. This leads to the simultaneous equations

$$4c_1 + c_2 = -5$$
$$2c_1 + c_2 = 1$$

Solving, we obtain $c_1 = -3$ and $c_2 = 7$, which yields the desired solution as

$$y(n) = (-3)2^{-n} + 7 + 2n + 3^{-n}, \qquad n \geq 0$$

Example 2.2-2: Find the general solution of the first-order difference equation

$$y(n) - 0.9y(n - 1) = 0.5 + (0.9)^{n-1}, \quad n \geq 0 \quad (2.2\text{-}14)$$

with $y(-1) = 5$.

Solution:

STEP 1. Substituting $y(n) = a^n$ in the homogeneous equation

$$y(n) - 0.9y(n - 1) = 0$$

we obtain $\qquad\qquad y_c(n) = c_1(0.9)^n \qquad\qquad\qquad (2.2\text{-}15)$

since we are dealing with a first-order difference equation.

STEP 2. From the forcing function in (2.2-14), the complementary solution in (2.2-15), and lines 1 and 3 of Table 2.2-1, it follows that

$$y_p(n) = c_2 n(0.9)^n + c_3$$

Substitution of $y(n) = y_p(n)$ in (2.2-14) results in

$$c_3 + c_2 n(0.9)^n - 0.9c_2(n - 1)(0.9)^{n-1} - 0.9c_3 = 0.5 + (0.9)^{n-1}$$

which leads to

$$0.1c_3 = 0.5$$

and $\qquad\qquad\qquad (0.9)^n c_2 = (0.9)^{n-1}$

Thus we have

$$c_3 = 5 \quad \text{and} \quad c_2 = \frac{10}{9}$$

which implies that

$$y_p(n) = \frac{10}{9} n(0.9)^n + 5 \qquad\qquad (2.2\text{-}16)$$

Combining (2.2-15) and (2.2-16), we get

$$y(n) = c_1(0.9)^n + \frac{10}{9} n(0.9)^n + 5 \qquad\qquad (2.2\text{-}17)$$

STEP 3. From (2.2-17) and the initial condition $y(-1) = 5$, it follows that $c_1 = \dfrac{10}{9}$. Hence the desired solution can be written as

$$y(n) = (n + 1)(0.9)^{n-1} + 5, \qquad n \geq 0$$

Example 2.2-3: Find the general solution of the second-order difference equation

$$y(n) - 1.8y(n - 1) + 0.81y(n - 2) = 2^{-n}, \qquad n \geq 0 \quad (2.2\text{-}18)$$

Leave the answer in terms of unknown constants, which one can evaluate if the initial conditions are given.

Solution:
STEP 1. With $y(n) = a^n$ substituted into the homogeneous counterpart of (2.2-18), we obtain

$$a^2 - 1.8a + 0.81 = 0$$

which results in the *repeated roots*

$$a_1 = a_2 = 0.9$$

Thus, as in the case of differential equations, we consider the complementary solution to be

$$y_c(n) = c_1(0.9)^n + c_2 n(0.9)^n \tag{2.2-19}$$

STEP 2. From the given forcing function in (2.2-18), $y_c(n)$ in (2.2-19), and line 3 of Table 2.2-1, it is clear that

$$y_p(n) = c_3 2^{-n} \tag{2.2-20}$$

Substitution of (2.2-20) in (2.2-18) leads to

$$c_3[1 - (3)(6) + (3)(24)]2^{-n} = 2^{-n}$$

which yields $c_3 = \dfrac{25}{16}$. Thus the desired solution is given by (2.2-19) and (2.2-20) to be

$$y(n) = c_1(0.9)^n + c_2 n(0.9)^n + \left(\frac{25}{16}\right)2^{-n}$$

where c_1 and c_2 can be evaluated if two initial conditions are specified.

Example 2.2-4: Find the particular solution for the first-order difference equation

$$y(n) - 0.5y(n - 1) = \sin\left(\frac{n\pi}{2}\right), \qquad n \geq 0 \qquad (2.2\text{-}21)$$

Solution: Since the forcing function is sinusoidal, we refer to line 5 of Table 2.2-1 and choose a particular solution of the form

$$y_p(n) = c_1 \sin\left(\frac{n\pi}{2}\right) + c_2 \cos\left(\frac{n\pi}{2}\right) \qquad (2.2\text{-}22)$$

Substitution of $y(n) = y_p(n)$ in (2.2-21) leads to

$$c_1 \sin\left(\frac{n\pi}{2}\right) + c_2 \cos\left(\frac{n\pi}{2}\right) - 0.5c_1 \sin\left[\frac{(n-1)\pi}{2}\right]$$
$$- 0.5c_2 \cos\left[\frac{(n-1)\pi}{2}\right] = \sin\left(\frac{n\pi}{2}\right) \qquad (2.2\text{-}23)$$

We now use the following identities:

$$\sin\left[\frac{(n-1)\pi}{2}\right] = \sin\left(\frac{n\pi}{2} - \frac{\pi}{2}\right) = -\cos\left(\frac{n\pi}{2}\right)$$
$$\cos\left[\frac{(n-1)\pi}{2}\right] = \cos\left(\frac{n\pi}{2} - \frac{\pi}{2}\right) = \sin\left(\frac{n\pi}{2}\right) \qquad (2.2\text{-}24)$$

Substituting (2.2-24) in (2.2-23), we obtain

$$(c_1 - 0.5c_2) \sin\left(\frac{n\pi}{2}\right) + (0.5c_1 + c_2) \cos\left(\frac{n\pi}{2}\right) = \sin\left(\frac{n\pi}{2}\right)$$

which yields the simultaneous equations

$$c_1 - 0.5c_2 = 1 \qquad (2.2\text{-}25)$$
$$0.5c_1 + c_2 = 0$$

The solution of (2.2-25) yields $c_1 = \frac{4}{5}$ and $c_2 = -\frac{2}{5}$. Hence the desired result is given by (2.2-22) to be

$$y_p(n) = \frac{4}{5} \sin\left(\frac{n\pi}{2}\right) - \frac{2}{5} \cos\left(\frac{n\pi}{2}\right), \quad n \geq 0$$

Example 2.2-5: In Example 2.2-3, suppose the forcing function is $(0.9)^n$ instead of 2^{-n}, $n \geq 0$. What would be the appropriate choice for the particular solution?

Solution: The complementary solution is given by (2.2-19) to be

$$y_c(n) = c_1(0.9)^n + c_2 n(0.9)^n$$

Since the forcing function is $(0.9)^n$, line 3 of Table 2.2-1 implies that a choice for the particular solution is $c_3(0.9)^n$. However, since this choice and the preceding $y_c(n)$ have a term in common, we must modify our choice according to the footnote of Table 2.2-1 to obtain $c_3 n(0.9)^n$. But this choice again has a term in common with $y_c(n)$. Thus we apply to the footnote of Table 2.2-1 once again to obtain

$$y_p(n) = c_3 n^2(0.9)^n \tag{2.2-26}$$

which has no more terms in common with $y_c(n)$.

Hence $y_p(n)$ in (2.2-26) is the appropriate choice for the particular solution for the difference equation in (2.2-18) when the forcing function is $(0.9)^n$; that is,

$$y(n) - 1.8y(n - 1) + 0.81y(n - 2) = (0.9)^n, \quad n \geq 0$$

2.3 SUMMARY

Our treatment of linear difference equations with constant coefficients in this chapter was confined to first- and second-order difference equations. Higher-order difference equations of this type will be considered in Chapters 3 and 5. Although our interest in such difference equations is restricted to DT systems as they relate to electrical engineering, they have a variety of applications in diverse areas such as economics, psychology, and sociology. The interested reader may refer to [3] for more details.

PROBLEMS

2–1 Solve the following second-order linear difference equations with constant coefficients.

(a) $2f(n - 2) - 3f(n - 1) + f(n) = 3$, $n \geq 0$, with $f(-2) = -3$ and $f(-1) = -2$.

(b) $f(n - 2) - 2f(n - 1) + f(n) = 1$, $n \geq 0$, with $f(-1) = -0.5$ and $f(-2) = 0$.
Hint: $f_c(n) = c_1 + c_2 n$, and $f_p(n) = c_3 n^2$ (why?).

(c) $y(n) - 0.8y(n - 1) = (0.8)^n$, $n \geq 0$, with $y(0) = 6$.

(d) $y(n) - y(n - 1) = 1 + (0.5)^n$, $n \geq 0$, with $y(0) = 1$.

(e) $y(n - 2) - 6y(n - 2) + 5y(n) = 1$, $n \geq 0$, with $y(-2) = 2$ and $y(-1) = 1$.

(f) $y(n) + y(n - 1) + y(n - 2) = n^2 + n + 1$, $n \geq 0$. The answer may be left in terms of two arbitrary constants.
Hint: $y_p(n) = c_3 n^2 + c_4 n + c_5$.

2–2 Given the following *nonlinear* difference equation:

$$y(n) = \frac{1}{2}\left[y(n - 1) + \frac{u(n)}{y(n - 1)}\right], \qquad n \geq 0$$

where $u(n) = \begin{cases} 0, & n < 0 \\ 2, & n \geq 0 \end{cases}$

(a) If $y(-1) = 1$, find $y(n)$, $0 \leq n \leq 4$.

(b) By inspecting the answers obtained in part (a), could you guess the value of $y(n)$ for large n?

2–3 Find the particular solution of the second-order difference equation

$$8y(n) - 6y(n - 1) + y(n - 2) = 5 \sin\left(\frac{n\pi}{2}\right), \qquad n \geq 0$$

2–4 Solve the second-order difference equation

$$y(n) + ky(n - 2) = 0, \qquad n \geq 0$$

for the two cases when $k = 1$ and $k = \frac{1}{4}$, with the initial conditions $y(0) = 1$ and $y(-1) = 0$.

REFERENCES

1. F. B. HILDEBRAND, *Methods of Applied Mathematics*, Prentice-Hall, Englewood Cliffs, N.J., 1963, Chap. 3.
2. J. A. CADZOW, *Discrete-time Systems*, Prentice-Hall, Englewood Cliffs, N.J., 1973, Chaps. 1 and 2.
3. S. GOLDBERG, *Difference Equations*, Wiley, New York, 1966 (third printing).

3

The Z-Transform

The purpose of this chapter is to develop the Z-transform (ZT) and discuss some of its important properties. The ZT is then used to solve linear difference equations with constant coefficients, which were introduced in Chapter 2. This material will be used in Chapter 5 in connection with the representation and frequency analysis of discrete-time (DT) systems. The ZT is a fundamental contribution to the area of DT systems; it is attributed to Zadeh and Ragazzini [1].

3.1 DEFINITIONS

Let $f(t)$ be a continuous-time (CT) signal that is passed through an ideal sampler as illustrated in Fig. 3.1-1, where the sampling rate is indicated as f_s samples per second (sps), and T is the sampling interval in seconds; that is, $f_s = 1/T$. We assume that $f(t) = 0$ for $t < 0$.

The output of the ideal sampler is the sampled signal $f^*(t)$, which was defined in (1.1-2). It can be regarded as the product of $f(t)$ and a train of unit impulse functions $\delta_T(t)$, as follows:

$$f^*(t) = f(t)\delta_T(t) \tag{3.1-1}$$

$$= f(t)\delta(t) + f(t)\delta(t - T) + f(t)\delta(t - 2T) + \cdots$$

From (1.7-1) and (3.1-1), there results

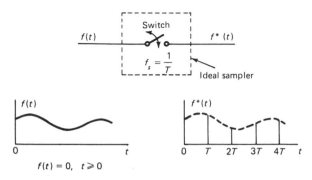

Fig. 3.1-1 On sampling a continuous-time signal to obtain the corresponding sampled signal.

$$F^*(s) = \int_0^\infty [f(t)\delta(t) + f(t)\delta(t - T) + f(t)\delta(t - 2T) + \cdots] e^{-st} dt$$

where $F^*(s)$ is the Laplace transform (LT) of the sampled signal $f^*(t)$.

Term-by-term integration of this expression for $F^*(s)$ and application of the impulse function property in (1.3-9b) lead to

$$F^*(s) = f(0) + f(T)e^{-Ts} + f(2T)e^{-2Ts} + \cdots$$

$$= \sum_{n=0}^\infty f(nT) e^{-nTs} \qquad (3.1\text{-}2)$$

We now define a complex variable z such that

$$z = e^{sT} \quad \text{or} \quad s = \frac{1}{T} \ln(z) \qquad (3.1\text{-}3)$$

Substitution of (3.1-3) in (3.1-2) results in

$$F(z) = \sum_{n=0}^\infty f(nT)z^{-n} \qquad (3.1\text{-}4)$$

where $F(z)$ is called the ZT of the sequence or DT signal, $f(nT)$.

The ZT $F(z)$ in (3.1-4) is said to converge outside the circle $|z| > R$, where R is called the radius of absolute convergence; that is, the *region of convergence* of $F(z)$ is given by $|z| > R$.

For convenience, we shall denote the sequence $f(nT)$ by $f(n)$, where the sampling interval T is implied. Hence the ZT of the sequence $f(n)$ will be denoted as

$$F(z) = Z\{f(n)\} = \sum_{n=0}^{\infty} f(n)z^{-n} \qquad (3.1\text{-}5)$$

where Z represents the ZT.

A fundamental property is that there is a *one-to-one correspondence* between a sequence $f(n)$ and its ZT. In other words, the ZT is a *unique* transformation. Hence, given $F(z)$, we can recover $f(n)$ via the inverse ZT (IZT), which we denote as

$$f(n) = Z^{-1}\{F(z)\} \qquad (3.1\text{-}6)$$

where Z^{-1} represents the IZT. Methods for finding the IZT will be considered in Section 3.5.

Fourier Transform Considerations

Equation (1.3-2) yields the Fourier transform (FT) of $f^*(t)$ in (3.1-1) to be

$$F^*(\omega) = \int_{-\infty}^{\infty} [f(t)\delta(t) + f(t)\delta(t - T) + f(t)\delta(t - 2T) + \cdots] e^{-j\omega t}dt$$

which simplifies via (1.3-9b) to yield

$$F^*(\omega) = f(0) + f(T)e^{-j\omega T} + f(2T)e^{-j2\omega T} + \cdots$$

Thus

$$F^*(\omega) = \sum_{n=0}^{\infty} f(n)e^{-jn\omega T} \qquad (3.1\text{-}7)$$

is the FT of the sampled signal $f^*(t)$.

Now, substituting $z = e^{j\omega T}$ in (3.1-5), we obtain

$$\bar{F}(\omega) = F(e^{j\omega T}) = \sum_{n=0}^{\infty} f(n)e^{-jn\omega T} \qquad (3.1\text{-}8)$$

where $\bar{F}(\omega)$ is defined as the FT of the sequence $f(n)$.

Combining (3.1-7) and (3.1-8), we get

$$\bar{F}(\omega) = F^*(\omega) = F(z)\big|_{z=e^{j\omega T}} \tag{3.1-9}$$

Equation (3.1-9) leads to the conclusion that the FTs of a sequence and the corresponding sampled signal are equivalent. Furthermore, they can be obtained from the ZT by replacing z by $e^{j\omega T}$, that is, by evaluating the ZT on the unit circle $|z| = 1$.

Next the functions $e^{-jn\omega T}$ are periodic with periods ω_s, where $\omega_s = 2\pi f_s$ and f_s is the sampling frequency. This can be readily shown by noting that

$$e^{jn(\omega \pm \omega_s)T} = e^{jn\omega T} \cdot e^{\pm jn\omega_s T}$$

$$= e^{jn\omega T}$$

since $\omega_s T = 2\pi$ and $e^{\pm jn2\pi} = 1$, for all n. Hence $F^*(\omega)$ and $\bar{F}(\omega)$ are also periodic functions, with period ω_s.

Also, the functions $e^{jn\omega T}$ are orthogonal over the period ω_s, since they satisfy the condition

$$\int_{-\omega_s/2}^{\omega_s/2} e^{-jn\omega T} e^{jm\omega T} \, d\omega = \begin{cases} 0, & m \neq n \\ \omega_s, & m = n \end{cases} \tag{3.1-10}$$

Using (3.1-8) and (3.1-10), it is straightforward to show that $f(n)$ can be recovered from $\bar{F}(\omega)$ as follows:

$$f(n) = \frac{1}{\omega_s} \int_{-\omega_s/2}^{\omega_s/2} \bar{F}(\omega)e^{jn\omega T} d\omega, \qquad n \geq 0 \tag{3.1-11}$$

Equation (3.1-11) is defined as the inverse Fourier transform (IFT) of $\bar{F}(\omega)$.

Thus, in summary, $\bar{F}(\omega)$ and $f(n)$ form the following FT pair:

$$\bar{F}(\omega) = \sum_{n=0}^{\infty} f(n)e^{-jn\omega T} \tag{3.1-12a}$$

and

$$f(n) = \frac{1}{\omega_s} \int_{-\omega_s/2}^{\omega_s/2} \bar{F}(\omega)e^{jn\omega T} d\omega \tag{3.1-12b}$$

3.2 Z-TRANSFORM OF A SET OF SEQUENCES

It is instructive to present some illustrative examples related to finding the ZT of simple sequences. These examples will be followed by a table of Z-transforms (ZTs) of some commonly used sequences.

Example 3.2-1: Given the unit sequence

$$f(n) = 1, \quad n \geq 0 \tag{3.2-1}$$

find $F(z)$.

Solution: Substitution of (3.2-1) in (3.1-5) leads to

$$F(z) = \sum_{n=0}^{\infty} z^{-n} = \frac{z}{z-1} \tag{3.2-2}$$

since $\sum_{n=0}^{\infty} z^{-n}$ is a complex-valued geometric series with common ratio z^{-1}. This series converges if $|z^{-1}| < 1$ or equivalently if $|z| > 1$. Thus the region of convergence of $F(z)$ in (3.2-2) is given by $|z| > 1$, as depicted in Fig. 3.2-1.

Example 3.2-2: Find the ZT of the sequence

$$f(n) = e^{-anT}, \quad n \geq 0 \tag{3.2-3}$$

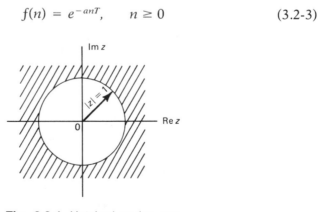

Fig. 3.2-1 Hatched region represents the region of convergence of $F(z)$ in (3.2-2).

for two cases: (1) a is real; (2) a is imaginary.

Solution:

CASE 1. From (3.2-3) and (3.1-5) we obtain

$$F(z) = \sum_{n=0}^{\infty} e^{-anT}z^{-n} = \sum_{n=0}^{\infty} (e^{-aT}z^{-1})^n \qquad (3.2\text{-}4)$$

If $u = e^{-aT}z^{-1}$, then (3.2-4) becomes

$$F(z) = \sum_{n=0}^{\infty} u^n = \frac{1}{1-u} \qquad (3.2\text{-}5)$$

since the infinite series in (3.2-5) is a geometric series with common ratio u, which converges if $|u| < 1$. Since $|u| = e^{-aT}|z^{-1}|$, the equivalent condition for convergence is that $|z| > e^{-aT}$. Thus

$$F(z) = \frac{z}{z - e^{-aT}}, \qquad |z| > e^{-aT} \qquad (3.2\text{-}6a)$$

CASE 2. Here the infinite series in (3.2-5) converges if $|z| > 1$ since the $u = e^{-aT}z^{-1}$ yields $|u| = |z^{-1}|$ when a is imaginary. Thus

$$F(z) = \frac{z}{z - e^{-aT}}, \qquad |z| > 1 \qquad (3.2\text{-}6b)$$

It is worthwhile noting that, while the ZTs in (3.2-6a) and (3.2-6b) are the same, the corresponding regions of convergence are different. Similarly, if a is a complex number, then it can be shown that (Problem 3-10)

$$F(z) = \frac{z}{z - e^{-aT}}, \qquad |z| > |e^{-aT}| \qquad (3.2\text{-}6c)$$

Example 3.2-3: Given the sinusoidal sequence

$$f(n) = \sin(n\omega T), \qquad n \geq 0 \qquad (3.2\text{-}7)$$

Find $F(z)$.

Solution: Substituting (3.2-7) in (3.1-5), we get

$$F(z) = \sum_{n=0}^{\infty} \sin (n\omega T)z^{-n} \tag{3.2-8}$$

Now since

$$\sin (n\omega T) = \frac{1}{2j} (e^{jn\omega T} - e^{-jn\omega T})$$

(3.2-8) can be written as

$$F(z) = \frac{1}{2j} \left(\sum_{n=0}^{\infty} e^{jn\omega T}z^{-n} - \sum_{n=0}^{\infty} e^{-jn\omega T}z^{-n} \right)$$

That is,

$$F(z) = \frac{1}{2j} (Z\{e^{jn\omega T}\} - Z\{e^{-jn\omega T}\}) \tag{3.2-9}$$

Next, from (3.2-6b), it follows that

$$Z\{e^{\pm jn\omega T}\} = \frac{z}{z - e^{\pm j\omega T}}, \qquad |z| > 1 \tag{3.2-10}$$

Substitution of (3.2-10) in (3.2-9) and subsequent simplification leads to

$$F(z) = \frac{z \sin (\omega T)}{z^2 - 2z \cos \omega T + 1}, \qquad |z| > 1$$

Similarly, one can find the ZTs of various other sequences. Some of the more commonly used sequences and their transforms are summarized in Table 3.2-1 along with the corresponding regions of convergence.

3.3 MAPPING PROPERTIES OF $z = e^{sT}$ TRANSFORMATION

The $z = e^{sT}$ transformation, which was introduced in (3.1-3), represents a mapping of regions in the s-plane to the z-plane, since both s and z

Table 3.2-1   Z-Transforms of Some Commonly Used Sequences

$f(n), n \geq 0$	$F(z)$	Region of Convergence
1. $f(0) = k = \text{const.}^{\dagger}$ $f(n) = 0, n = 1, 2, \cdots$	k	$\|z\| > 0$
2. $f(m) = k = \text{const.}$ $f(n) = 0, n \neq m$	kz^{-m}	$\|z\| > 0$
3. $k = \text{constant}^{\ddagger}$	$\dfrac{kz}{z - 1}$	$\|z\| > 1$
4. kn	$\dfrac{kz}{(z - 1)^2}$	$\|z\| > 1$
5. kn^2	$\dfrac{kz(z + 1)}{(z - 1)^3}$	$\|z\| > 1$
6. ke^{-anT}; a is complex	$\dfrac{kz}{z - e^{-aT}}$	$\|z\| > \|e^{-aT}\|$
7. kne^{-anT}; a is complex	$\dfrac{kze^{-aT}}{(z - e^{-aT})^2}$	$\|z\| > \|e^{-aT}\|$
8. $\sin(\omega_0 nT)$	$\dfrac{z \sin \omega_0 T}{z^2 - 2z \cos \omega_0 T + 1}$	$\|z\| > 1$
9. $\cos(\omega_0 nT)$	$\dfrac{z(z - \cos \omega_0 T)}{z^2 - 2z \cos \omega_0 T + 1}$	$\|z\| > 1$
10. $e^{-anT} \sin(\omega_0 nT)$	$\dfrac{ze^{-aT} \sin \omega_0 T}{z^2 - 2e^{-aT}z \cos \omega_0 T + e^{-2aT}}$	$\|z\| > e^{-aT}$
11. $e^{-anT} \cos(\omega_0 nT)$	$\dfrac{ze^{-aT}(ze^{aT} - \cos \omega_0 T)}{z^2 - 2e^{-aT}z \cos \omega_0 T + e^{-2aT}}$	$\|z\| > e^{-aT}$
12. α^n, where α is a constant	$\dfrac{z}{z - \alpha}$	$\|z\| > \alpha$
13. $n\alpha^n$	$\dfrac{\alpha z}{(z - \alpha)^2}$	$\|z\| > \alpha$
14. $\sinh(\omega_0 nT)$	$\dfrac{z \sinh \omega_0 T}{z^2 - 2z \cosh \omega_0 T + 1}$	$\|z\| > \cosh \omega_0 T$
15. $\cosh(\omega_0 nT)$	$\dfrac{z(z - \cosh \omega_0 T)}{z^2 - 2z \cosh \omega_0 T + 1}$	$\|z\| > \sinh \omega_0 T$

†Sometimes called the *impulse sequence*, and the *the unit impulse sequence* when $k = 1$.
‡Sometimes called the *step sequence*, and the *unit step sequence* when $k = 1$.

are complex variables. With $s = \sigma + j\omega$, the transformation $z = e^{sT}$ yields

$$z = e^{\sigma T} \cdot e^{j\omega T} \tag{3.3-1}$$

Representing z as a complex number, we obtain

$$z = |z| \angle \theta \tag{3.3-2}$$

where $|z| = e^{\sigma T}$

$\theta = \omega T = 2\pi(\omega/\omega_s)$, since $\omega_s = 2\pi/T$

The behavior of z in (3.3-1) is now examined for three cases.

CASE 1. $\sigma = 0$. Equation (3.3-1) simplifies to yield

$$z = e^{j2\pi(\omega/\omega_s)}$$

from which we obtain the following table:

ω	$|z|$	θ
0	1	0
$\dfrac{\omega_s}{4}$	1	$\dfrac{\pi}{2}$
$\dfrac{\omega_s}{2}$	1	π
$\dfrac{3\omega_s}{4}$	1	$\dfrac{3\pi}{2}$
ω_s	1	2π

From this table it is apparent that the portion of the $j\omega$-axis between $\omega = 0$ and ω_s in the s-plane is mapped on the circumference of the unit circle in the z-plane, as illustrated in Fig. 3.3-1. Again as ω increases from ω_s to $2\omega_s$, another counterclockwise encirclement of the unit circle results. Thus we conclude that, as ω varies from 0 to ∞, there is an infinite number of encirclements of the unit circle in the *counterclockwise* direction. Similarly, there is an infinite number of encirclements of the unit circle in the *clockwise* direction as ω varies from 0 to $-\infty$.

CASE 2. $\sigma < 0$. Equation (3.3-1) implies that $|z| < 1$. Hence from case 1 it follows that each strip of width ω_s in the left-half s-plane is

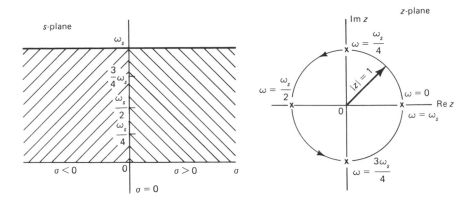

Fig. 3.3-1 Relation between the s-plane and z-plane.

mapped *into* the unit circle; for example, see hatched region (///) in Fig. 3.3-1. This mapping occurs in the form of concentric circles in the z-plane as σ is varied from 0^- to $-\infty$. We note that $\sigma = 0^-$ and $\sigma = -\infty$ correspond to circles of radii 1^- and 0, respectively, in the z-plane.

CASE 3. $\sigma > 0$. Here (3.3-1) implies that $|z| > 1$. Thus from case 1 it is apparent that each strip of width ω_s in the right-half s-plane is mapped *outside* the unit circle; for example, see hatched region (\\\) in Fig. 3.3-1. As in case 2, this mapping occurs in concentric circles in the z-plane as σ is varied from 0^+ to ∞. We observe that $\sigma = 0^+$ and $\sigma = \infty$ correspond to circles of radii 1^+ and ∞, respectively, in the z-plane.

From the preceding discussion we conclude that the $j\omega$-axis of the s-plane corresponds to the circumference of the unit circle, while the left- and right-half s-planes correspond to regions of the z-plane that lie inside and outside the unit circle, respectively. We note that this mapping is *not* one-to-one since there is more than one point on the s-plane that corresponds to a single point on the unit circle.

3.4 SOME PROPERTIES OF THE Z-TRANSFORM

Linearity

The ZT is a linear transformation; that is, if

$$f(n) = \sum_{i=1}^{N} f_i(n)$$

then
$$F(z) = \sum_{i=1}^{N} F_i(z) \qquad (3.4\text{-}1)$$

where $F(z) = Z\{f(n)\}$ and $F_i(z) = Z\{f_i(n)\}$

Multiplication by an Exponential Sequence

If $g(n) = e^{-anT} f(n)$, $n \geq 0$, then it can be shown that (Problem 3-2)

$$G(z) = F(ze^{aT}) \qquad (3.4\text{-}2)$$

where $G(z) = Z\{g(n)\}$.

Shift Theorem

If
$$g(n) = f(n - k)u(n - k), \qquad n \geq 0 \qquad (3.4\text{-}3)$$

where $u(n - k) = \begin{cases} 0, & n < k \\ 1, & n \geq k \end{cases}$

for a specified positive integer (constant) k, then the shift theorem states that

$$G(z) = z^{-k}F(z) \qquad (3.4\text{-}4)$$

where $G(z) = Z\{g(n)\}$ and $F(z) = Z\{f(n)\}$

To prove (3.4-4), we first observe that $g(n)$ represents a shifted version of $f(n)$. The amount of shift is k, as illustrated in Fig. 3.4-1. Thus the ZT of $g(n)$ is given by

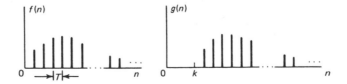

Fig. 3.4-1 Graphical implication of (3.4-3).

$$G(z) = \sum_{n=0}^{\infty} g(n)z^{-n} = \sum_{n=k}^{\infty} f(n - k)z^{-n}$$

$$= z^{-k} \sum_{n=k}^{\infty} f(n - k)z^{-(n-k)}$$

$$= z^{-k}[f(0) + f(1)z^{-1} + f(2)z^{-2} + \cdots]$$

That is,

$$G(z) = z^{-k}F(z)$$

which is the desired result; see (3.4-4).

Initial Value Theorem

This theorem enables us to find the initial value $f(0)$ from a given $F(z)$ via the relation

$$f(0) = \lim_{z \to \infty} F(z)$$

The proof readily follows from definition of $F(z)$; that is,

$$F(z) = \sum_{n=0}^{\infty} f(n)z^{-n} = f(0) + f(1)z^{-1} + f(2)z^{-2} + \cdots$$

Final Value Theorem

Using this theorem we can find the value of $f(n)$ as n tends to infinity, since it states that

$$\lim_{n \to \infty} f(n) = \lim_{z \to 1} \left[\frac{z - 1}{z} \right] F(z) \tag{3.4-5}$$

where $F(z)$ is assumed to have all its poles *within* the unit circle. The proof of this theorem is quite tedious. The interested reader may refer to [2] for the same.

Multiplication by *n*

This theorem states that if

$$g(n) = nf(n), \qquad n \geq 0$$

then
$$G(z) = -z\,\frac{dF(z)}{dz}$$
(3.4-6)

where $G(z) = Z\{g(n)\}$ and $F(z) = Z\{f(n)\}$

The proof of this theorem is left as an exercise; see Problem 3-3.

Periodic Sequence Theorem

The essence of this theorem is that the ZT of an N-periodic sequence can be expressed in terms of the ZT of its first period. That is, if

$$f(n) = \underbrace{\{f(0)\ f(1)\ \cdots\ f(N-1)}_{\text{First period}}\ f(0)\ f(1)\ \cdots\}$$
(3.4-7)

is the N-periodic sequence, and

$$f_1(n) = \begin{cases} f(n), & 0 \le n \le N-1 \\ 0, & \text{elsewhere} \end{cases}$$

denotes the sequence in the first period of the sequence in (3.4-7), then

$$F(z) = \frac{F_1(z)}{1 - z^{-N}}, \qquad |z^N| > 1$$
(3.4-8)

where $F_1(z) = Z\{f_1(n)\}$.

To prove (3.4-8) we note that the periodic property enables us to write

$$F(z) = \sum_{n=0}^{\infty} f(n)z^{-n}$$

$$= \sum_{n=0}^{\infty} f_1(n)\,z^{-n} + \sum_{n=0}^{\infty} f_1(n-N)z^{-n}$$
(3.4-9)

$$+ \sum_{n=0}^{\infty} f_1(n-2N)z^{-n} + \cdots$$

Application of the shift theorem in (3.4-4) to (3.4-9) leads to

$$F(z) = F_1(z) + z^{-N}F_1(z) + z^{-2N}F_1(z) + \cdots$$
(3.4-10)

$$= F_1(z)[1 + z^{-N} + z^{-2N} + \cdots]$$

Equation (3.4-10) involves an infinite geometric series with common ratio z^{-N}, which converges if $|z^{-N}| < 1$. Thus

$$F(z) = \frac{F_1(z)}{1 - z^{-N}}, \qquad |z^N| > 1$$

which is (3.4-8).

Another useful property of the ZT is known as *Parseval's theorem*, and is left as an exercise; see Problems 3-15 and 3-16.

3.5 INVERSE Z-TRANSFORM

As mentioned in Section 3.1, the ZT $F(z)$ of a given sequence $f(n)$ is unique. Hence the ZT and its inverse in (3.1-5) and (3.1-6), respectively, form a transform pair; that is,

$$F(z) = Z\{f(n)\}, \qquad \text{the forward transform}$$

and $\qquad f(n) = Z^{-1}\{F(z)\}, \qquad \text{the inverse transform}$

Thus far we have addressed the problem of obtaining $F(z)$ for a given $f(n)$. In this section we address the reverse problem of obtaining $f(n)$ from a given $F(z)$. There are three methods for doing so:

1. Power series method
2. Partial fraction expansion method
3. Residue method

We will discuss these methods individually. The partial fraction expansion is restricted to the case when $F(z)$ is in the form of a rational function, while the other two methods apply to a larger class of functions.

Power Series Method

By means of a straightforward long-division process, a given $F(z)$ is expressed in the form of power series to obtain

$$F(z) = a_0 + a_1 z^{-1} + a_2 z^{-2} + \cdots \qquad (3.5\text{-}1)$$

Again, from the definition of the ZT, we have

$$F(z) = \sum_{n=0}^{\infty} f(n)z^{-n} = f(0) + f(1)z^{-1} + f(2)z^{-2} + \cdots \quad (3.5\text{-}2)$$

Examination of (3.5-1) and (3.5-2) shows that there is a one-to-one correspondence between the power series coefficients a_n and the desired sequence values $f(n)$. Hence $f(n)$ is obtained via the relation

$$f(n) = a_n, \quad n \ge 0 \quad (3.5\text{-}3)$$

To illustrate, let

$$F(z) = \frac{z^2 + z}{z^3 - 3z^2 + 3z - 1} = \frac{N(z)}{D(z)} \quad (3.5\text{-}4)$$

Using long division, $N(z)$ is divided by $D(z)$ to obtain the power series

$$F(z) = z^{-1} + 4z^{-2} + 9z^{-3} + 16z^{-4} + \cdots \quad (3.5\text{-}5)$$

From (3.5-1) and (3.5-5) it follows that

$$a_0 = 0, \quad a_1 = 1, \quad a_2 = 4, \quad a_3 = 9, \text{etc.}$$

Thus (3.5-3) yields

$$f(0) = 0, \quad f(1) = 1, \quad f(2) = 4, \quad f(3) = 9, \text{etc.}$$

which *suggests* that

$$f(n) = n^2$$

is the sequence whose ZT is $F(z)$ in (3.5-4).

A computer program for finding the IZT using this method is given in Appendix 3.1. A shortcoming of this approach is that, in general, it does not yield a general closed-form solution for $f(n)$.

Partial Fraction Expansion (PFE) Method

The approach entertained here parallels that used to obtain $f(t)$ from its LT $F(s)$, as discussed in Section 1.9. In this case we work with the PFE of the function $F(z)/z$ instead of $F(s)$, as illustrated by the following examples.

Example 3.5-1: If $F(z) = z/(3z^2 - 4z + 1)$, find $f(n)$.

Solution: We first obtain the PFE of $F(z)/z$. To do this, we write

$$\frac{F(z)}{z} = \frac{1}{(z-1)(3z-1)} = \frac{A}{z-1} + \frac{B}{3z-1} \qquad (3.5\text{-}6)$$

where $A = (z-1)\left.\frac{F(z)}{z}\right|_{z=1} = \frac{1}{2}$

$\quad B = (3z-1)\left.\frac{F(z)}{z}\right|_{z=1/3} = -\frac{3}{2}$

Equation (3.5-6) now yields

$$F(z) = \frac{1}{2}\left[\frac{z}{z-1} - \frac{z}{z-\frac{1}{3}}\right]$$

Next the IZT of each of the terms $z/(z-1)$ and $z/\left(z-\frac{1}{3}\right)$ is found by refering to Table 3.2-1. This process yields

$$Z^{-1}\left\{\frac{z}{z-1}\right\} = 1 \quad \text{and} \quad Z^{-1}\left\{\frac{z}{z-\frac{1}{3}}\right\} = \left(\frac{1}{3}\right)^n$$

Thus the desired result is

$$f(n) = \frac{1}{2}(1 - 3^{-n}), \qquad n \geq 0$$

From this example it is clear that there are basically two steps involved in the PFE method:

STEP 1. Obtain the PFE of $F(z)/z$.

STEP 2. Find the IZT of the terms that appear in the PFE obtained in step 1 by referring to Table 3.2-1.

Example 3.5-2: If $F(z) = \dfrac{2z^2}{(z+1)(z+2)^2}$, find $f(n)$.

Solution:

STEP 1. $\dfrac{F(z)}{z} = \dfrac{2z}{(z + 1)(z + 2)^2} = \dfrac{A}{z + 1} + \dfrac{B}{z + 2} + \dfrac{C}{(z + 2)^2}$

where $A = (z + 1)\dfrac{F(z)}{z}\Big|_{z = -1} = -2$

$B = \dfrac{d}{dz}\left\{(z + 2)^2\dfrac{F(z)}{z}\right\}_{z = -2} = 2$

$C = (z + 2)^2\dfrac{F(z)}{z}\Big|_{z = -2} = 4$

which results in

$$F(z) = \dfrac{-2z}{z + 1} + \dfrac{2z}{z + 2} + \dfrac{4z}{(z + 2)^2} \tag{3.5-7}$$

or $f(n) = Z^{-1}\left\{\dfrac{-2z}{z + 1}\right\} + Z^{-1}\left\{\dfrac{2z}{z + 2}\right\} + Z^{-1}\left\{\dfrac{4z}{(z + 2)^2}\right\}$

STEP 2. From Table 3.2-1 we obtain

$$Z^{-1}\left\{\dfrac{z}{z + 1}\right\} = (-1)^n$$

$$Z^{-1}\left\{\dfrac{z}{z + 2}\right\} = (-2)^n$$

and

$$Z^{-1}\left\{\dfrac{4z}{(z + 2)^2}\right\} = (-2)Z^{-1}\left\{\dfrac{-2z}{(z + 2)^2}\right\} \tag{3.5-8}$$

$$= (-2)n(-2)^n = n(-2)^{n + 1}$$

From (3.5-7) and (3.5-8) it follows that

$$f(n) = -2(-1)^n + 2(-2)^n + n(-2)^{n + 1}, \quad n \geq 0$$

Example 3.5-3: Given: $F(z) = \dfrac{z^2 + z}{(z - 1)^2}$

Determine $f(n)$.

Solution:

STEP 1. $\dfrac{F(z)}{z} = \dfrac{z+1}{(z-1)^2} = \dfrac{A}{z-1} + \dfrac{B}{(z-1)^2}$

where $A = \dfrac{d}{dz}\left\{(z-1)^2\dfrac{F(z)}{z}\right\}_{z=1} = 1$

$B = (z-1)^2\dfrac{F(z)}{z}\bigg|_{z=1} = 2$

Hence

$$F(z) = \dfrac{z}{z-1} + \dfrac{2z}{(z-1)^2}$$

which implies that

$$f(n) = Z^{-1}\left\{\dfrac{z}{z-1}\right\} + Z^{-1}\left\{\dfrac{2z}{(z-1)^2}\right\} \qquad (3.5\text{-}9)$$

STEP 2. Referring to Table 3.2-1, we have

$$Z^{-1}\left\{\dfrac{z}{z-1}\right\} = 1 \qquad (3.5\text{-}10)$$

and

$$Z^{-1}\left\{\dfrac{2z}{(z-1)^2}\right\} = 2n$$

Substitution of (3.5-10) in (3.5-9) yields

$$f(n) = (1+2n), \qquad n \geq 0$$

Residue Method

From the definition of the ZT of a given sequence $f(m)$, we have

$$F(z) = \sum_{m=0}^{\infty} f(m)z^{-m}, \qquad |z| > R_1 \qquad (3.5\text{-}11)$$

where $|z| > R_1$ specifies the region of convergence. We will first show that we can recover $f(m)$ from $F(z)$ using the following *inverse integral formula:*

$$f(n) = \frac{1}{2\pi j} \oint_C z^{n-1} F(z) \, dz, \qquad n \geq 0 \qquad (3.5\text{-}12)$$

where C is any simple closed curve enclosing $|z| = R_1$, and $\oint_C$ denotes contour integration along C in the counterclockwise direction.

PROOF. Multiplying (3.5-11) by z^{n-1} and integrating both sides around the contour C, it follows that

$$\frac{1}{2\pi j} \oint_C z^{n-1} F(z) \, dz = \sum_{m=0}^{\infty} f(m) \frac{1}{2\pi j} \oint_C z^{n-m} \frac{dz}{z} \qquad (3.5\text{-}13)$$

Without loss of generatity, we can let C be a circle of radius R, where $R > R_1$. Then, making the change of variables

$$z = Re^{j\theta}$$

results in

$$dz = jz \, d\theta$$

Hence we have

$$\frac{1}{2\pi j} \oint_C z^{n-m} \frac{dz}{z} = \frac{R^{n-m}}{2\pi} \int_0^{2\pi} e^{j(n-m)\theta} \, d\theta$$

$$= \begin{cases} 1, & m = n \\ 0, & \text{otherwise} \end{cases}$$

Substitution of this result in (3.5-13) leads to

$$\frac{1}{2\pi j} \oint_C z^{n-1} F(z) \, dz = f(n), \qquad n \geq 0$$

which is the desired result in (3.5-12).

The inversion integral in (3.5-12) can be easily evaluated using Cauchy's *residue theorem*, which is a fundamental result in the subject area of complex variables. Before stating the same, it is instructive to present a brief discussion related to the concept of residues. To this end, we consider the $F(z)$ that was introduced in Example 3.5-2; that is,

$$F(z) = \frac{2z^2}{(z + 1)(z + 2)^2}$$

Writing $F(z)$ in the form of a PFE, we obtain

$$F(z) = \frac{A}{z + 1} + \frac{B_1}{z + 2} + \frac{B_2}{(z + 2)^2}$$

with respect to which we make the following comments:

1. The coefficient A is the residue of $F(z)$ corresponding to the *simple* pole $z = -1$.

2. The coefficient B_1 is the residue of $F(z)$ corresponding to the *multiple* (second-order) pole at $z = -2$.

3. The coefficient B_2 is *not* a residue.

It is important to note that, regardless of the order of a multiple pole, there is only *one* residue. It is the coefficient B_1 of the term $B_1/(z - a)$ in the PFE of a given $F(z)$ that has an mth order pole at $z = a$.

RESIDUE THEOREM. Given an $F(z)$, $|z| > R_1$, the corresponding IZT can be found by evaluating the integral

$$f(n) = \frac{1}{2\pi j} \oint_C z^{n-1} F(z) \, dz, \qquad n \geq 0$$

= sum of the residues of $G(z) = z^{n-1}F(z)$ corresponding to the poles of $G(z)$ that lie inside a simple closed curve C that encloses $|z| = R_1$.　　　　(3.5-14)

The residue of $G(z)$ at a given pole at $z = a$, say $R_{z=a}$, can be calculated using the convenient formula

$$R_{z=a} = \frac{d^{m-1}}{dz^{m-1}} \left[\frac{(z - a)^m}{(m - 1)!} G(z) \right]\Bigg|_{z=a}, \qquad m \geq 1 \qquad (3.5\text{-}15a)$$

where m is the order of the pole at $z = a$. For example, if $m = 1$ and $m = 2$, we obtain the familiar expressions

$$R_{z=a} = (z - a)G(z)\Big|_{z=a} \qquad (3.5\text{-}15b)$$

and
$$R_{z=a} = \frac{d}{dz}[(z - a)^2 G(z)]\Big|_{z=a} \tag{3.5-15c}$$

which are associated with simple and second-order poles.

Example 3.5-4: Find the IZT of

$$F(z) = \frac{1}{(z - 1)(z - 0.5)}$$

Solution: The given $F(z)$ implies that

$$G(z) = \frac{z^{n-1}}{(z - 1)(z - 0.5)} \tag{3.5-16}$$

We note that $G(z)$ has a simple pole at $z = 0$ when $n = 0$, and no pole at $z = 0$ for $n \geq 1$. Thus the cases $n = 0$ and $n \geq 1$ will be considered separately.

CASE 1. $n = 0$. Equation (3.5-16) becomes

$$G(z) = \frac{1}{z(z - 1)(z - 0.5)}$$

Thus the residue theorem yields

$$f(0) = R_{z=0} + R_{z=1} + R_{z=0.5} \tag{3.5-17}$$

Evaluating these residues using (3.5-15b), we get

$$R_{z=0} = zG(z)\Big|_{z=0} = 2$$

$$R_{z=1} = (z - 1)G(z)\Big|_{z=1} = 2 \tag{3.5-18}$$

and
$$R_{z=0.5} = (z - 0.5)G(z)\Big|_{z=0.5} = -4$$

Substitution of (3.5-18) in (3.5-17) leads to

$$f(0) = 0 \tag{3.5-19}$$

CASE 2. $n \geq 1$. Here the residue theorem is applied using $G(z)$ in (3.5-16) to obtain

$$f(n) = R_{z=1} + R_{z=0.5}, \qquad n \geq 1$$

where $R_{z=1} = (z - 1)G(z)\Big|_{z=1} = 2$

$$R_{z=0.5} = (z - 0.5)G(z)\Big|_{z=0.5} = -2(0.5)^{n-1}$$

Thus

$$f(n) = 2 - 2(0.5)^{n-1}, \qquad n \geq 1 \qquad (3.5\text{-}20)$$

Combining (3.5-19) and (3.5-20), we get the desired solution:

$$f(n) = \begin{cases} 0, & n = 0 \\ 2[1 - (0.5)^{n-1}], & n \geq 1 \end{cases}$$

Example 3.5-5: Find $f(n)$ for the case when

$$F(z) = \frac{z^2 + z}{(z - 1)^2}$$

Solution: Here $G(z) = z^{n-1}F(z)$ yields

$$G(z) = \frac{z^{n+1} + z^n}{(z - 1)^2}$$

and from the residue theorem we have

$$f(n) = R_{z=1} \qquad (3.5\text{-}21)$$

Since $F(z)$ has a second-order pole, the corresponding residue is given by [see (3.5-15c)]

$$R_{z=1} = \frac{d}{dz}[(z - 1)^2 G(z)]\Big|_{z=1} = (2n + 1) \qquad (3.5\text{-}22)$$

Thus (3.5-21) yields the IZT to be

$$f(n) = (2n + 1), \qquad n \geq 0$$

3.6 SOLUTION OF DIFFERENCE EQUATIONS

We are now in a position to consider an important application of ZTs, the solution of linear difference equations with constant coefficients. The approach involves the development of an expression for the ZT of sequences of the form $f(n - m)$, where m is fixed and $n \geq 0$. To illustrate, consider the case when $m = 1$. Then the definition of the ZT in (3.1-5) yields

$$Z\{f(n - 1)\} = \sum_{n=0}^{\infty} f(n - 1)z^{-n}$$

$$= f(-1) + f(0)z^{-1} + f(1)z^{-2} + f(2)z^{-3} + \cdots$$

$$= f(-1) + z^{-1}[f(0) + f(1)z^{-1} + f(2)z^{-2} + \cdots]$$

which implies that

$$Z\{f(n - 1)\} = z^{-1}F(z) + f(-1) \tag{3.6-1}$$

Similarly, we obtain (see Problem 3-8)

$$Z\{f(n - 2)\} = z^{-2}F(z) + z^{-1}f(-1) + f(-2) \tag{3.6-2a}$$

and $Z\{f(n - 3)\} = z^{-3}F(z) + z^{-2}f(-1) + z^{-1}f(-2) + f(-3)$ (3.6-2b)

Using the principle of mathematical induction, it can be shown that (3.6-1) and (3.6-2) are special cases of general result that

$$Z\{f(n - m)\} = z^{-m}\{F(z) + \sum_{r=1}^{m} f(-r)z^r\} \tag{3.6-3}$$

for a specified m and $n \geq 0$.

Next we demonstrate via illustrative examples that (3.6-3) can be used to solve difference equations.

Example 3.6-1: Solve the second-order difference equation

$$2f(n - 2) - 3f(n - 1) + f(n) = 3^{n-2}, \qquad n \geq 0 \tag{3.6-4}$$

where $f(-2) = -\dfrac{4}{9}$ and $f(-1) = -\dfrac{1}{3}$

Solution: Taking the ZT of both sides of (3.6-4), we obtain

$$2Z\{f(n - 2)\} - 3Z\{f(n - 1)\} + Z\{f(n)\} = 3^{-2}Z\{3^n\}$$

Thus (3.6-1), (3.6-2), and Table 3.2-1 yield

$$2[z^{-2}F(z) + z^{-1}f(-1) + f(-2)]$$
$$- 3[z^{-1}F(z) + f(-1)] + F(z) = 3^{-2}\frac{z}{z - 3} \quad (3.6\text{-}5)$$

Substitution of $f(-2) = -\dfrac{4}{9}$ and $f(-1) = -\dfrac{1}{3}$ in (3.6-5) and subsequent simplification leads to

$$\frac{(z - 1)(z - 2)}{z^2} F(z) = \frac{2}{3z} - \frac{1}{9} + \frac{z}{9(z - 3)} = \frac{(z - 2)}{z(z - 3)}$$

That is,

$$F(z) = \frac{z}{(z - 1)(z - 3)} \quad (3.6\text{-}6)$$

To obtain $f(n)$ from $F(z)$, we can employ the PFE or residue methods. Use of the PFE method leads to

$$\frac{F(z)}{z} = \frac{A}{z - 1} + \frac{B}{z - 3}$$

where $\quad A = (z - 1)\dfrac{F(z)}{z} \bigg|_{z=1} = -\dfrac{1}{2}$

$$B = (z - 3)\frac{F(z)}{z} \bigg|_{z=3} = \frac{1}{2}$$

Thus

$$F(z) = -\frac{1}{2}\frac{z}{z - 1} + \frac{1}{2}\frac{z}{z - 3}$$

or $\quad f(n) = -\dfrac{1}{2}Z^{-1}\left\{\dfrac{z}{z - 1}\right\} + \dfrac{1}{2}Z^{-1}\left\{\dfrac{z}{z - 3}\right\}$ $\quad (3.6\text{-}7)$

From (3.6-7) and Table 3.2-1 we obtain

$$f(n) = \frac{1}{2}(3^n - 1), \qquad n \geq 0$$

as the solution of (3.6-4).

Example 3.6-2: Find $f(n)$ corresponding to the difference equation

$$f(n - 2) - 2f(n - 1) + f(n) = 1, \qquad n \geq 0 \qquad (3.6\text{-}8)$$

with initial conditions $f(-1) = -0.5$ and $f(-2) = 0$.

Solution: From (3.6-8) we have

$$Z\{f(n - 2)\} - 2Z\{f(n - 1)\} + Z\{f(n)\} = Z\{1\} \qquad (3.6\text{-}9)$$

The ZT of the left-hand side of (3.6-9) is obtained via (3.6-1) and (3.6-2), and Table 3.2-1 yields $Z\{1\}$. Thus (3.6-9) leads to

$$z^{-2}F(z) + z^{-1}f(-1) + f(-2) - 2[z^{-1}F(z) + f(-1)] + F(z) = \frac{z}{z - 1}$$

Substituting $f(-1) = -0.5$ and $f(-2) = 0$ in the preceding equation and simplifying the resulting expression, we obtain

$$F(z) = \frac{z^2}{(z - 1)^3} + \frac{0.5z}{(z - 1)^2} = \frac{1.5z^2 - 0.5z}{(z - 1)^3}$$

which yields

$$f(n) = Z^{-1}\left\{ \frac{1.5z^2 - 0.5z}{(z - 1)^3} \right\} \qquad (3.6\text{-}10)$$

The IZT in (3.6-10) is conveniently evaluated by resorting to the residue method, since

$$f(n) = R_{z=1} \qquad (3.6\text{-}11)$$

where $R_{z=1}$ is the residue of

$$G(z) = \frac{1.5z^{n+1} - 0.5z^n}{(z - 1)^3}$$

at $z = 1$. This residue is evaluated using (3.5-15a) with $m = 3$ to obtain

$$R_{z=1} = \frac{1}{2!} \frac{d^2}{dz^2} [1.5z^{n+1} - 0.5z^n] \bigg|_{z=1}$$

$$= 0.5n^2 + n$$

Thus (3.6-11) yields the solution of (3.6-8) to be

$$f(n) = 0.5n^2 + n, \quad n \geq 0$$

Example 3.6-3: Given the difference equation

$$y(n) + b^2 y(n - 2) = 0, \quad n \geq 0 \tag{3.6-12}$$

where $|b| < 1$ and the initial conditions are $y(-1) = 0$ and $y(-2) = -1$. Show that

$$y(n) = b^{n+2} \cos\left(\frac{n\pi}{2}\right) \tag{3.6-13}$$

Solution: The ZT of each term in (3.6-12) yields

$$Z\{y(n)\} + b^2 Z\{y(n - 2)\} = 0 \tag{3.6-14}$$

From (3.6-2) and (3.6-14) it follows that

$$Y(z) + b^2[z^{-2}Y(z) + z^{-1}y(-1) + y(-2)] = 0 \tag{3.6-15}$$

Substituting $y(-1) = 0$ and $y(-2) = -1$ in (3.6-15), there results

$$Y(z)(1 + b^2 z^{-2}) = b^2$$

which yields

$$Y(z) = \frac{b^2 z^2}{z^2 + b^2} \tag{3.6-16}$$

To find $y(n)$ we employ the residue method and hence use $Y(z)$ in (3.6-16) to form

$$G(z) = \frac{b^2 z^{n+1}}{z^2 + b^2} = \frac{b^2 z^{n+1}}{(z - jb)(z + jb)} \tag{3.6-17}$$

Thus

$$y(n) = R_{z=jb} + R_{z=-jb} \tag{3.6-18}$$

Evaluating $R_{z=jb}$, we obtain

$$R_{z=jb} = (z - jb)G(z)\Big|_{z=jb}$$

where $G(z)$ is given by (3.6-17). Thus

$$R_{z=jb} = \frac{b^2 b^{n+1} j^{n+1}}{2jb}$$

$$= \frac{b^{n+2} j^n}{2}$$

which yields

$$R_{z=jb} = \frac{b^{n+2} e^{jn\pi/2}}{2} \tag{3.6-19}$$

since $j = e^{j\pi/2}$.

Similarly, one obtains

$$R_{z=-jb} = \frac{b^{n+2} e^{-jn\pi/2}}{2} \tag{3.6-20}$$

Substitution of (3.6-19) and (3.6-20) in (3.6-18) results in

$$y(n) = b^{n+2} \left[\frac{e^{jn\pi/2} + e^{-jn\pi/2}}{2} \right] \tag{3.6-21}$$

Applying the identity

$$\cos(\theta) = \frac{e^{j\theta} + e^{-j\theta}}{2}$$

to (3.6-21), we obtain the desired result in (3.6-13); that is,

$$y(n) = b^{n+2} \cos(n\pi/2), \qquad n \geq 0$$

3.7 TWO-SIDED Z-TRANSFORM

Up to this point we have discussed the ZT in the context of sequences $f(n)$ that are *zero* for $n < 0$. Such sequences are said to be *causal* and are of practical interest. From a theoretical viewpoint, however, it is useful to consider some aspects of the ZT of *noncausal* sequences; that is, sequences $f(n)$ that may not be zero for $n < 0$. To this end,

$$F(z) = Z\{f(n)\} = \sum_{n=-\infty}^{\infty} f(n)z^{-n} \tag{3.7-1}$$

is defined as the *two-sided* ZT of a given sequence $f(n)$. In the event that $f(n)$ is causal, we note the ZT definitions in (3.7-1) and (3.1-5) are equivalent. Thus the definition of the ZT in (3.1-5) is known as the *one-sided* ZT.

From the discussion in Section 3.2, it is apparent that $F(z)$ in (3.7-1) converges everywhere *outside* a circle of radius R_1 if $f(n) = 0$ for $n < 0$. Similarly, it can be shown that $F(z)$ in (3.7-1) converges everywhere *inside* a circle of radius R_2 if $f(n) = 0$ for $n \geq 0$. Thus the region of convergence for the two-sided ZT in (3.7-1) is the annular region $R_1 < |z| < R_2$.

Starting with (3.7 1), it is straightforward to derive the following *inversion integral formula* for the two-sided ZT:

$$f(n) = \frac{1}{2\pi j} \oint_C z^{n-1}F(z)\, dz, \qquad |n| < \infty \tag{3.7-2}$$

where C is a simple closed path separating $|z_1| = R_1$ from $|z_2| = R_2$. The steps involved in deriving (3.7-2) from (3.7-1) parallel those used in obtaining the inversion integral formula in (3.5-12), starting with the one-sided ZT in (3.5-11).

Again, the inversion integral formula in (3.7-2) can be easily evaluated via Cauchy's residue theorem, according to which we have

$$f(n) = \sum_k R_{z=a_k}, \qquad n \geq 0 \tag{3.7-3a}$$

$$= -\sum_k R_{z=b_k}, \qquad n < 0 \tag{3.7-3b}$$

In (3.7-3), $R_{z=a_k}$ is the residue of $G(z) = z^{n-1}F(z)$ at the pole $z = a_k$ *inside* C, while $R_{z=b_k}$ is the residue of $G(z)$ at the pole $z = b_k$ *outside* C.

Example 3.7-1: If

$$f(n) = \begin{cases} 1, & n < 0 \\ (0.5)^n, & n \geq 0 \end{cases} \tag{3.7-4}$$

find $F(z)$.

Solution: Substitution of (3.7-4) in (3.7-1) results in

$$F(z) = \sum_{n=-\infty}^{-1} z^{-n} + \sum_{n=0}^{\infty} (0.5z^{-1})^n \tag{3.7-5}$$

Recognizing that

$$\sum_{n=-\infty}^{-1} z^{-n} = \sum_{n=1}^{\infty} z^n = \sum_{n=0}^{\infty} z^n - 1$$

(3.7-5) can be written as

$$F(z) = F_1(z) + F_2(z) \tag{3.7-6}$$

where $\quad F_1(z) = \sum_{n=0}^{\infty} z^n - 1 = \dfrac{z}{1-z}, \quad |z| < 1$

and $\quad F_2(z) = \sum_{n=0}^{\infty} (0.5z^{-1})^n = \dfrac{1}{1-0.5z^{-1}}, \quad |z| > 0.5$

Thus $\qquad\qquad F(z) = \dfrac{z}{1-z} + \dfrac{1}{1-0.5z^{-1}}$

which yields the desired result

$$F(z) = \frac{-0.5z}{(z-0.5)(z-1)}, \quad 0.5 < |z| < 1 \tag{3.7-7}$$

FT CONSIDERATIONS. We recall that the FT of a causal sequence is obtained from its ZT via the substitution $z = e^{j\omega T}$; see (3.1-9). Likewise, for the two-sided ZT, we have the relation

$$\bar{F}(\omega) = F(z)\Big|_{z=e^{j\omega T}} \tag{3.7-8}$$

$$= \sum_{n=-\infty}^{\infty} f(n)e^{-jn\omega T}$$

where $\bar{F}(\omega)$ is defined as the FT of the noncausal sequence $f(n)$. Again, applying the property in (3.1-10) to (3.7-8), we obtain the following definition for the IFT of $\bar{F}(\omega)$:

$$f(n) = \frac{1}{\omega_s} \int_{-\omega_s/2}^{\omega_s/2} \bar{F}(\omega) e^{jn\omega T} \, d\omega, \qquad |n| < \infty \qquad (3.7\text{-}9)$$

Example 3.7-2: Given $\bar{F}(\omega) = 1/(1 + a^2 - 2a \cos \omega T)$, where $0 < a < 1$. Find $F(z)$. If the region of convergence of $F(z)$ is $a < |z| < 1/a$, find $f(n)$ using (3.7-3).

Solution: Substituting $2 \cos \omega T = (e^{j\omega T} + e^{-j\omega T})$ in the expression for $\bar{F}(\omega)$, we obtain

$$\bar{F}(\omega) = \frac{1}{1 + a^2 - ae^{j\omega T} - ae^{-j\omega T}} \qquad (3.7\text{-}10)$$

Next, with $e^{j\omega T} = z$, (3.7-10) yields

$$F(z) = \frac{1}{1 + a^2 - az - az^{-1}} \qquad (3.7\text{-}11)$$

$$= \frac{1}{(a - z)(a - z^{-1})}$$

To find $f(n)$, we write $F(z)$ in (3.7-11) as

$$F(z) = \frac{-z/a}{(z - a)(z - 1/a)}, \qquad a < |z| < 1/a \qquad (3.7\text{-}12)$$

From (3.7-12) it is clear that $F(z)$ has two poles at $z = a$ and $z = 1/a$, respectively. Thus, to apply (3.7-3), we may choose a closed path C as indicated in Fig. 3.7-1. For convenience, C is chosen to be a circle.

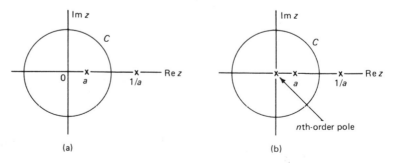

(a)   (b)

Fig. 3.7-1 Plots of the poles of $G(z)$ in Example 3.7-2 for $n \geq 0$ and $n \leq 0$, respectively.

Next we form

$$G(z) = z^{n-1}F(z) \tag{3.7-13}$$

$$= \frac{-z^n/a}{(z-a)(z-1/a)}$$

and consider two cases.

CASE 1. $n \geq 0$. From (3.7-13) it follows that C encloses only the pole $z = a$; see Fig. 3.7-1a. Hence (3.7-3a) yields

$$f(n) = R_{z=a} = (z-a)G(z) \Big|_{z=a} \tag{3.7-14}$$

or $$f(n) = \frac{a^n}{1-a^2}, \qquad n \geq 0$$

CASE 2. $n < 0$. Here $G(z)$ in (3.7-13) has an nth-order pole at $z = 0$, in addition to the poles at $z = a$ and $z = 1/a$; see Fig. 3.7-1b. Thus, using (3.7-3b), we obtain

$$f(n) = -R_{z-1/a} \tag{3.7-15}$$

where $$R_{z=1/a} = (z - \frac{1}{a})G(z) \Big|_{z=1/a}$$

$$= -\frac{a^{-n}}{1-a^2}$$

Thus (3.7-15) yields

$$f(n) = \frac{a^{-n}}{1-a^2}, \qquad n < 0 \tag{3.7-16}$$

Combining (3.7-14) and (3.7-16), we obtain the desired sequence as

$$f(n) = \frac{a^{|n|}}{1-a^2}, \qquad |n| < \infty \tag{3.7-17}$$

3.8 SUMMARY

We have introduced the ZT and discussed some of its important properties. It was also shown that the ZT provides a convenient means for solving linear difference equations with constant coefficients, which were introduced in Chapter 2. As such, the ZT plays a key role in obtaining transfer functions to represent DT systems. This aspect of the ZT will be studied in Chapter 5, and its relevance to the frequency analysis of sequences will be discussed in Chapter 4.

PROBLEMS

3–1   Using Table 3.2-1, find $G(z)$ for the sequence

$$g(n) = 2e^{-6n} - 2e^{-3n} + 24ne^{-6n}$$

3–2   If $g(n) = e^{-anT}f(n)$, $n \geq 0$, show that

$$G(z) = Z\{g(n)\} = F(ze^{aT})$$

3–3   If $g(n) = nf(n)$, $n \geq 0$, show that

$$G(z) = -z\frac{dF(z)}{dz}$$

where $G(z) = Z\{g(n)\}$ and $F(z) = Z\{f(n)\}$.

3–4   Given the 6-periodic sequence

$$f(n) = \{1, 1, 1, -1, -1, -1, 1, 1, \cdots\}$$

show that

$$F(z) = \frac{z(z^2 + z + 1)}{z^3 + 1}$$

3–5   Given:

$$F(z) = \frac{0.165z^3 + 0.268z^2}{z^3 - 1.702z^2 + 0.837z - 0.135}$$

Find $f(n)$, $0 \leq n \leq 3$.

3–6   Use the PFE method to find the IZT of the following ZTs:

(a) $F(z) = \dfrac{z}{(z - e^{-a})(z - e^{-b})}$, where a and b are positive constants.

(b) $F(z) = \dfrac{z^2}{(z - 1)(z - 0.8)}$.

3–7   Use the residue method to find the IZT of the following ZTs:

(a) $F(z) = \dfrac{z^2 + 3z}{(z - 0.5)^3}$.

(b) $F(z) = \dfrac{1}{(z + 1)^2(z - 0.5)}$.

3–8   Prove the results in (3.6-2).

3–9   Solve the following difference equations using the ZT approach.

(a) $f(n - 2) - \dfrac{3}{2} f(n - 1) + \dfrac{1}{2} f(n) = 1 + 3^{-n}$, $n \geq 0$, with $f(-2) = 0$
and $f(-1) = 1$.

(b) $f(n) - 0.9f(n - 1) = 0.5 + (0.9)^{n-1}$, $n \geq 0$, with $f(-1) = 5$.

(c) $f(n) - 0.5f(n - 1) = \sin(n\pi/2)$, $n \geq 0$, with $f(-1) = 0$.
Hint: To find the ZT of $\sin(n\pi/2)$, use line 8 of Table 3.2-1 with $\omega_0 T = \pi/2$.

3–10   Given the sequence $f(n) = e^{-anT}$, $n \geq 0$, where a is a complex number, show that

$$F(z) = \frac{z}{z - e^{-aT}}, \qquad |z| > |e^{-aT}|$$

3–11   Given the difference equation

$$f(n - 2) - f(n - 1) + f(n) = 0, \qquad n \geq 0$$

with initial conditions $f(-1) = 1$ and $f(-2) = 0$.
(a) Show that

$$F(z) = \frac{z(z - 1)}{(z - z_1)(z - \tilde{z}_1)}$$

where $z_1 = e^{-j\theta}$, with $\theta = \tan^{-1}(\sqrt{3})$, and $\tilde{z}_1$ denotes its complex conjugate.
(b) Use the $F(z)$ of part (a) and the residue method to show that

$$f(n) = \frac{\sin[(n + 1)\theta] - \sin(n\theta)}{\sin(\theta)}, \qquad n \geq 0$$

3–12   Given the difference equation

$$y(n) = 2 \cos(\omega_0 T)y(n - 1) - y(n - 2), \qquad n \geq 0$$

with $y(-1) = -\sin \omega_0 T$ and $y(-2) = -\sin (2\omega_0 T)$. Use the ZT method to show that its solution is given by

$$y(n) = \sin (n\omega_0 T), \qquad n \geq 0$$

Note: The preceding difference equation represents a DT sine-wave generator. Difference equations to generate square and triangular waveforms are also available; see [5].

3–13   It can be shown that

$$F(z) = \frac{b_0 + b_1 z^{-1} + b_2 z^{-2} + b_3 z^{-3} + \cdots}{1 + a_1 z^{-1} + a_2 z^{-2} + a_3 z^{-3} + \cdots}$$
$$= c_0 + c_1 z^{-1} + c_2 z^{-2} + \cdots$$

where the c_i can be computed recursively as follows:

$$c_n = b_n - \sum_{i=1}^{n} c_{n-i} a_i, \qquad n \geq 1$$

with $c_0 = b_0$. Write $F(z)$ in Problem 3-5 as a ratio of polynomials of z^{-1}, and then use this recursion to verify that the same values of $f(n)$, $0 \leq n \leq 3$, are obtained.

Comment: The preceding recursive relation can be used as an algorithm to obtain the IZT via the long-division method; see Appendix 3.1 for the related computer program.

3–14   Given:

$$F(z) = \frac{z(b - 1)}{(z - b)(z - 1)}, \qquad b < |z| < 1$$

where $0 < b < 1$. Use (3.7-3) to show that

$$f(n) = \begin{cases} 1, & n < 0 \\ b^n, & n \geq 0 \end{cases}$$

3–15   Let $F(z)$ be the one-sided ZT whose poles lie within the unit circle—that is,

$$F(z) = \sum_{n=0}^{\infty} f(n) z^{-n}, \quad |z| > 1$$

Show that

$$\sum_{n=0}^{\infty} f^2(n) = \frac{1}{2\pi j} \oint_{C_u} z^{-1} F(z) F(z^{-1}) \, dz \qquad \text{(P3.15-1)}$$

where $\oint_{C_u}$ denotes the integral around the unit circle in the counter-clockwise direction. This result is known as *Parseval's Theorem* for the one-sided ZT, where $\sum_{n=0}^{\infty} f^2(n)$ denotes the total energy in the sequence $f(n)$.

$$\textit{Hint: } F(z) F(z^{-1}) = \sum_{n=0}^{\infty} f^2(n) + \sum_{\substack{m=0 \\ m \neq n}}^{\infty} \sum_{n=0}^{\infty} f(m) f(n) z^{m-n}$$

3–16   With $z = e^{j\theta}$ where $\theta = \omega T$, show that Parseval's theorem in (P3.15-1) can also be expressed as

$$\sum_{n=0}^{\infty} f^2(n) = \frac{1}{2\pi} \int_0^{2\pi} |F(e^{j\theta})|^2 \, d\theta \qquad \text{(P3.16-1)}$$

APPENDIX 3.1[†] COMPUTER PROGRAM TO EVALUATE THE INVERSE Z-TRANSFORM

A subroutine that implements the long-division algorithm discussed in Problem 3-13 is given in this appendix.

```
C*****************************************************************************
C
C              THE OUTPUT SEQUENCE f(n) CAN BE EVALUATED FOR
C              A MAXIMUM OF 1000 POINTS. F(Z) SHOULD BE
C              OF THE FORM P(Z)/Q(Z). THE POLYNOMIALS P(Z)
C              AND Q(Z) CONTAN ONLY NEGETIVE POWERS
C              OF Z. THE MAXIMUM NUMBER OF COEFFS. IS 10.
C              IT CAN BE INCREASED BY CHANGING THE SIZE OF
C              THE ARRAYS A AND B.
C
C*****************************************************************************
C
        DIMENSION A(0:11)/12*0.0/,B(0:11)/12*0.0/,C(0:1010)
        ACCEPT ' # OF NUMERATOR COEFF. ? -> ',N
        N=N-1
```

---

†Program developed by D. Haran.

```
        DO 10 I=0,N
        K=-I
        WRITE(10,1) K
1       FORMAT(/,' INPUT COEFF. OF Z** ',I2,Z)
        ACCEPT ' -> ',A(I)
        WRITE(10,2) K,A(I)
2       FORMAT(' THE COEFF. Z** ',I2,'=',G16.8)
10      CONTINUE
        TYPE
        ACCEPT ' # OF DENOMINATOR COEFF. ? -> ',M
        M=M-1
        DO 20 J=0,M
        L=-J
        WRITE(10,1) L
        ACCEPT '-> ',B(J)
        WRITE(10,2) L,B(J)
20      CONTINUE
        TYPE
        ACCEPT ' # OF POINTS IN f(n) ',LASTN
        C(0)=A(0)/B(0)
        DO 30 I=1,LASTN
        CONVOL=0.0
        K=I
        IF(I.GT.10)K=11
        DO 40 J=1,K
        CONVOL=CONVOL+C(I-J)*B(J)
40      CONTINUE
        BNOTCI=A(K)-CONVOL
        C(I)=BNOTCI/B(0)
30      CONTINUE
        WRITE(10,70)
70      FORMAT(//,' ********** THE SEQUENCE f(n) IS **********')
        DO 50 J=0,LASTN
        WRITE(10,60) J,C(J)
50      CONTINUE
60      FORMAT(5X,' f(',I3,')=',G16.8)
        STOP
        END
```

REFERENCES

1. J. R. RAGAZZINI and L. A. ZADEH, "Analysis of Sampled-Data Systems," *Trans. AIEE*, Vol. 71, Part II, 1952, pp. 225–234.

2. Y. H. KU, *Transient Circuit Analysis*, Van Nostrand Reinhold, New York, 1961, Chap. 11.

3. J. A. CADZOW, *Discrete-Time Systems*, Prentice-Hall, Englewood Cliffs, N.J., Chap. 5.

4. W. D. STANLEY, *Digital Signal Processing*, Reston, Reston, Va., 1975, Chap. 4.

5. D. G. CHILDERS and A. DURLING, *Digital Filtering and Signal Processing*, West, St. Paul, Minn., 1975, pp. 226–235.

4

Fourier
Representation of
Sequences

This chapter is concerned with the Fourier representation of sequences (DT signals). It will be shown that a sequence can be represented in terms of the discrete Fourier transform (DFT), which can be computed efficiently via an algorithm called the fast Fourier transform. Several DFT properties that play an important role with respect to DT signals and systems are also developed.

4.1 FUNDAMENTALS

The sampled signal $x^*(t)$ in the following figure is obtained by sampling a CT signal $x(t)$ at a rate of f_s sps. The sampling interval is T. In accordance with (3.1-1), we can express $x^*(t)$ as

$$x^*(t) = x(t)p_\delta(t) \qquad (4.1\text{-}1)$$

where $\quad p_\delta(t) = \sum_{m=-\infty}^{\infty} \delta(t - mT)$

Since $p_\delta(t)$ is a periodic function with period T, its FS expansion is given by (1.2-1) and (1.2-4) to be

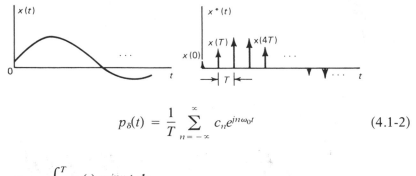

$$p_\delta(t) = \frac{1}{T} \sum_{n=-\infty}^{\infty} c_n e^{jn\omega_0 t} \qquad (4.1\text{-}2)$$

where $c_n = \int_0^T p_\delta(t) e^{-jn\omega_0 t}\, dt$

with $\omega_0 = 2\pi/T = \omega_s$ radians/second. Since $p_\delta(t)$ involves the impulse function, $\delta(t)$ in the interval $0 \le t < T$, c_n in (4.1-2) can be simplified to obtain

$$c_n = [e^{-jn\omega_0 t}]_{t=0} \qquad (4.1\text{-}3)$$

$$= 1, \quad \text{for all } n$$

Substitution of (4.1-3) in (4.1-2) leads to

$$p_\delta(t) = \frac{1}{T} \sum_{n=-\infty}^{\infty} e^{jn\omega_0 t} \qquad (4.1\text{-}4)$$

Combining (4.1-1) and (4.1-4), and noting that $\omega_0 = \omega_s$, we get

$$x^*(t) = \frac{1}{T} \sum_{n=-\infty}^{\infty} x(t) e^{jn\omega_s t} \qquad (4.1\text{-}5)$$

Let $X(\omega)$ and $X^*(\omega)$ denote the FTs of $x(t)$ and $x^*(t)$, respectively. Then, by definition of the FT in (1.3-2), we have

$$X(\omega) = \int_{-\infty}^{\infty} x(t) e^{-j\omega t}\, dt \qquad (4.1\text{-}6)$$

and

$$X^*(\omega) = \int_{-\infty}^{\infty} x^*(t) e^{-j\omega t}\, dt \qquad (4.1\text{-}7)$$

Substitution of (4.1-5) in (4.1-7) leads to

$$X^*(\omega) = \frac{1}{T} \int_{-\infty}^{\infty} \left\{ \sum_{n=-\infty}^{\infty} x(t) e^{jn\omega_s t} \right\} e^{-j\omega t}\, dt \qquad (4.1\text{-}8)$$

Interchanging the order of integration and summation in (4.1-8), we obtain

$$X^*(\omega) = \frac{1}{T} \sum_{n=-\infty}^{\infty} \left[\int_{-\infty}^{\infty} x(t) e^{-j(\omega - n\omega_s)t} \, dt \right] \qquad (4.1\text{-}9)$$

If ω in (4.1-6) is replaced by $(\omega - n\omega_s)$, then we have

$$X(\omega - n\omega_s) = \int_{-\infty}^{\infty} x(t) e^{-j(\omega - n\omega_s)t} \, dt \qquad (4.1\text{-}10)$$

Equations (4.1-9) and (4.1-10) yield

$$X^*(\omega) = \frac{1}{T} \sum_{n=-\infty}^{\infty} X(\omega - n\omega_s) \qquad (4.1\text{-}11a)$$

or

$$X^*(f) = \frac{1}{T} \sum_{n=-\infty}^{\infty} X(f - nf_s) \qquad (4.1\text{-}11b)$$

Next, from (3.1-9) we know that the FTs of a sequence and the corresponding sampled signal are equivalent; that is, $\bar{X}(\omega) = X^*(\omega)$, where $\bar{X}(\omega)$ is the FT of the sequence $x(mT)$ obtained by sampling the CT signal $x(t)$. Hence, corresponding to (4.1-11b), we have

$$\bar{X}(f) = \frac{1}{T} \sum_{n=-\infty}^{\infty} X(f - nf_s) \qquad (4.1\text{-}12)$$

Equation (4.1-12) is a fundamental result, which states that $\bar{X}(f)$ is a *periodic function* with period $f_s = 1/T$ or $\omega_s = 2\pi/T$. Furthermore, $\bar{X}(f)$ is directly proportional to $X(f)$ in each period, where $X(f)$ is the FT of the CT signal $x(t)$. This important aspect is illustrated in Fig. 4.1-1, which shows an example of an $X(f)$ and the corresponding $\bar{X}(f)$. Here $X(f)$ is said to be *band limited* in the sense that it is zero beyond some frequency B, which is called the *bandwidth* of the CT signal $x(t)$. We note that the bandwidth of a signal is the maximum frequency component it contains.

4.2 SAMPLING THEOREM

From Fig. 4.1-1, it is apparent that the CT signal $x(t)$ can be recovered from the sequence $x(mT)$ by passing it through an *ideal* low-pass filter whose bandwidth is B Hz. This is because an ideal low-pass filter has magnitude and phase responses as shown in Fig. 4.1-2, where it is noted

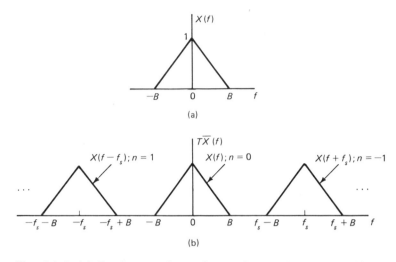

(a)

(b)

Fig. 4.1-1 (a) Fourier transform of a continuous-time signal $x(t)$;
(b) Fourier transform of the sequence $x(mT)$ obtained by sampling
$x(t)$ at $f_s = 1/T$ sps.

that all frequencies beyond B Hz are completely removed, while those
in the range $|f| \leq B$ are passed through without any attenuation. Fur-
thermore, all frequencies in the range $|f| \leq B$ are subject to exactly the
same delay, since the phase function is a *linear* function of f; that is,

$$T_d(f) = -\frac{d}{df} \text{ (phase function)}$$

$$= c, \qquad |f| \leq B$$

where $T_d(f)$ is the time-delay function, which was defined in (1.11-17).

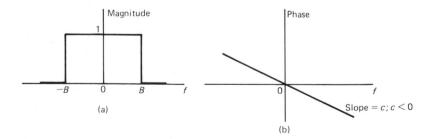

Fig. 4.1-2 Magnitude and phase responses of an ideal low-pass filter with
bandwidth B hertz.

Next, from Fig. 4.1-1b it is clear that, if *exact reconstruction* of $x(t)$ is to be achieved, there must be no overlap between adjacent components of $\bar{X}(f)$. For example, there must be no overlap between $X(f)$ and $X(f - f_s)$ or $X(f + f_s)$. As such, we observe that the following condition has to be satisfied:

$$f_s - B \geq B \tag{4.2-1}$$

or
$$f_s \geq 2B$$

where f_s is in hertz or samples per second. That is, the sampling frequency must be at least twice as large as the bandwidth B of the CT signal $x(t)$. This fundamental result is known as the *sampling theorem*. It was developed by Shannon in 1948 [1]. In essence, it establishes a *minimum* rate for sampling a given band-limited CT signal so that it can be *uniquely* determined by its sampled values. This is because the minimum sampling rate preserves all the information in a given signal; any higher rate of sampling also preserves this information. Again, the frequency $f_s/2$ is defined as the *Nyquist frequency* and will be denoted by f_N.

4.3 ALIASING

A natural question that one might ask as a sequel to the discussion of the sampling theorem is the following: what happens if f_s is *less than* 2B sps? In other words, what happens if the requirement in (4.2-1) is not satisfied? We now show that a phenomenon known as *aliasing* occurs if the minimum sampling frequency requirement is violated.

We consider two frequencies f_1 and f_2 such that $f_1 = 1$ Hz and $f_2 = 4$ Hz, and suppose the sampling frequency $f_s = 5$ sps, rather than *at least* 8 sps in accordance with the sampling theorem. Then, given a set of sampled values $x(mT)$, $0 \leq m \leq 4$, as shown in Fig. 4.3-1, one *cannot* determine whether they came from $\sin(2\pi f_1 t)$ or $\sin(2\pi f_2 t)$. Therefore, f_1 and f_2 are said to be *aliases* of one another, in the sense that they cannot be distinguished by their equally spaced sampled values. In general, f_2 is related to f_1 as follows [2]:

$$f_2 = 2kf_N \pm f_1, \qquad k \geq 1 \tag{4.3-1}$$

where f_N is the Nyquist frequency.

For example, if $f_1 = 1$ and $f_s = 5$, then the set of aliases f_2 corresponding to f_1 is given by (4.3-1) to be

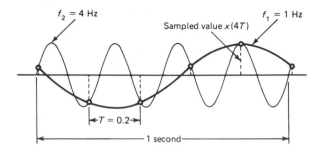

Fig. 4.3-1 Aliasing [2].

$$f_2 = 2kf_N \pm 1$$

$$= (5k \pm 1), \qquad k = 1, 2, 3, \cdots$$

(4.3-2)

The information provided by (4.3-2) is conveniently displayed in the form of an *aliasing diagram*, as shown in Fig. 4.3-2.

From Fig. 4.3-2 it is apparent that, when aliasing occurs, noise frequencies in $x(t)$ that are higher than f_N will *fold over* into the region $0 \le f \le f_N$. As such, the Nyquist frequency f_N is also known as the *folding frequency*.

We conclude our discussion related to aliasing with an illustrative example.

Example 4.3-1: Suppose that CT sinusoid

$$x(t) = \cos \lfloor (2\pi)(20)t \rfloor, \qquad t \ge 0$$

(4.3-3)

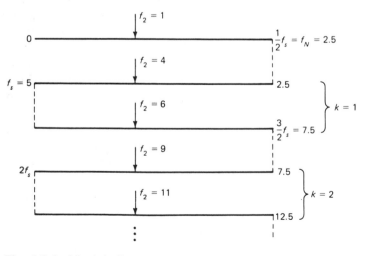

Fig. 4.3-2 Aliasing diagram.

is sampled at 30 sps to obtain the sampled signal

$$x^*(t) = \sum_{m=0}^{\infty} \cos\left[(2\pi)(20)mT\right]\delta(t - mT)$$

where $T = \dfrac{1}{30}$ second. If $x^*(t)$ is passed through an ideal low-pass filter whose bandwidth is 25 Hz, find the frequencies present in the resulting output $x'(t)$.

Solution: The bandwidth of the sinusoid $x(t)$ in (4.3-3) is 20 Hz. The sampling theorem thus requires f_s to be at least 40 sps. However, since $f_s = 30$ sps, aliasing occurs, as depicted in the following diagram.

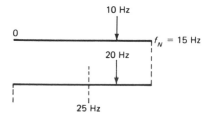

From this aliasing diagram it is clear that the 20-Hz component folds over into the (0 to f_N) range at the 10-Hz point. Thus $x'(t)$ will consist of two sinusoids with frequencies of 10 and 20 Hz. One can come to the same conclusion by examining $X(f)$ and $X^*(f)$ that are sketched in Fig. 4.3-3, which readily follow from (4.1-11b). We note that if f_s were 40 sps or higher, then $x'(t)$ would consist of a single 20-Hz sinusoid, as is the case with the input $x(t)$; that is, no aliasing occurs.

4.4 FOURIER SERIES EXPANSION OF PERIODIC SAMPLED SIGNALS

Let $x_p(t)$ be a CT periodic signal whose period is L seconds; see Fig. 4.4-1. We can represent $x_p(t)$ in the form of a FS using (1.2-1) and (1.2-4), as follows:

$$x_p(t) = \frac{1}{L} \sum_{n=-\infty}^{\infty} d_n e^{jn\omega_0 t} \tag{4.4-1}$$

where $\quad d_n = \displaystyle\int_0^L x_p(t) e^{-jn\omega_0 t}\, dt$

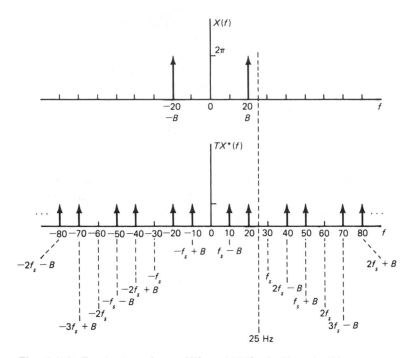

Fig. 4.3-3 Fourier transforms $X(f)$ and $X^*(f)$ of $x(t)$ and $x^*(t)$, respectively, defined in Example 4.3-1.

is the nth FS coefficient, and $\omega_0 = 2\pi/L$ is the fundamental radian frequency.

Next we assume that $x_p(t)$ is band limited. Then its FS expansion in (4.4-1) has a finite number of harmonics; thus

$$d_n = 0, \qquad |n| > M \qquad (4.4\text{-}2)$$

where M is a positive integer. This means that the bandwidth, B, of $x_p(t)$ is given by

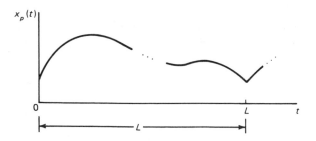

Fig. 4.4-1 Periodic signal $x_p(t)$.

$$B = M f_0 = \frac{M}{L} \tag{4.4-3}$$

where $f_0 = 1/L$ is the fundamental frequency in hertz.

If we now wish to sample $x_p(t)$, then the minimum sampling frequency is $2M/L$ sps, according to the sampling theorem requirement in (4.2-1). Hence we let

$$f_s = \frac{2M}{L} \text{ sps} \tag{4.4-4}$$

where f_s denotes the sampling frequency. This results in the $2M$-periodic sequence

$$x(m) = x(mT), \qquad 0 \le m \le 2M - 1 \tag{4.4-5}$$

over an interval of length L.

The corresponding sampled signal, $x_p^*(t)$, is

$$x_p^*(t) = \sum_{m=0}^{2M-1} x(m)\delta(t - mT) \tag{4.4-6}$$

Our objective is to find the FS expansion of $x_p^*(t)$, which is depicted in Fig. 4.4-2. Thus, corresponding to (4.4-1), we have

$$x_p^*(t) = \frac{1}{L} \sum_{n=-M}^{M} c_n e^{jn\omega_0 t} \tag{4.4-7}$$

where $\quad c_n = \int_0^L x_p^*(t) e^{-jn\omega_0 t} dt, \qquad |n| \le M$

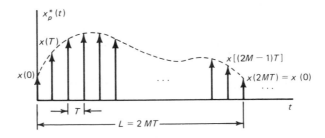

Fig. 4.4-2 Periodic DT signal $x^*{}_p(t)$.

To evaluate the c_n in (4.4-7), we substitute (4.4-6) in the expression for c_n. This results in

$$c_n = \int_0^L \left[\sum_{m=0}^{2M-1} x(m)\delta(t - mT) \right] e^{-jn\omega_0 t} \, dt \qquad (4.4\text{-}8)$$

Interchanging the order of integration and summation in (4.4-8), there results

$$c_n = \sum_{m=0}^{2M-1} x(m) \left[\int_0^L e^{-jn\omega_0 t} \, \delta(t - mT) \, dt \right]$$

which yields

$$c_n = \sum_{m=0}^{2M-1} x(m) e^{-jnm\omega_0 T} \qquad (4.4\text{-}9)$$

since

$$\int_0^L e^{-jn\omega_0 t} \delta(t - mT) \, dt = \left[e^{-jn\omega_0 t} \right]_{t=mT}$$

With $L = 2MT$ and $\omega_0 = 2\pi/L$, (4.4-9) becomes

$$c_n = \sum_{m=0}^{2M-1} x(m) e^{-jnm\pi/M}, \qquad |n| \le M \qquad (4.4\text{-}10)$$

which yields the FS coefficients required to obtain $x_p^*(t)$ in (4.4-7). In (4.4-10) we note that the Mth harmonic corresponds to the Nyquist (folding) frequency, f_N. This is because $Mf_0 = f_N$ with $f_0 = 1/L$ and $L = 2MT$.

4.5 DISCRETE FOURIER TRANSFORM

The discrete Fourier transform (DFT) is a Fourier representation of a given sequence $x(m)$, $0 \le m \le N - 1$, and is defined as

$$X(n) = \sum_{m=0}^{N-1} x(m) W^{nm}, \qquad 0 \le n \le N - 1 \qquad (4.5\text{-}1)$$

where N is finite,

$$W = e^{-j2\pi/N}$$

and $X(n)$ is the nth DFT coefficient. We note that W is the Nth root of unity, since $e^{-j2\pi} = 1$. Again, since the W^{nm} are N-periodic, the DFT coefficients are also N-periodic.

It is easy to show that (Problem 4-6) the DFT is directly related to the FS of the periodic sampled signal $x_p^*(t)$ in (4.4-6) as follows:

$$c_0 = X(0)$$

$$c_k = X(k)$$

$$c_{-k} = X(2M - k), \qquad 1 \le k < M$$

and

$$c_{\pm M} = X(\pm M)$$

(4.5-2)

where the c_i are given by (4.4-10).

The DFT can be manipulated to obtain a very efficient algorithm to compute it. This algorithm is used widely in a variety of disciplines; it is known as the *fast Fourier transform*, details of which will be discussed in Sections 4.6 through 4.8.

The exponential functions W^{nm} in (4.5-1) are orthogonal, and hence have the property

$$\sum_{n=0}^{N-1} W^{(m-k)n} = \begin{cases} N, & \text{if } m = k \\ 0, & \text{otherwise} \end{cases}$$

(4.5-3)

Combining (4.5-1) and (4.5-3), one can show that (Problem 4-5)

$$x(m) = \frac{1}{N} \sum_{n=0}^{N-1} X(n)W^{-nm}, \qquad 0 \le m \le N - 1$$

(4.5-4)

Equation (4.5-4) implies that, if the DFT coefficients $X(n)$ are known, then $x(m)$ can be recovered *uniquely*. Thus (4.5-4) is defined as the inverse discrete Fourier transform (IDFT).

COMMENT. Note that the scalar $1/N$ that appears in the IDFT in (4.5-4) does not appear in the DFT. However, if we had chosen to define the DFT with the scalar $1/N$, then it would not have appeared in the IDFT. Clearly, both forms of these definitions are equivalent.

We now discuss some properties of the DFT via examples.

Example 4.5-1: If $x(m)$, $0 \leq m \leq 2M - 1$, is a *real* sequence, show that

$$X(M + k) = \tilde{X}(M - k), \qquad 0 \leq k \leq M \qquad (4.5\text{-}5)$$

where $\tilde{X}(n)$ denotes the complex conjugate of $X(n)$.

Solution: By definition of the DFT, we have

$$X(n) = \sum_{m=0}^{2M-1} x(m)W^{nm}, \qquad 0 \leq n \leq 2M - 1$$

where $W = e^{-j2\pi/2M}$

Thus

$$X(M + k) = \sum_{m=0}^{2M-1} x(m)W^{(M+k)m}$$

$$= \sum_{m=0}^{2M-1} x(m)W^{-(M-k)m}W^{2Mm} \qquad (4.5\text{-}6)$$

Now $W = e^{-j2\pi/2M}$ implies that

$$W^{2Mm} = (e^{-j2\pi})^m \equiv 1 \qquad \text{for all } m$$

since $e^{-j2\pi} = 1$. Hence (4.5-6) simplifies to yield

$$X(M + k) = \sum_{m=0}^{2M-1} x(m)W^{-(M-k)m} \qquad (4.5\text{-}7)$$

The definition of the DFT in (4.5-1) implies that

$$\tilde{X}(n) = \sum_{m=0}^{2M-1} x(m)W^{-nm} \qquad (4.5\text{-}8)$$

since $\tilde{W}^{nm} = W^{-nm}$. Equation (4.5-8) implies that the right-hand side of (4.5-7) is $\tilde{X}(M - k)$. Thus

$$X(M + k) = \tilde{X}(M - k)$$

which is the desired result in (4.5-5).

We shall refer to (4.5-5) as the *complex-conjugate* property of the DFT. It states that only the first $(M + 1)$ DFT coefficients of a real data sequence with $2M$ points are independent. The rest can be obtained from these as illustrated in the following.

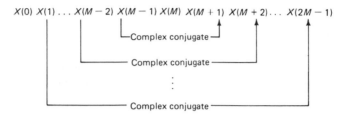

Example 4.5-2: If $x(m)$, $0 \leq m \leq 2M - 1$, is a real sequence, show that $X(0)$ and $X(M)$ are real numbers.

Solution: From the DFT definition in (4.5-1) we have

$$X(n) = \sum_{m=0}^{2M-1} x(m)W^{nm}, \qquad 0 \leq n \leq 2M - 1$$

which yields

$$X(0) = \sum_{m=0}^{2M-1} x(m) \tag{4.5-9a}$$

and

$$X(M) = \sum_{m=0}^{2M-1} x(m)W^{Mm} \tag{4.5-9b}$$

From (4.5-9a) it is clear that $X(0)$ is a real number if the data sequence is real, and is the dc component of the sequence.

Next, since $W = e^{-j2\pi/M}$, we have $W^M = e^{-j\pi} = -1$. Thus (4.5-9b) becomes

$$X(M) = \sum_{m=0}^{2M-1} x(m)(-1)^m$$

which is also a real number.

Example 4.5-3: Develop power, amplitude, and phase spectra for the DFT. For convenience, assume that $x(m)$, $0 \leq m \leq 2M - 1$, is real.

Solution: To define a power spectrum, we derive Parseval's theorem for the DFT. To this end, we take the complex conjugate of both sides of the DFT definition in (4.5-1) to obtain

$$\widetilde{X}(n) = \sum_{m=0}^{2M-1} x(m)W^{-nm}, \qquad 0 \le n \le 2M - 1$$

where the symbol $\sim$ denotes complex conjugate, and $\widetilde{W}^{nm} = W^{-nm}$. The product of $X(n)$ in (4.5-1) and the preceding $\widetilde{X}(n)$ leads to

$$|X(n)|^2 = \sum_{p=0}^{2M-1} \sum_{q=0}^{2M-1} x(p)x(q)W^{+np}W^{-nq}$$

Summing both sides of this equation over n, we get

$$\sum_{n=0}^{2M-1} |X(n)|^2 = \sum_{p=0}^{2M-1} \sum_{q=0}^{2M-1} x(p)x(q)\left[\sum_{n=0}^{2M-1} W^{+np}W^{-nq}\right]$$

which simplifies due to the orthogonality property in (4.5-3) to yield

$$\sum_{n=0}^{2M-1} |X(n)|^2 = 2M \sum_{p=0}^{2M-1} x^2(p)$$

That is,

$$\sum_{m=0}^{2M-1} x^2(m) = \frac{1}{2M} \sum_{n=0}^{2M-1} |X(n)|^2 \tag{4.5-10}$$

Next, from the complex-conjugate property in (4.5-5) it follows that (4.5-10) can be expressed as

$$\sum_{m=0}^{2M-1} x^2(m) = \frac{1}{2M}\left[X^2(0) + 2\sum_{n=1}^{M-1} |X(n)|^2 + X^2(M)\right] \tag{4.5-11}$$

or $\dfrac{1}{N}\displaystyle\sum_{m=0}^{N-1} x^2(m) = \dfrac{1}{N^2}\left[X^2(0) + 2\sum_{n=1}^{M-1} |X(n)|^2 + X^2(M)\right]$

with $N = 2M$.

We refer to (4.5-11) as Parseval's theorem for the DFT. If $x(m)$ is obtained by sampling a periodic voltage or current signal across a 1-Ω

resistance, then the left-hand side of (4.5-11) represents the average power in the sequence. Thus we define the *DFT power spectrum* of a real $x(m)$ as

$$P_0 = \frac{X^2(0)}{N^2}$$

$$P_n = \frac{2|X(n)|^2}{N^2}, \qquad 1 \le n < M \qquad (4.5\text{-}12)$$

and

$$P_M = \frac{X^2(M)}{N^2}$$

where P_0 is the dc power, and P_M is the power in the Mth harmonic, that is, the Nyquist frequency.

Next, since $X(n)$ is a complex number, we can express it in polar form as

$$X(n) = |X(n)|\, e^{j\psi_n}$$

which leads to the definitions

$$p_n^x = |X(n)| \qquad (4.5\text{-}13\text{a})$$

as the *DFT amplitude* spectrum, and

$$\psi_n^x = \tan^{-1}\left\{\frac{\text{Im}[X(n)]}{\text{Re}[X(n)]}\right\} \qquad (4.5\text{-}13\text{b})$$

as the *DFT phase spectrum.*

Example 4.5-4: Corresponding to a given N-periodic sequence $x(m)$, $0 \le m \le N - 1$, let $z(m)$ be a shifted sequence whose elements are obtained from $x(m)$ as follows:

$$z(m) = x(m - h), \qquad 0 \le h \le N - 1$$

where h is the amount by which $x(m)$ is shifted to the *right*. Show that

$$Z(n) = W^{nh}X(n), \qquad 0 \le n \le N - 1 \qquad (4.5\text{-}14)$$

Solution: The DFT of $z(m)$ is given by

$$Z(n) = \sum_{m=0}^{N-1} z(m)W^{nm}$$

which yields

$$Z(n) = \sum_{m=0}^{N-1} x(m - h)W^{nm}$$

Let $m - h = \gamma$. Then we have

$$Z(n) = W^{nh} \left\{ \sum_{\gamma=-h}^{-h+N-1} x(\gamma)W^{n\gamma} \right\} \qquad (4.5\text{-}15)$$

Now, since $x(\gamma)$ and $W^{n\gamma}$, $0 \le \gamma \le N - 1$, are both N-periodic sequences, their product is also N-periodic, and hence (see Problem 4-8)

$$\sum_{\gamma=-h}^{-h+N-1} x(\gamma)W^{n\gamma} \equiv \sum_{\gamma=0}^{N-1} x(\gamma)W^{n\gamma} \qquad (4.5\text{-}16)$$

Combining (4.5-15) and (4.5-16), we get

$$Z(n) = W^{nh} \left\{ \sum_{\gamma=0}^{N-1} x(\gamma)W^{n\gamma} \right\}$$

That is,

$$Z(n) = W^{nh}X(n), \qquad 0 \le n \le N - 1$$

which is (4.5-14).

The result is (4.5-14) is known as a DFT *shift theorem*. It states that the DFT coefficients of a shifted N-periodic sequence are related to those of the original sequence via the scalar W^{nh}, where h is the size of the shift.

Example 4.5-5: If $x(m)$ and $y(m)$ are real N-periodic sequences, their convolution sequence $z(m)$ is defined as

$$z(m) = \sum_{h=0}^{N-1} x(h)y(m - h), \qquad 0 \le m \le N - 1 \qquad (4.5\text{-}17)$$

where $0 \le h \le N - 1$ is a shift parameter. Show that

$$Z(n) = X(n)Y(n), \qquad 0 \le n \le N - 1 \qquad (4.5\text{-}18)$$

Solution: The DFT of $z(m)$ is given by

$$Z(n) = \sum_{m=0}^{N-1} z(m)W^{nm}$$

which upon substituting for $z(m)$ in (4.5-17) yields

$$Z(n) = \sum_{m=0}^{N-1} \sum_{h=0}^{N-1} x(h)y(m - h)W^{nm} \tag{4.5-19}$$

Interchanging the order of summation in (4.5-19), we obtain

$$Z(n) = \sum_{h=0}^{N-1} x(h) \left\{ \sum_{m=0}^{N-1} y(m - h)W^{nm} \right\} \tag{4.5-20}$$

Thus from the shift theorem in (4.5-14), with $z(m) = y(m - h)$, and (4.5-20) we have

$$Z(n) = Y(n) \left\{ \sum_{h=0}^{N-1} x(h)W^{nh} \right\}$$

$$= X(n)Y(n)$$

which is (4.5-18).

Equations (4.5-17) and (4.5-18) represent *discrete convolution in time*. It implies that convolution in the DT domain is equivalent to multiplication in the DFT domain.

COMMENT. A corollary to the preceding result is the following, which represents *discrete convolution in frequency:*
If

$$z(m) = x(m)y(m), \qquad 0 \le m \le N - 1 \tag{4.5-21}$$

then

$$Z(n) = \frac{1}{N} \sum_{h=0}^{N-1} X(h)Y(n - h), \qquad 0 \le n \le N - 1 \tag{4.5-22}$$

for $0 \le h \le N - 1$.

From (4.5-21) and (4.5-22), it follows that multiplication in the

DT domain is equivalent to convolution in the DFT domain. The proof related to this result is left as an exercise; see Problem 4-13.

Example 4.5-6: Consider the sequence $x(m)$, $0 \le m \le N - 1$, which is obtained by sampling a band-limited signal $x(t)$. Let the sampling interval be T. Find the relation between its FT, $\bar{X}(\omega)$, and its DFT coefficients $X(n)$.

Solution: By definition, the DFT is given by

$$X(n) = \sum_{m=0}^{N-1} x(m)W^{nm}, \qquad 0 \le n \le N - 1 \qquad (4.5\text{-}23)$$

where $W = e^{-j2\pi/N}$

Again, (3.1-8) yields the FT of $x(m)$ to be

$$\bar{X}(\omega) = \sum_{m=0}^{N-1} x(m)e^{-jm\omega T} \qquad (4.5\text{-}24)$$

where $\bar{X}(\omega) = X(e^{j\omega T})$ yields the FT for *all* ω.

Now suppose we desire $\bar{X}(\omega)$ at only a set of values that are $\omega_0 = 2\pi/NT$ radians/second apart; that is, the spacing between the FT values is the fundamental radian frequency. Then ω in (4.5-24) is replaced by $n\omega_0$ to obtain

$$\bar{X}(n\omega_0) = \sum_{m=0}^{N-1} x(m)e^{-jnm\omega_0 T}, \qquad 0 \le n \le N - 1 \qquad (4.5\text{-}25)$$

Again, since $\omega_0 = 2\pi/NT$, (4.5-25) becomes

$$\bar{X}(n\omega_0) = \sum_{m=0}^{N-1} x(m)e^{-j2\pi nm/N} \qquad (4.5\text{-}26)$$

or

$$\bar{X}(n\omega_0) = \sum_{m=0}^{N-1} x(m)W^{nm}, \qquad 0 \le n \le N - 1$$

since $W = e^{-j2\pi/N}$.

Comparing (4.5-23) and (4.5-26), we get the relation

$$X(n) = \bar{X}(n\omega_0), \qquad 0 \le n \le N - 1 \qquad (4.5\text{-}27)$$

Equation (4.5-27) implies that the DFT coefficients yield the FT of a given sequence at a set of points that are $\omega_0 = 2\pi/NT$ radians/second apart.

We also note that (3.1-9) and (4.5-27) relate the DFT and the ZT as follows:

$$X(n) = X(z) \big|_{z = e^{jn\omega_0 T}} \qquad (4.5\text{-}28)$$

Thus the DFT of a sequence yields its ZT *on the unit circle* at a set of points that are $\omega_0 T = 2\pi/N$ radians apart.

4.6 FAST FOURIER TRANSFORM

As mentioned in the previous section, the fast Fourier transform (FFT) is a very efficient algorithm for computing the DFT. The need for such an algorithm is apparent by examining the number of arithmetic operations that are required for a direct computation of the DFT via (4.5-1). To this end, we write (4.5-1) in the form

$$
\begin{aligned}
X(n) &= \sum_{m=0}^{N-1} x(m) W_N^{nm} \\
&= [x(0) + x(1)\, W_N^n + \cdots + x(N-1)\, W_N^{n(N-1)}]
\end{aligned}
\qquad (4.6\text{-}1)
$$

where N is the total number of points in the data sequence $x(m)$, $0 \le n \le N - 1$, and $W_N = e^{-j2\pi/N}$.

From (4.6-1) it follows that to compute each $X(n)$ we need approximately N complex multiplications and N complex additions. Since there are N DFT coefficients, this implies that a total of approximately N^2 complex multiplications and N^2 complex additions is required. As such, the number of arithmetic operations required to obtain the DFT by this type of direct computation becomes very large for large values of N. Thus computational procedures that reduce the number of multiplications and additions are of considerable interest.

There is also another serious drawback in using the direct computation approach. It concerns the multipliers W_N^{nm} in (4.6-1). Computing and storing them becomes a formidable (if not impossible) task for large values of N.

Most approaches that are used to improve the efficiency of the computation of the DFT exploit the fact that there is a great deal of redundancy in the values taken by W_N and its powers. For example, different powers of W_N have the same value, as illustrated in Fig. 4.6-1

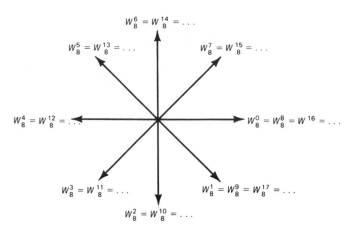

Fig. 4.6-1 Some powers of W$_8$ that are equivalent.

for $N = 8$. Also $W^{nm} = \widetilde{W}^{nm}$, and $W^{nm} = -\widetilde{W}^{nm}$ for certain combinations of m and n. Thus by carrying out the computations in an organized way, arithmetic operations associated with powers of W_N that have the same value can be avoided. The first major contribution in the regard was made in 1965 by Cooley and Tukey [3]. The publication of their paper led to the development of a variety of efficient algorithms, which are collectively referred to as "the FFT."

Several texts that provide a detailed study of the FFT are available [4 to 10]. For our purposes, it is sufficient to derive some commonly used FFT algorithms for the case when N is a power of 2; that is, $N = 2, 4, 8, 16, 32$, etc. Clearly, there is no loss in generality in restricting out attention to the case when N is a power of 2, as long as N is large enough to satisfy the sampling theorem requirement in (4.2-1).

Algorithm Development

The first step consists of dividing the data sequence into two halves, as follows:

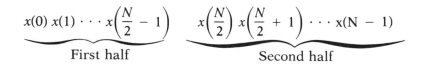

Then $X(n)$ in (4.6-1) can be expressed as the sum of two components to obtain

$$X(n) = \left[\sum_{m=0}^{(N/2)-1} x(m)W_N^{nm} + \sum_{m=N/2}^{N-1} x(m)W_N^{nm} \right]$$

(4.6-2)

$$\text{or } X(n) = \left[\sum_{m=0}^{(N/2)-1} x(m)W_N^{nm} + W_N^{(N/2)n} \sum_{m=0}^{(N/2)-1} x\left(m + \frac{N}{2}\right) W_N^{nm} \right]$$

In (4.6-2) we note that $W^{(N/2)n} = (-1)^n$. Hence it simplifies to yield

$$X(n) = \sum_{m=0}^{(N/2)-1} \left[x(m) + (-1)^n x\left(m + \frac{N}{2}\right) \right] W_N^{nm}$$

(4.6-3)

We now consider even and odd values of the index n in (4.6-3) separately, and let $X(2k)$ and $X(2k + 1)$ represent the even-numbered and odd-numbered DFT coefficients, respectively. As such, (4.6-3) can be split into two equations as follows:

$$X(2k) = \sum_{m=0}^{(N/2)-1} \left[x(m) + x\left(m + \frac{N}{2}\right) \right] W_N^{2km}$$

(4.6-4)

$$\text{and} \quad X(2k + 1) = \sum_{m=0}^{(N/2)-1} \left[x(m) - x\left(m + \frac{N}{2}\right) \right] W_N^m W_N^{2km}$$

where $0 \le k \le (N/2) - 1$.

Next we note that the terms W_N^{2km} can equivalently be written as $W_{N/2}^{km}$, since $W_N = e^{-j2\pi/N}$. Thus (4.6-4) yields

$$X(2k) = \sum_{m=0}^{(N/2)-1} \left[x(m) + x\left(m + \frac{N}{2}\right) \right] W_{N/2}^{km}$$

(4.6-5)

$$\text{and} \quad X(2k + 1) = \sum_{m=0}^{(N/2)-1} \left[x(m) - x\left(m + \frac{N}{2}\right) \right] W_N^m W_{N/2}^{km}$$

Now let

$$g(m) = x(m) + x\left(m + \frac{N}{2}\right)$$

(4.6-6)

$$\text{and} \quad h(m) = x(m) - x\left(m + \frac{N}{2}\right), \quad 0 \le m \le (N/2) - 1$$

Then from (4.6-5) and (4.6-6) it follows that the N-point DFT in (4.6-1) can now be computed via two $N/2$-point DFTs as follows:

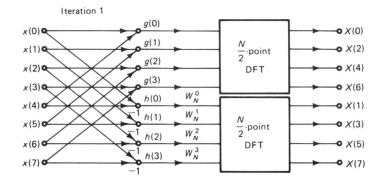

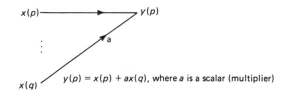

Fig. 4.6-2 Flow graph illustrating decomposition of an N-point discrete Fourier transform into two $N/2$ discrete Fourier transforms for $N = 8$.

1. Form the sequences $g(m)$ and $h(m)$, $0 \leq m \leq (N/2) - 1$ as defined in (4.6-6).

2. Compute the sequence $h(m)W_N^m$, $0 \leq m \leq (N/2) - 1$ using $h(m)$ from step 1.

3. Compute $N/2$-point DFTs of the sequences obtained in steps 1 and 2 to obtain the even- and odd-numbered DFT coefficients, respectively.

This procedure is illustrated for the case $N = 8$ in Fig. 4.6-2 and is referred to as *iteration 1*. We shall refer to this figure as a *flow graph*, which uses the following notation:

Since N is a power of 2, $N/2$ is even, and hence each of the $N/2$-point DFTs in (4.6-5) can be computed via two $N/4$-point DFTs by merely repeating the process. This results in moving from Fig. 4.6-2 to Fig. 4.6-3. We note that $X(0)$, $X(2)$, $X(4)$, and $X(6)$ in Fig. 4.6-2 are reordered in Fig. 4.6-3 as $X(0)$, $X(4)$, $X(2)$, and $X(6)$, respectively. This is because $X(0)$, $X(2)$, $X(4)$, and $X(6)$ are considered to be the 0th, 1st, 2nd, and 3rd DFT coefficients in a 4-point DFT. Rearranging them as even- and odd-numbered coefficients results in the ordering 0th, 2nd, 1st, and 3rd, which yields $X(0)$, $X(4)$, $X(2)$, and $X(6)$ as shown in Fig. 4.6-3. Similarly, the coefficients $X(1)$, $X(3)$, $X(5)$, and $X(7)$ in Fig. 4.6-2 appear in the order $X(1)$, $X(5)$, $X(3)$, and $X(7)$, respectively, in Fig. 4.6-3.

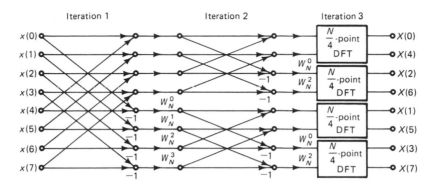

Fig. 4.6-3 Flow graph illustrating decomposition of an N-point discrete Fourier transform into four $N/4$-point discrete Fourier transforms for $N = 8$.

It is easily seen that, by repeating the process associated with (4.6-2), we will finally end up with a set of 2-point DFTs, since N is a power of 2. This aspect is evident from Fig. 4.6-3, where the $N/4$-point DFTs under iteration 3 are 2-point DFTs since $N = 8$. From the definition of the DFT in (4.6-1) it follows that $W_2 = -1$, and hence a 2-point DFT can be computed by merely adding and subtracting the input points, as illustrated in Fig. 4.6-4. Because of the appearance of the flow graph in Fig. 4.6-4, it is occasionally called a *butterfly* computation.

Replacing the $N/4$-point DFTs in Fig. 4.6-3 by butterflies, we obtain the flow graph in Fig. 4.6-5, which we shall refer to as an *FFT flow graph*.

We make the following observations with respect to the FFT flow graph in Fig. 4.6-5.

1. The number of iterations is 3, which equals $\log_2 8$. In general, when $N = 2^u$, there are $u = \log_2 N$ iterations.

2. In the rth iteration, $1 \le r \le u$, the multipliers are

$$W_N^{(2^{r-1})s}, \qquad 0 \le s \le 2^{u-r} - 1 \qquad (4.6-7)$$

where $u = \log_2 N$.

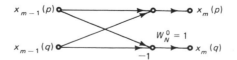

Fig. 4.6-4 Flow graph for computing a 2-point discrete Fourier transform.

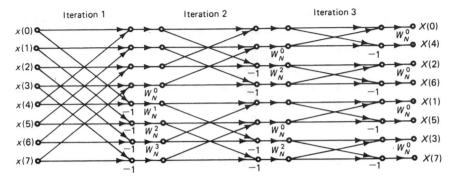

Fig. 4.6-5 Flow graph illustrating complete decomposition of an 8-point discrete Fourier transform for $N = 8$.

3. The order in which the DFT coefficients are obtained is 0, 4, 2, 6, 1, 5, 3, 7. This ordering is related to the normal ordering 0, 1, 2, 3, 4, 5, 6, 7 via a *bit-reversal* process, as illustrated in Fig. 4.6-6. From Fig. 4.6-6 it follows that there is a one-to-one correspondence between the members of the sequences

$$S_8 = \{0, 1, 2, 3, 4, 5, 6, 7\}$$

and

$$\overleftarrow{S}_8 = \{0, 4, 2, 6, 1, 5, 3, 7\}$$

Each member of $\overleftarrow{S}_8$ is the decimal number equivalent of the bit string obtained by reversing the 3-bit representation of its counterpart in S_8. The bit-reversal process (algorithm) in Fig. 4.6-6 can be used for any value of N that is a power of 2.

4. The output of the rth iteration can be stored in the same locations as those of its input. To see this, let us consider the output points $g(0) =$

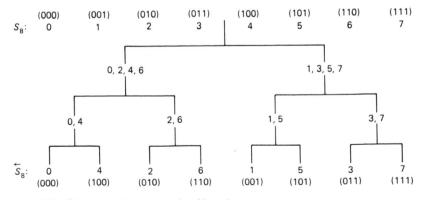

Fig. 4.6-6 Bit-reversal process for $N = 8$.

$x(0) + x(4)$ and $h(0) = x(0) - x(4)$ in Fig. 4.6-2. It is apparent that $g(0)$ and $h(0)$ can be stored in two temporary locations, $T1$ and $T2$, by computing $T1 = x(0) + x(4)$ and $T2 = x(0) - x(4)$. Then, since $x(0)$ and $x(4)$ are no longer needed for any further computations in iteration 1 (see Fig. 4.6-2), the contents of $T1$ and $T2$ can be stored back in the locations of $x(0)$ and $x(4)$. That is, $g(0)$ and $h(0)$ are stored in the locations of $x(0)$ and $x(4)$, respectively. Similarly, the locations of $x(1)$ and $x(5)$ are used to store $g(1)$ and $h(1)$, and so on. The same procedure can be used for the outputs of iterations 2 and 3 in Fig. 4.6-5. This convenient feature of the FFT flow graph in Fig. 4.6-5 is known as the *in-place* property. As such, essentially no additional storage is required as the data are being processed. The total number of storage locations required is $2N$, assuming complex data sequences. In addition, $2N$ storage locations are required for the bit-reversal process.

4.7 INVERSE FAST FOURIER TRANSFORM

The algorithm development approach we introduced in the previous section in connection with the FFT can easily be used to develop an efficient algorithm to compute the IDFT. This is apparent from similarities present in the DFT and IDFT definitions given in (4.5-1) and (4.5-4), respectively. As in the case of the DFT, we let $N = 2^u$ and $W_N = e^{-j2\pi/N}$. This causes the IDFT in (4.5-4) to become

$$x(m) = \frac{1}{N} \sum_{n=0}^{N-1} X(n)\, W_N^{-nm}, \qquad 0 \le m \le N - 1 \qquad (4.7\text{-}1)$$

where $W_N = e^{-j2\pi/N}$.

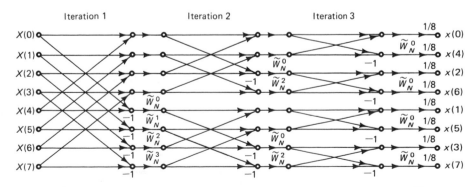

Fig. 4.7-1 Inverse fast Fourier transform flow graph for $N = 8$, where $\widetilde{W}_N$ is the complex conjugate of W_N.

Comparing (4.7-1) with the DFT definition in (4.6-1), we see that an FFT algorithm can be used to compute the IDFT by accounting for the term $1/N$ in (4.7-1) and using powers of W_N^{-1}, instead of powers of W_N. Thus, corresponding to the FFT flow graph in Fig. 4.6-5, we obtain the *IFFT flow graph* in Fig. 4.7-1 for $N = 8$, where $\tilde{W}_N$ denotes the complex conjugate of W_N. Here we note that the DFT coefficients are in normal order while the data sequence is in bit-reversed order at the output.

From Figs. 4.6-5 and 4.7-1 it is clear that the FFT and IFFT flow graphs are essentially identical. Hence the advantages we cited in the previous section for the FFT are also valid for the IFFT when the latter is compared to a direct computation of the IDFT.

4.8 OTHER FORMS OF THE FAST FOURIER TRANSFORM

The FFT algorithm discussed in the last two sections is said to belong to a class of *decimation-in-frequency* algorithms. This is because such algorithms are based upon the decomposition of the pertinent *output* sequence into smaller and smaller subsequences. The output sequence for the DFT computation is $X(n)$, while that for the IDFT computation is $x(m)$.

Alternately, FFT algorithms can be derived by forming smaller and smaller subsequences of the *input* sequence. The input sequence is $x(m)$, for the DFT computation and $X(n)$ for the IDFT computation. The class of FFT algorithms so obtained is called *decimation-in-time* algorithms. To illustrate, we outline the development of one such algorithm for computing the DFT

$$X(n) = \sum_{m=0}^{N-1} x(m)W_N^{nm}, \qquad 0 \le n \le N - 1 \qquad (4.8\text{-}1)$$

where $W_N = e^{-j2\pi/N}$, and N is an integer power of 2.

First, we divide the input sequence $x(m)$ into two subsequences consisting of even- and odd-numbered members, respectively, as follows:

$$\underbrace{x(0)\ x(2)\ \cdots\ x(N/2)}_{\text{even-numbered}} \qquad \underbrace{x(1)\ x(3)\ \cdots\ x(N-1)}_{\text{odd-numbered}}$$

Then (4.8-1) becomes

$$X(n) = \sum_{m=0}^{(N/2)-1} x(2m)W_N^{2nm} + \sum_{m=0}^{(N/2)-1} x(2m+1)W_N^{n(2m+1)}$$

$$= \sum_{m=0}^{(N/2)-1} x(2m)W_N^{2nm} + W_N^n \sum_{m=0}^{(N/2)-1} x(2m+1)W_N^{2nm}$$

Since $W_N^2 = W_{N/2}$, the preceding equation can be written as

$$X(n) = \sum_{m=0}^{(N/2)-1} x(2m)W_{N/2}^{nm} + W_N^n \sum_{m=0}^{(N/2)-1} x(2m+1)W_{N/2}^{nm} \qquad (4.8\text{-}2)$$

In (4.8-2) we recognize that each of the two summations is an $N/2$-point DFT. Thus we define

$$a(n) = \sum_{m=0}^{(N/2)-1} x(2m)W_{N/2}^{nm}$$

and

$$b(n) = \sum_{m=0}^{(N/2)-1} x(2m+1)W_{N/2}^{nm}$$

where $a(n)$ and $b(n)$ are DFT coefficients. Since $a(n)$ and $b(n)$ pertain to $N/2$-point DFTs, they are $N/2$-periodic; that is,

$$a\left(k + \frac{N}{2}\right) = a(k)$$

$$(4.8\text{-}3)$$

and

$$b\left(k + \frac{N}{2}\right) = b(k), \qquad 0 \le k \le \frac{N}{2} - 1$$

Substitution of (4.8-3) in (4.8-2) leads to

$$X(n) = a(n) + W_N^n b(n), \qquad 0 \le n \le N - 1 \qquad (4.8\text{-}4)$$

Equation (4.8-4) implies that the DFT coefficients $X(n)$, $0 \le n \le N - 1$, can be obtained by a *linear combination of two (N/2)-point DFTs*. Again, we note that W_N^n has the property

$$W_N^{(N/2+k)} = -W_N^k, \qquad 0 \le k \le \frac{N}{2} - 1 \qquad (4.8\text{-}5)$$

For example, let us now consider the case $N = 8$. Then (4.8-3) and (4.8-5) imply that (4.8-4) yields the following set of equations:

$$X(n) = a(n) + W_8^n b(n), \qquad 0 \le n \le 3$$

and

$$X(4 + n) = a(n) - W_8^n b(n), \qquad 0 \le n \le 3$$

$$(4.8\text{-}6)$$

Equation (4.8-6) represents the computations indicated in Fig. 4.8-1 under iteration 1.

Next, each 4-point DFT in Fig. 4.8-1 is decomposed into two 2-point DFTs by repeating the preceding decomposition process. This results in four 2-point DFTs, as shown in Fig. 4.8-2 under iteration 2. Then the 2-point DFTs are computed via the four butterflies (iteration 3) shown in Fig. 4.8-3, which represents a FFT algorithm based on the decimation-in-time approach for $N = 8$. Similarly, one can derive the corresponding IFFT algorithm; see Problem 4-12.

The FFT has a variety of applications; see [6, 16, 17]. Many of these applications involve the notions of *spectral estimation* and *fast convolution*. We shall discuss some aspects of spectral estimation in Section 4.9. Since DT convolution concepts will be studied in some detail in Chapter 5, a discussion of fast convolution will be undertaken toward the end of that chapter.

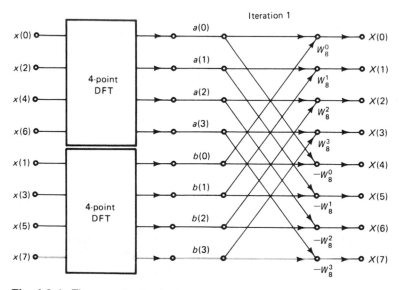

Fig. 4.8-1 Flow graph of a decimation-in-time computation of an 8-point discrete Fourier transform into two 4-point discrete Fourier transforms.

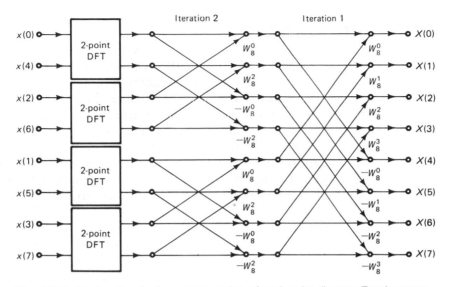

Fig. 4.8-2 Decimation-in-time computation of an 8-point discrete Fourier transform into four 2-point discrete Fourier transforms.

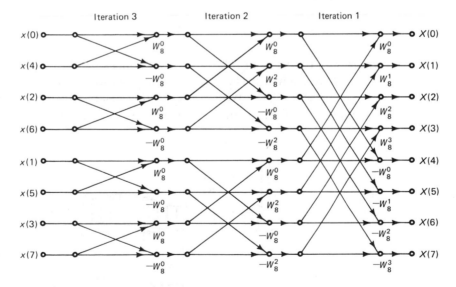

Fig. 4.8-3 Flow graph of a decimation-in-time fast Fourier transform to compute an 8-point discrete Fourier transform.

4.9 SPECTRAL ESTIMATION

The power density spectrum (PDS) defined in (1.6-25) is a very useful concept in the analysis of random signals, since it provides a meaningful measure for the distribution of the average power in such signals. The term *spectral estimation* refers to the process of estimating the PDS.

The subject of spectral estimation has received a greal deal of attention; for example, see [19 to 21]. An in-depth study of this subject matter requires a good background in the analysis of random signals, and hence is beyond the scope of this book. Thus our attention will be restricted to some computational aspects of estimating the PDS via the FFT.

Consider a sequence whose PDS is desired and whose length is indefinite. In practice, one typically has access to only a finite portion of it, say $x(n)$. Thus we can only estimate the actual PDS of the entire sequence using $x(n)$. If $\bar{G}_x(\omega) = G(e^{j\omega T})$ is the actual PDS, where T is the sampling interval, then let $\hat{G}_x(\omega)$ denote its estimate.

One way of computing $\hat{G}_x(\omega)$ consists of decomposing $x(m)$ into subsequences $x_\ell(m)$ of length M samples each. These subsequences are spaced D samples apart, where $D = M/2$, as depicted in Fig. 4.9-1. We note there is a 50% overlap between successive subsequences. It is apparent that the relation between $x_\ell(m)$ and $x(m)$ is

$$x_\ell(m) = x[m + (\ell - 1)D], \qquad 1 \le \ell \le K \qquad (4.9\text{-}1)$$

where K is the total number of subsequences used to estimate the PDS.

Next, each $x_\ell(m)$ is multiplied by a window sequence $w(m)$, which is a DT version of the window functions discussed in Section 1.5. This results in "windowed subsequences" $x_\ell'(m)$, which are given by

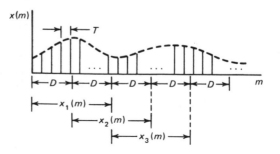

Fig. 4.9-1 Subsequences used to estimate the power density spectrum.

$$x_\ell'(m) = x_\ell(m)w(m), \qquad 0 \le m \le M - 1 \qquad (4.9\text{-}2)$$

for $1 \le \ell \le K$.

Computing the FFT of $x_\ell'(m)$ in (4.9-2), we get

$$X_\ell'(n) = \sum_{m=0}^{M-1} x_\ell'(m)W_M^{nm}, \qquad 0 \le n \le M - 1 \qquad (4.9\text{-}3)$$

where $W_M = e^{-j2\pi/M}$, and $X_\ell'(n)$ is the nth DFT coefficient of the ℓth subsequence.

We now compute a quantity $Q_\ell(f_n)$, which is called a *periodogram*, and is defined as

$$Q_\ell(f_n) = \frac{1}{MU} |X_\ell'(n)|^2 \qquad 0 \le n \le M - 1 \qquad (4.9\text{-}4)$$

where $f_n = \dfrac{n}{MT}, \qquad 0 \le n \le M - 1$

$U = \dfrac{1}{M} \sum_{m=0}^{M-1} w^2(m) = $ average power in the window sequence

The DFT coefficients $|X_\ell'(n)|^2$ in (4.9-4) are normalized by U to account for the loss of power caused by the windowing process, and f_n is the frequency associated with the nth DFT component. If $x(n)$ is real, then $f_{M/2}$ is the Nyquist frequency.

The desired PDS is then computed as

$$\hat{G}(f_n) = \frac{1}{K} \sum_{\ell=1}^{K} Q_\ell(f_n) = \frac{1}{KMU} \sum_{\ell=1}^{K} |X_\ell'(n)|^2 \qquad (4.9\text{-}5)$$

for $0 \le n \le M - 1$.

From (4.9-5) it is clear that the PDS estimate so obtained is a weighted sum of the periodograms of each of the individual subsequences.

A comprehensive study of the statistical properties of the PDS estimate in (4.9-5) has been done; for example, see [20]. He has shown that the 50% overlap between successive subsequences (see Fig. 4.9-1) helps to improve certain statistical properties of this estimate.

Leakage Phenomenon

The reason for introducing the window sequence $w(n)$ in (4.9-2) is to combat a phenomenon known as *leakage*. This phenomenon is best introduced by means of a simple example, which we present in the sequel.

Let $x(t)$ be a unit amplitude sine wave with frequency 32 Hz, which is sampled at 128 sps. Then the corresponding DFT amplitude spectrum p_n^x defined in (4.5-13a), is as shown in Fig. 4.9-2a. This spectrum was obtained using 128 data points. For convenience, only the spectrum points p_n^x, $22 \leq n \leq 42$, are displayed in Fig. 4.9-2a. The only nonzero term that occurs in p_n^x is at $n = 32$, which corresponds to 32 Hz. This is because the related fundamental frequency f_0 is given by

$$f_0 = \frac{1}{NT} = 1 \text{ Hz}$$

where N is the number of data points and T is the sampling interval. It is clear that an amplitude spectrum with a single nonzero spectral component is what one would expect for a pure sinusoid.

Now let us consider another sine wave $y(t)$ which frequency is slightly different from that of $x(t)$ just considered. For example, suppose $y(t)$ is a 31.5-Hz unit amplitude sine wave, which is also sampled at 128 sps. Then its DFT amplitude spectrum p_n^y obtained using 128 data points is as shown in Fig. 4.9-2b. Unlike p_n^x in Fig. 4.9-2a, we observed that p_n^y has several spectral components that are symmetric about 31.5 Hz. The reason why p_n^y does not have a single spectral component is that the interval NT ($= 1$ sec) does *not* contain an integral number of cycles of the 31.5-Hz sine wave. On the other hand, there is an integral number of cycles of a 32-Hz sine wave in the 1 sec interval. The result is the components of $y(t)$ that are not periodic over the interval NT tend to "leak out" into portions of the spectrum that are adjacent to the correct spectral value, which in this case is 31.5 Hz.

From the preceding discussion it follows that leakage will occur for any given signal which has components that are not periodic over the interval NT. This is because the DFT is a FS expansion which assumes that all components in $x(t)$ are periodic over NT, and hence have an integral number of cycles over that interval. The effect of not having an integral number of cycles over NT is to produce discontinuities in the periodic extension of a signal whose FS is desired. This aspect is illustrated in Fig. 4.9-3 for sine waves with frequencies of 1 Hz and 1.25 Hz, where L is the period for the FS expansion. Convergence problems that arise from such discontinuities can be discussed in terms of the *Gibbs phenomenon*. We shall have more to say about the Gibbs phenomenon in Chapter 7, in connection with designing a class of DT filters.

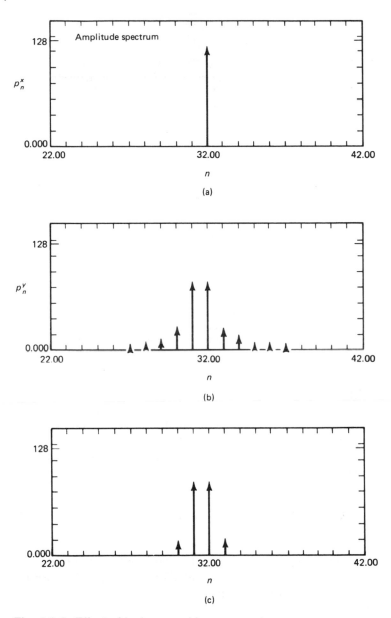

Fig. 4.9-2 Effect of leakage problem on spectra.

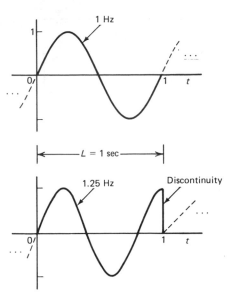

Fig. 4.9-3 Effect of having a noninte-gral number of cycles in a period to ob-tain a Fourier series expansion.

Smoothing

To understand how windowing reduces leakage, we note that (4.9-2) involves the *product* of two sequences $x_\ell(m)$ and $w(m)$. This product is equivalent to a convolution of their DFTs, as indicated by (4.5-22). Thus we have

$$X'_\ell(n) = \frac{1}{M} \sum_{k=0}^{M-1} W(k)X_\ell(n-k), \qquad 0 \le n \le M - 1 \qquad (4.9\text{-}6a)$$

where $X_\ell(k)$ and $W(k)$ are the kth DFT coefficients of $x_\ell(m)$ and $w(m)$, respectively. Since $X_\ell(k)$ and $W(k)$ are both M-periodic sequences, (4.9-6a) can also be written as

$$X'_\ell(n) = \frac{1}{M} \sum_{k=-1}^{M-2} W(k)X_\ell(n-k), \qquad 0 \le n \le M - 1 \qquad (4.9\text{-}6b)$$

This is because summation over $k = -1$ through $M - 2$ covers a full period.

As an illustration, let us consider the case when $w(m)$ is a DT version of a hanning window, which was introduced in Chapter 1 (see Table 1.5-1). It is defined as

$$w_{ha}(m) = \frac{1}{2}\left(1 - \cos\frac{2\pi m}{M}\right), \qquad 0 \le m < M \qquad (4.9\text{-}7)$$

It can be shown that (Problem 4-14) the DFT of $w_{ha}(m)$ given by

$$W_{ha}(k) = \begin{cases} -M/4, & k = -1 \\ M/2, & k = 0 \\ -M/4, & k = 1 \\ 0, & \text{elsewhere} \end{cases} \qquad (4.9\text{-}8)$$

Substitution of $W_{ha}(k)$ in (4.9-8) for $W(k)$ in (4.9-6b) leads to

$$X'_\ell(n) = -0.25X_\ell(n + 1) + 0.5X_\ell(n) - 0.25X_\ell(n - 1) \qquad (4.9\text{-}9)$$

Equation (4.9-9) implies that the effect of using the hanning window is to produce a set of "smoothed" DFT coefficients $X'_\ell(n)$, which are obtained as a weighted average of the DFT coefficient $X_\ell(n)$ and its adjacent neighbors $X_\ell(n - 1)$ and $X_\ell(n + 1)$. It is this *smoothing* process that reduces leakage, as demonstrated by the DFT amplitude spectrum in Fig. 4.9-2c. This spectrum was obtained by first multiplying the sequence $y(m)$ by $w_{ha}(m)$; then computing the DFT spectrum of the product sequences, and finally normalizing the result to account for power loss due to windowing. If U_{ha} denotes the average power in the hanning window $w_{ha}(m)$, from Parseval's theorem in (4.5-11) and the DFT coefficients in (4.9-8), we obtain [with $W_{ha}(-1) = W_{ha}(M-1)$]

$$U_{ha} = 2\left(\frac{1}{4}\right)^2 + \left(\frac{1}{2}\right)^2 = \frac{3}{8} \qquad (4.9\text{-}10)$$

Since Fig. 4.9-2c displays the amplitude spectrum rather than the power spectrum, we normalize the DFT amplitude spectrum of the windowed sequence by $\sqrt{\frac{3}{8}} \approx 0.61$. Comparing the spectra in Figs. 4.9-2b and 4.9-2c, we see that windowing has the desirable effect of reducing leakage by narrowing the spread of the spectrum about 31.5 Hz.

Other window sequences besides hanning can be used to reduce leakage. These include the DT counterparts of the triangular, Hamming, and Kaiser windows that were considered in Chapter 1 (Table 1.5-1). We list these in Table 4.9-1 along with the hanning window sequence for values of m from 0 through $M - 1$.

Table 4.9-1   Additional Window Sequences for Smoothing[†]

1. Triangular

$$w_{tr}(m) = \begin{cases} \dfrac{2m}{M-1}, & 0 \le m \le \dfrac{M-1}{2} \\[3mm] 2 - \dfrac{2m}{M-1}, & \dfrac{M-1}{2} \le m \le M-1 \end{cases}$$

2. Hanning

$$w_{ha}(m) = \frac{1}{2}[1 - \cos\left(\frac{2\pi m}{M-1}\right)], \qquad 0 \le m \le M-1$$

3. Hamming

$$w_H(m) = 0.54 - 0.46 \cos\left(\frac{2\pi m}{M-1}\right), \qquad 0 \le m \le M-1$$

4. Kaiser

$$w_K(m) = \frac{I_0\left[\theta\sqrt{\left(\frac{M-1}{2}\right)^2 - \left[m - \left(\frac{M-1}{2}\right)\right]^2}\right]}{I_0\left[\theta\left(\frac{M-1}{2}\right)\right]}$$

where I_0 is the modified Bessel function of the first kind and zero order; the smoothing effect can be adjusted by varying the parameter θ

[†]This table is derived from [4].

Before presenting an illustrative example, it is instructive to verify that the rectangular window sequence $w_r(m)$ produces no smoothing whatsoever. This window sequence is defined as

$$w_r(m) = 1, \qquad 0 \le m \le M-1 \tag{4.9-11}$$

It can be shown that (Problem 4-15) the DFT of $w_r(m)$ is given by

$$W_r(k) = \begin{cases} M, & k = 0 \\ 0, & \text{elsewhere} \end{cases} \tag{4.9-12}$$

Substitution of (4.9-12) in (4.9-6a) or (4.9-6b) leads to

$$X'_\ell(n) \equiv X_\ell(n)$$

indicating that no smoothing is realized.

Because of the smoothing process that occurs via appropriate windowing, this procedure for estimating the PDS is sometimes called the *method of modified periodograms*.

Illustrative Example

We now use the method of modified periodograms to estimate the PDS of the signal displayed in Fig. 4.9-4a.[†] This signal represents helicopter noise and was acquired using an omnidirectional microphone. The microphone was fixed on a short platform and the helicopter was a few hundred feet above the microphone. The CT signal so obtained was sampled at 256 sps to obtain a sequence of 2048 points, of which 2000 are plotted in Fig. 4.9-4a. This sequence appears as a CT signal in Fig. 4.9-4a because of interpolation that is introduced by the plotting device.

The data sequence was processed with 50% overlap as illustrated in Fig. 4.9-1, and each of the subsequences $x_\ell(m)$ consisted of 256 points; that is, $M = 256$ in (4.9-2). In addition, the hanning window sequence defined in Table 4.9-1 was used with $M = 256$. Thus the PDS obtained by taking 256-point FFTs consisted of 129 ($= M/2 + 1$) points, since the data sequence is real. The set of frequencies associated with the PDS points are given by f_n in (4.9-4) to be

$$f_n = \frac{n}{MT} = n, \qquad 0 \le n \le 128 \qquad (4.9\text{-}13)$$

since $M = 256$ and $T = 1/256$ (i.e., the PDS points are 1 Hz apart). It is apparent from (4.9-13) that the spacing between the spectrum points depends on M, which is the FFT size. This spacing is usually called the *frequency resolution* of the PDS.

The resulting PDS is displayed in Fig. 4.9-4b on a decibel scale, where dB equals 10 times the logarithm of each PDS value. From this PDS it is clear that the helicopter noise shown in Fig. 4.9-4a is made up primarily of three sinusoidal components with frequencies of 20, 40, and 60 Hz, respectively. It is apparent that the 40- and 60-Hz components are the second and third harmonics of the predominant 20-Hz component. Similarly, spectral estimation can be used as a powerful tool to analyze the frequency content of complex time signals pertaining to a variety of applications.

[†]Courtesy Sandia Laboratories, Albuquerque, New Mexico.

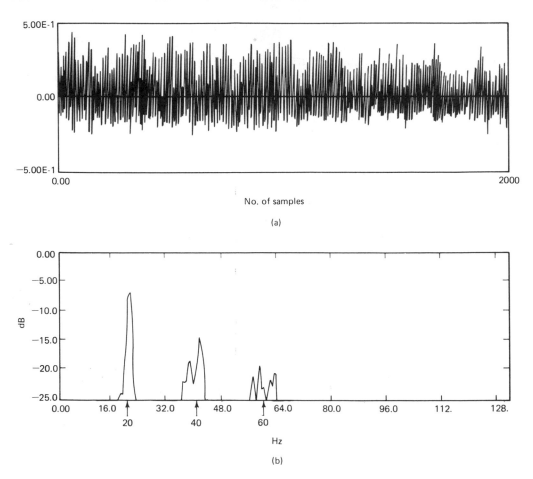

No. of samples

(a)

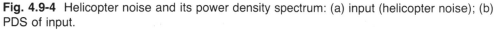

(b)

Fig. 4.9-4 Helicopter noise and its power density spectrum: (a) input (helicopter noise); (b) PDS of input.

4.10 SUMMARY

Starting with the basic definition of the FS for CT signals, we derived the DFT that enables a Fourier analysis of sequences. The important transition from CT signals to corresponding sequences was carried out via Shannon's sampling theorem.

Efficient FFT and IFFT algorithms for computing the DFT and IDFT, respectively, were developed. The relevance of such algorithms to the notions of decimation-in-time and decimation-in-frequency was introduced. It was shown that FFT algorithms result in substantial savings in computations and storage requirements compared to the direct computation of the DFT. The reduction in the number of computations has a significant impact on the related execution time. For example, it has been reported that using an IBM 7094 digital computer, an FFT algorithm required 5 seconds to compute all 8192 DFT coefficients, while the time taken for direct computation was of the order of half an hour [11]. In addition, it has been shown that, compared to a direct computation of the DFT, the FFT reduces the roundoff error by a factor of approximately $N/(\log_2 N)$ [12].

A method for estimating the PDS was introduced. The related leakage phenomenon was considered, and the role of windowing (smoothing) to reduce it was explained.

Our discussion related to the FFT and IFFT was restricted to the case when N is a power of 2, since this is the case that is most widely used in practice. The case when N is not necessarily a power of 2 is discussed elsewhere [13 to 15]. Finally, the interested reader may consult [4], [6], [16], and [17] for a variety of applications of the FFT.

PROBLEMS

4–1   The FT $X(f)$ of a CT signal $x(t)$ is shown in the following sketch. If $x(t)$ is sampled at 8000 sps, sketch $\bar{X}(f)$, the FT of the corresponding sequence $x(n)$.

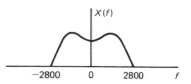

4–2   Repeat Example 4.3-1 for a 60-Hz sinusoid that is sampled at 100 sps and passed through an ideal low pass filter whose bandwidth is 50 Hz. Use the aliasing diagram. Verify the answer so obtained by sketching $X(f)$ and $X^*(f)$.

4–3   Using an aliasing diagram, find the first five aliasing frequencies of f = 2000 Hz for the case when a CT signal of bandwidth 4000 Hz is sampled at 7000 sps.

4–4   The ZT of the sequence $x(n) = (0.6)^n$, $n \geq 0$, is given as

$$X(z) = \frac{z}{z - 0.6}$$

and the related sampling interval T is 0.002 sec. If $\bar{X}(f)$ denotes the FT of this sequence, evaluate $|\bar{X}(f)|$ at $f = 80$ Hz. *Hint:* Use equation (3.1-9).

4–5   Using the definition of the DFT in (4.5-1) and the orthogonal property in (4.5-3), show that the IDFT is as given in (4.5-4).

4–6   Using (4.4-10) and (4.5-1), derive (4.5-2).

4–7   The first five DFT coefficients of a real 8-periodic sequence are $X(0) = 0.5, X(1) = 2 + j, X(2) = 3 + 2j, X(3) = j$, and $X(4) = 3$. Find $X(5), X(6)$, and $X(7)$.

4–8   Verify that (4.5-16) is true for $N = 4$ and $h = 1$.

4–9   The DFT coefficients of a 4-periodic sequence $x(m)$ are $X(0) = 5, X(1) = (2 + j), X(2) = -5$, and $X(3) = (2 - j)$. Use the IDFT to find $x(m)$.

4–10 Use the bit-reversal process (algorithm) illustrated in Fig. 4.6-6 to find the bit-reversed sequence corresponding to the normal ordered sequence 0, 1, 2, 3, . . . , 14, 15, which corresponds to the case $N = 16$.

4–11 Given the 8-periodic sequence $x(0) = 1, x(1) = 2, x(2) = 1, x(3) = 1, x(4) = 3, x(5) = 2, x(6) = 1$, and $x(7) = 2$. Use the FFT flow graph in Fig. 4.6-5 to compute the DFT coefficients $X(n), 0 \leq n \leq 7$.

4–12 Derive the decimation-in-time IFFT corresponding to the flow graph shown in Fig. 4.8-3.

4–13 Derive the discrete convolution in frequency result given by (4.5-21) and (4.5-22).

4–14 Show that the DFT of $w_{ha}(m)$ is (4.9-7) is as given by (4.9-8). *Hint:* Write $\cos(2\pi m/M)$ in $w_{ha}(m)$ as $\frac{1}{2}(W^m + W^{-m})$, where $W = e^{-j2\pi/M}$, and then take the DFT; also use the orthogonality property in (4.5-3).

4–15 Show that $W_r(k), 0 \leq k \leq M - 1$, is as given by (4.9-12) for the rectangular window sequence in (4.9-11).

4–16 The sequence $x(m), 0 \leq m \leq 63$, given in Table P4-16 is obtained by sampling a signal $x(t)$ at 960 sps. The duration of the signal is approximately 1/15 sec. Use the FFT program in Appendix 4.1 to compute the DFT power spectrum. By inspection of this spectrum, determine the maximum frequency present in $x(t)$; that is, the bandwidth of $x(t)$.

4–17 Use the FFT program in Appendix 4.1 to compute the DFT amplitude spectrum of a 31.5-Hz sinusoid of unit amplitude, which is sampled at 128 sps. Use a 128-point FFT, and the triangular, Hamming, and Kaiser window sequences given in Table 4.9-1. Compare the spectra so obtained with those obtained by the rectangular and hanning window sequences displayed in Figs. 4.9-2b and 4.9-2c, respectively.

Table P4-16

m	$x(m)$	m	$x(m)$
0	17.00	41	0.53
1	21.36	42	−0.53
2	10.15	43	0.00
3	0.00	44	1.00
4	1.00	45	0.26
5	5.00	46	−0.82
6	3.30	47	0.00
7	0.00	48	1.00
8	1.00	49	−0.10
9	3.11	50	−1.22
10	1.87	51	0.00
11	0.00	52	1.00
12	1.00	53	−0.67
13	2.35	54	−1.87
14	1.22	55	0.00
15	0.00	56	1.00
16	1.00	57	−1.79
17	1.91	58	−3.30
18	0.82	59	0.00
19	0.00	60	1.00
20	1.00	61	−5.74
21	1.60	62	−10.15
22	0.53	63	0.00
23	0.00		
24	1.00		
25	1.36		
26	0.30		
27	0.00		
28	1.00		
29	1.15		
30	0.10		
31	0.00		
32	1.00		
33	0.95		
34	−0.10		
35	0.00		
36	1.00		
37	0.75		
38	−0.30		
39	0.00		
40	1.00		

APPENDIX 4.1[†] A FAST FOURIER TRANSFORM SUBROUTINE

In this appendix we list a computer program that implements a decimation-in-frequency FFT algorithm. Appendix 4.1 of [7] was used as a guide for writing this program.

```
C**************************************************************************
C
C          CALLING SEQUENCE
C
C                  CALL FFT(X,N,INV)
C
C          PURPOSE
C
C                  FAST FOURIER TRANSFORMATION
C
C          ROUTINE(S) CALLED BY THIS ROUTINE
C
C                  NONE
C
C          ARGUMENT(S) REQUIRED FROM THE CALLING ROUTINE
C
C                  X        -         COMPLEX VECTOR TO BE TRANSFORMED
C                  N        -         NUMBER OF POINTS TO BE TRANSFORMED
C                                     (MUST BE A POWER OF TWO)
C                  INV      -         INV=0 -> FORWARD TRANSFORM
C                                     INV=1 -> INVERSE TRANSFORM
C
C          ARGUMENT(S) SUPPLIED  TO  THE CALLING ROUTINE
C
C                  X        -         COMPLEX TRANSFORMED VECTOR
C
C**************************************************************************
C
C          NOTE 1: This subroutine makes no checks on the validity
C                  of the data supplied by the calling routine.
C
C          NOTE 2: Argument(s) supplied by the calling routine are
C                  not modified by this subroutine.
C
C**************************************************************************
C
          SUBROUTINE FFT(X,N,INV)
          COMPLEX X(1),W,T
          ITER=0
          IREM=N
10        IREM=IREM/2
          IF (IREM.EQ.0) GO TO 20
          ITER=ITER+1
          GO TO 10
20        CONTINUE
          S=-1
          IF (INV.EQ.1) S=1
          NXP2=N
          DO 50 IT=1,ITER
          NXP=NXP2
```

[†]Program written by Fred Ratcliffe.

```
NXP2=NXP/2
WPWR=3.141592/FLOAT(NXP2)
DO 40 M=1,NXP2
ARG=FLOAT(M-1)*WPWR
W=CMPLX(COS(ARG),S*SIN(ARG))
DO 40 MXP=NXP,N,NXP
J1=MXP-NXP+M
J2=J1+NXP2
T=X(J1)-X(J2)
X(J1)=X(J1)+X(J2)
40      X(J2)=T*W
50      CONTINUE
N2=N/2
N1=N-1
J=1
DO 65 I=1,N1
IF(I.GE.J) GO TO 55
T=X(J)
X(J)=X(I)
X(I)=T
55      K=N2
60      IF(K.GE.J) GO TO 65
J=J-K
K=K/2
GO TO 60
65      J=J+K
IF (INV.EQ.1) GO TO 75
DO 70 I=1,N
70      X(I)=X(I)/FLOAT(N)
75      CONTINUE
RETURN
END
```

REFERENCES

1. C. E. SHANNON, "A Mathematical Theory of Communication," *Bell System Tech. J.*, Vol. 27, 1948, pp. 379–423.

2. R. B. BLACKMAN and J. W. TUKEY, *The Measurement of Power Spectra*, Dover, New York, 1958.

3. J. W. COOLEY and R. W. TUKEY, "An Algorithm for Machine Computation of Complex Fourier Series," *Mathematics of Computation*, Vol. 19, 1965, pp. 297–301.

4. A. V. OPPENHEIM and R. SCHAFER, *Digital Signal Processing*, Prentice-Hall, Englewood Cliffs, N.J., 1975.

5. E. O. BRIGHAM, *The Fast Fourier Transform*, Prentice-Hall, Englewood Cliffs, N.J., 1974.

6. L. R. RABINER and B. GOLD, *Theory and Application of Digital Signal Processing*, Prentice-Hall, Englewood Cliffs, N.J., 1975.

7. N. AHMED and K. R. RAO, *Orthogonal Transforms for Digital Signal Processing*, Springer-Verlag, New York/Berlin, 1975.

8. W. D. STANLEY, *Digital Signal Processing*, Reston, Reston, Va., 1975.

9. S. D. STEARNS, *Digital Signal Analysis*, Hayden, Rochelle Park, N.J., 1975.

10. B. GOLD and C. RADER, *Digital Processing of Signals*, McGraw-Hill, New York, 1969.

11.   W. T. Cochran et al., "What is the Fast Fourier Transform?" *Proc. IEEE*, Vol. 55, 1967, pp. 1664–1674.

12.   T. Kaneko, "Accumulation of Round-off Errors in Fast Fourier Transforms," *J. ACM*, Vol. 17, 1970, pp. 637–654.

13.   G. Sande, "Arbitrary Radix One-Dimensional Fast Fourier Transform Subroutines," University of Chicago, Chicago, 1968.

14.   R. C. Singleton, "An Algorithm for Computing the Mixed Radix Fast Fourier Transform," *IEEE Trans. Acoustics, Speech and Signal Processing*, Vol. AU-17, 1969, pp. 99–103.

15.   C. Rader, "Discrete Fourier Transforms When the Number of Data Samples Is Prime," *Proc. IEEE*, Vol. 56, 1968, pp. 1107–1108.

16.   J. W. Cooley, P. A. W. Lewis, and P. D. Welch, "The Fast Fourier Transform and Its Applications," IBM Res. Paper, RC-1743, 1967, IBM Watson Research Center, Yorktown Heights, N.Y.

17.   A. V. Oppenheim, ed., *Applications of Digital Signal Processing*, Prentice-Hall, Englewood Cliffs, N.J., 1978.

18.   G. D. Bergland, "A Guided Tour of the Fast Fourier Transform," *IEEE Spectrum*, Vol. 6, July 1969, pp. 41–52.

19.   C. Bingham, M. D. Godfrey, and J. W. Tukey, "Modern Techniques of Power Spectral Estimation," *IEEE Trans. Audio and Electroacoust.*, Vol. AU-15, 1967, pp. 56–65.

20.   P. D. Welch, "The Use of the Fast Fourier Transform for the Estimation of Power Spectra," *IEEE Trans. Audio and Electroacoust.*, Vol. AU-15, 1967, pp. 70–73.

21.   P. I. Richards, "Computing Reliable Power Spectra," *IEEE Spectrum*, Vol. 14, 1967, pp. 83–90.

22.   A. Papoulis, *The Fourier Integral and Its Applications*, McGraw-Hill, New York, 1962, pp. 29–32.

23.   R. W. Hamming, *Digital Filters*, Prentice-Hall, Englewood Cliffs, N.J., 1977, Chap. 5.

5

Discrete-Time System Transfer Functions

From Chapter 1 we recall that differential equations and the LT enable us to represent CT systems in terms of transfer functions. We now show that difference equations and the ZT enable us to do so for DT systems. As such, we will be able to discuss DT systems in the context of familiar frequency-domain concepts, such as magnitude response, phase response, bandwidth, and the like.

5.1 FUNDAMENTALS

Let $x(n)$ and $y(n)$ be real input and output sequences, respectively, of the linear, time-invariant DT system shown in Fig. 5.1-1. This system is represented by a set of coefficients a_i, $0 \leq i \leq k$, and b_j, $1 \leq j \leq m$. The input-output relationship of the system in Fig. 5.1-1 is given by the following difference equation, which represents *DT convolution:*

$$y(n) = a_0 x(n) + a_1 x(n - 1) + \cdots + a_k x(n - k) \tag{5.1-1}$$
$$- b_1 y(n - 1) - b_2 y(n - 2) - \cdots - b_m y(n - m)$$

where n is the time index.

In (5.1-1) we observe that the output at time n is obtained as a weighted sum of the input at time n, and a set of past k input and m output values (samples), respectively.

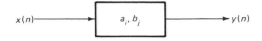

Fig. 5.1-1 Discrete-time system represented by coefficients a_i, b_j.

Assuming that the system is initially relaxed,[†] the ZT of (5.1-1) yields [see (3.6-3)]

$$Y(z) = a_0 X(z) + a_1 z^{-1} X(z) + \cdots + a_k z^{-k} X(z) \tag{5.1-2}$$
$$- b_1 z^{-1} Y(z) - b_2 z^{-2} Y(z) - \cdots - b_m z^{-m} Y(z)$$

where $X(z)$ and $Y(z)$ denote the ZTs of the input and output sequences, respectively.

Collecting terms in (5.1-2), we obtain

$$\frac{Y(z)}{X(z)} = H(z) = \frac{a_0 + a_z z^{-1} + a_2 z^{-2} + \cdots + a_k z^{-k}}{1 + b_1 z^{-1} + b_2 z^{-2} + \cdots + b_m z^{-m}} \tag{5.1-3}$$

where $H(z)$ is defined as the *transfer function* of the DT system. We note that $H(z)$ is a ratio of two polynomials of z^{-1}. Hence the a_i and b_j in (5.1-3) are referred to as the numerator and denominator coefficients of the system transfer function.

Without loss of generality, we let $m = k$ in (5.1-3) to obtain

$$H(z) = \frac{a_0 + a_1 z^{-1} + a_2 z^{-2} + \cdots + a_k z^{-k}}{1 + b_1 z^{-1} + b_2 z^{-2} + \cdots + b_k z^{-k}} \tag{5.1-4a}$$

which can be written as

$$H(z) = \frac{a_0 z^k + a_1 z^{k-1} + a_2 z^{k-2} + \cdots + a_k}{z^k + b_1 z^{k-1} + b_2 z^{k-2} + \cdots + b_k} \tag{5.1-4b}$$

Factoring the numerator and denominator polynomials of $H(z)$ in (5.1-4b), we obtain

$$H(z) = \frac{a_0(z - z_1)(z - z_2) \cdots (z - z_k)}{(z - p_1)(z - p_2) \cdots (z - p_k)} \tag{5.1-5}$$

where z_i and p_i denote the *i*th *zero* and *pole*, respectively, of $H(z)$.

[†]That is, all initial conditions are zero.

The *impulse response* of the preceding system is the output that results when the input is the unit impulse sequence $x(n) = 1$, for $n = 0$, and zero elsewhere. Since the ZT of such an input equals unity (see line 1 of Table 3.2-1), (5.1-3) implies that the corresponding output is $H(z)$. Thus the impulse response is obtained by simply evaluating the IZT of the transfer function $H(z)$. That is,

$$h(n) = Z^{-1}\{H(z)\} \qquad (5.1\text{-}6)$$

Next, let us combine (5.1-3) and (5.1-5) to obtain

$$Y(z) = \frac{a_0(z - z_1)(z - z_2) \cdots (z - z_k)}{(z - p_1)(z - p_2) \cdots (z - p_k)} X(z) = H(z)X(z) \qquad (5.1\text{-}7)$$

where $X(z)$ is in the form of a rational function. Then the portion of the output

$$y(n) = Z^{-1}\{Y(z)\} \qquad (5.1\text{-}8)$$

that is due to the *poles of $X(z)$* is called the *forced response* of the system; that portion due to p_i, the *poles of $H(z)$* is called the *natural response*. If a system has all its poles within the unit circle, then its natural response dies down as $n \to \infty$, and hence is referred to as the *transient response*. Also, if the input to such systems is periodic, then the corresponding forced response is called the *steady-state response*.

The preceding system is *stable* if the poles of $H(z)$ in (5.1-7) lie within the unit circle; that is,

$$|p_i| < 1, \qquad 1 \le i \le k$$

in which case

$$\lim_{n \to \infty} \{h(n)\} = 0 \qquad (5.1\text{-}9)$$

For example, if

$$H(z) = \frac{z}{z - b}$$

then (see Table 3.2-1)

$$h(n) = b^n, \qquad n \ge 0$$

and the system is stable if $|b| < 1$.

Conversely, if $|p_i| > 1$ for *any* $0 \leq i \leq k$, then the system is *unstable* since

$$\lim_{n \to \infty} \{h(n)\} \to \infty \tag{5.1-10}$$

A system is also unstable if $H(z)$ has multiple-order pole(s) on the unit circle; for example, if $H(z) = z/(z - 1)^2$, then $h(n) = n$, which is unstable. Finally, a system is said to be *marginally stable* or *oscillatory bounded* if

$$\lim_{n \to \infty} \{h(n)\} = c \tag{5.1-11}$$

where c is a nonzero constant. It is easy to verify that first-order poles of such a system lie *on* the unit circle, as is the case, for example, with

$$H(z) = \frac{z}{z + 1}$$

whose impulse response oscillates between ± 1, since (see Table 3.2-1)

$$h(n) = (-1)^n, \qquad n \geq 0$$

In conclusion, we introduce some terminology that is usually used to classify DT systems into two groups, which result from the following considerations:

1. *All* the b_j in (5.1-1) are zero; that is,

$$y(n) = \sum_{i=0}^{k} h(i)x(n - i), \qquad n \geq 0 \tag{5.1-12}$$

where $h(i) = a_i$ is the ith impulse response coefficient, and the upper summation index k is a *finite constant*. This means that the number of impulse response coefficients, k, is finite. Such systems are thus called *nonrecursive* or *finite impulse response* (FIR) systems.

2. *At least* one $b_j \neq 0$ in (5.1-1). Such systems are called *recursive* or *infinite impulse response* (IIR) systems, since their impulse responses consist of an *infinite* number of coefficients. To illustrate, let $k = m = 1$ in (5.1-1) to obtain

$$y(n) = a_0 x(n) + a_1 x(n - 1) - b_1 y(n - 1) \tag{5.1-13}$$

where $b_1 \neq 0$.

Then (5.1-13) implies that

$$y(n - 1) = a_0x(n - 1) + a_1x(n - 2) - b_1y(n - 2) \quad (5.1\text{-}14)$$

Substitution for $y(n - 1)$ in (5.1-13) using (5.1-14) leads to

$$y(n) = a_0x(n) + (a_1 - a_0b_1)x(n - 1) - a_1b_1x(n - 2) + b_1^2y(n - 2)$$
$$(5.1\text{-}15)$$

Again, from (5.1-13) we have

$$y(n - 2) = a_0x(n - 2) + a_1x(n - 3) - b_1y(n - 3) \quad (5.1\text{-}16)$$

Next, substituting for $y(n - 2)$ in (5.1-15) using (5.1-16), we obtain

$$y(n) = \sum_{i=0}^{3} h(i)x(n - i) - b_1^3y(n - 3) \quad (5.1\text{-}17)$$

where $\quad h(0) = a_0$

$\qquad h(1) = a_1 - a_0b_1$

$\qquad h(2) = -a_1b_1 + a_0b_1^2$

$\qquad h(3) = a_1b_1^2$

By repeating this process, it is apparent that $y(n)$ can be expressed in terms of an *infinite* number of impulse response coefficients $h(i)$, which are due to b_1 in (5.1-13) being nonzero. Hence IIR systems can be represented as

$$y(n) = \sum_{i=0}^{n} h(i)x(n - i), \quad n \geq 0 \quad (5.1\text{-}18)$$

where the upper summation index n is a *variable.*

Finally, from (5.1-4b) we see that FIR systems have *no poles*, except at $z = 0$, since $b_j = 0$, $1 \leq j \leq k$. On the other hand, IIR systems can have poles anywhere in the z-plane, but confined to the unit circle if an IIR system is to be stable.

We shall study methods of designing a class of IIR and FIR systems in Chapters 6 and 7, respectively.

5.2 MORE ON DISCRETE-TIME CONVOLUTION

The DT convolution relationships in (5.1-12) and (5.1-18) can be conveniently represented in the form of a block diagram as indicated in Fig. 5.2-1. It is instructive to consider a graphical interpretation of the same. To this end, we consider an FIR system that consists of four impulse response coefficients $h(0)$, $h(1)$, $h(2)$, and $h(3)$. Then from (5.1-12) it follows that

$$y(n) = \sum_{i=0}^{3} h(i)x(n - i), \qquad n \geq 0 \qquad (5.2\text{-}1)$$

which yields

$$y(0) = h(0)x(0)$$
$$y(1) = h(0)x(1) + h(1)x(0)$$
$$y(2) = h(0)x(2) + h(1)x(1) + h(2)x(0)$$

and

$$y(n) = h(0)x(n) + h(1)x(n - 1) + h(2)x(n - 2) + h(3)x(n - 3)$$

$$(5.2\text{-}2)$$
$$\text{for } n \geq 3$$

The geometrical interpretation of (5.2-2) is apparent from Figure 5.2-2a. It is clear that DT convolution represents a step-by-step process of multiplication and addition, followed by shifting. We also note that the input sequence is "turned around," as is the case when two CT functions $h(t)$ and $x(t)$ are convolved to obtain the resulting function $y(t)$; for example, see Fig. 1.8-2. This aspect is illustrated again in Fig. 5.2-2b.

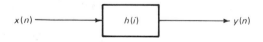

Fig. 5.2-1 Block diagram representation of discrete-time convolution in terms of impulse response values.

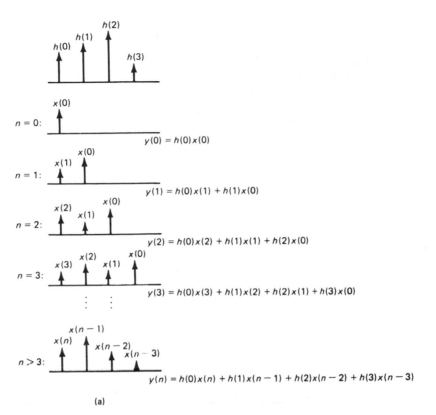

(a)

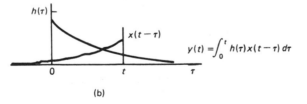

(b)

Fig. 5.2-2 Graphical interpretation of convolution: (a) DT case; (b) CT case.

Fundamental Theorem

Convolution in the DT domain is equivalent to multiplication in the ZT domain. That is, if

$$y(n) = \sum_{i=0}^{n} x(i)h(n - i), \qquad n \geq 0 \tag{5.2-3}$$

then
$$Y(z) = H(z)X(z) \tag{5.2-4}$$

We note that this theorem parallels that related to CT systems, discussed in Section 1.7. It states that if $x(t)$ and $h(t)$ are convolved to produce $y(t)$, then $Y(s) = H(s)X(s)$, where s is the LT variable.

PROOF. By definition of the ZT, we have

$$Y(z) = \sum_{n=0}^{\infty} y(n)z^{-n} \tag{5.2-5}$$

Substitution of (5.2-3) in (5.2-5) leads to

$$Y(z) = \sum_{n=0}^{\infty} \left\{ \sum_{i=0}^{n} x(i)h(n - i) \right\} z^{-n}$$

$$= \sum_{n=0}^{\infty} \sum_{i=0}^{\infty} x(i)h(n - i)z^{-n} \tag{5.2-6}$$

since $h(n - i) = 0$ for $i > n$, which means that the system is assumed to be causal.

Interchanging the order of summation in (5.2-6), we obtain

$$Y(z) = \sum_{i=0}^{\infty} \sum_{n=0}^{\infty} x(i)h(n - i)z^{-n} \tag{5.2-7}$$

With $(n - i) = k$, (5.2-7) becomes

$$Y(z) = \sum_{i=0}^{\infty} \sum_{k=-i}^{\infty} x(i)h(k)z^{-(k + i)}$$

$$= \left\{ \sum_{i=0}^{\infty} x(i)z^{-i} \right\} \left\{ \sum_{k=0}^{\infty} h(k)z^{-k} \right\} \tag{5.2-8}$$

realizing that $h(k) = 0$, $k < 0$. Thus (5.2-8) yields

$$Y(z) = H(z)X(z)$$

which is the desired result.

5.3 ILLUSTRATIVE EXAMPLES

Example 5.3-1: The following difference equation is known as the first-order trapezoidal integration rule in numerical analysis:

$$y(n) = \frac{1}{2} [x(n) + x(n - 1)] + y(n - 1), \qquad n \geq 0 \qquad (5.3\text{-}1)$$

Find:

(a) The transfer function.

(b) The impulse response.

(c) The output if the input is the unit step sequence $x(n) = 1, n \geq 0$.

Solution:

(a) The ZT of (5.3-1) yields

$$Y(z) = \frac{1}{2}[X(z) + z^{-1}X(z)] + z^{-1}Y(z)$$

which results in the transfer function

$$H(z) = \frac{Y(z)}{X(z)} = 0.5\left(\frac{z + 1}{z - 1}\right) \qquad (5.3\text{-}2)$$

(b) To obtain $h(n)$, we need to evaluate the IZT of $H(z)$ in (5.3-2). We do so by means of the residue method, and hence form

$$G(z) = z^{n-1}H(z) \qquad (5.3\text{-}3)$$

$$= 0.5\, z^{n-1}\left(\frac{z + 1}{z - 1}\right)$$

CASE 1. $n = 0$ results in two poles at $z = 0$ and $z = 1$, since

$$G(z) = \frac{0.5(z + 1)}{z(z - 1)}$$

Thus

$$h(0) = R_{z=0} + R_{z=1}$$

where $R_{z=0} = \dfrac{0.5(z+1)}{z-1}\bigg|_{z=0} = -0.5$

$R_{z=1} = \dfrac{0.5(z+1)}{z}\bigg|_{z=1} = 1$

which implies that

$$h(0) = 0.5 \qquad (5.3\text{-}4)$$

CASE 2. $n \geq 1$ causes $G(z)$ in (5.3-3) to have only one pole at $z = 1$. Evaluating the corresponding residue, we obtain

$$h(n) = 1, \qquad n \geq 1 \qquad (5.3\text{-}5)$$

Combining (5.3-4) and (5.3-5) we get the impulse response displayed in the following sketch.

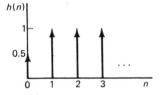

(c) The ZT of the unit step sequence is given in Table 3.2-1 as

$$X(z) = \frac{z}{z-1}$$

Thus the ZT of the corresponding output is

$$Y(z) = H(z)X(z)$$
$$= \frac{0.5(z^2 + z)}{(z-1)^2}$$

To find $y(n)$, we obtain the IZT of $Y(z)$ using the residue method. To this end, we form

$$G(z) = z^{n-1}Y(z)$$
$$= 0.5\frac{(z^{n+1} + z^n)}{(z-1)^2}$$

Thus
$$y(n) = R_{z=1}$$

where $R_{z=1}$ is the residue of $G(z)$ at $z = 1$. It is evaluated using (3.5-15c) to obtain the desired result

$$y(n) = \left\{ 0.5 \frac{d}{dz} [z^{n+1} + z^n] \right\}\Big|_{z=1}$$
$$= 0.5(2n + 1), \quad n \geq 0$$
(5.3-6)

From (5.3-6) it is apparent that

$$y(0) = 0.5, \quad y(1) = 1.5, \quad y(2) = 2.5, \quad y(3) = 3.5, \quad \text{etc.}$$

which demonstrates that the trapezoidal integration rule (5.3-1) does indeed integrate the input unit-step sequence. We also observe that this process represents a marginally stable system since the corresponding $H(z)$ in (5.3-2) has a pole on the unit circle; that is, $z = 1$.

Example 5.3-2: Consider the second-order IIR system

$$y(n) = K_1 y(n - 1) + K_2 y(n - 2) + x(n), \quad n \geq 0 \quad (5.3-7)$$

where K_1 and K_2 are constants.

(a) Evaluate $H(z)$.

(b) Find $h(n)$, $n \geq 0$ for the following cases:
(1) $K_1^2/4 + K_2 < 0$
(2) $K_1^2/4 + K_2 > 0$
(3) $K_1^2/4 + K_2 = 0$
Discuss stability conditions related to these cases.

Solution:
  (a) Assuming zero initial conditions, the ZT of (5.3-7) yields

$$Y(z) = K_1 z^{-1} Y(z) + K_2 z^{-2} Y(z) + X(z)$$

from which we obtain

$$H(z) = \frac{Y(z)}{X(z)} = \frac{z^2}{z^2 - K_1 z - K_2} \quad (5.3-8)$$

(b) From (5.3-8) it follows that $H(z)$ has two roots, z_1 and z_2, where

$$z_1 = \frac{K_1 + \sqrt{K_1^2 + 4K_2}}{2}$$

$$= \frac{K_1}{2} + \sqrt{(K_1^2/4) + K_2}$$

and
$$z_2 = \frac{K_1}{2} - \sqrt{(K_1^2/4) + K_2} \qquad (5.3\text{-}9)$$

CASE 1.

$$\frac{K_1^2}{4} + K_2 < 0 \qquad (5.3\text{-}10)$$

From (5.3-9) and (5.3-10) it is clear that z_1 and z_2 are complex conjugate roots. Hence we let

$$z_1 = \gamma e^{j\omega_0 T} \qquad (5.3\text{-}11a)$$

and
$$z_2 = \gamma e^{-j\omega_0 T}$$

where γ and $\omega_0 T$ specify the location of the roots in the unit circle, as illustrated.

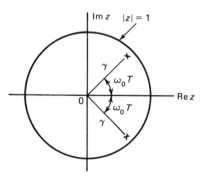

To find γ and ω_0, we use the fact that

$$z^2 - K_1 z - K_2 = (z - z_1)(z - z_2)$$

$$= (z - \gamma e^{j\omega_0 T})(z - \gamma e^{-j\omega_0 T})$$

That is,

$$z^2 - K_1 z - K_2 = z^2 - 2\gamma \cos(\omega_0 T) z + \gamma^2$$

which implies that

$$\gamma^2 = -K_2 \qquad (5.3\text{-}11b)$$

and

$$\omega_0 = \frac{1}{T} \cos^{-1} \frac{K_1}{2\gamma}$$

From (5.3-11b) it is clear that γ and ω_0 are readily obtained from the given difference equation in (5.3-7) since K_1 and K_2 are specified. Using the residue method we know that

$$h(n) = Z^{-1}\{H(z)\} \qquad (5.3\text{-}12)$$

$$= R_{z=z_1} + R_{z=z_2}$$

where $R_{z=z_1}$ and $R_{z=z_2}$ are the residues of

$$G(z) = \frac{z^{n+1}}{(z - z_1)(z - z_2)}$$

at the poles $z = z_1$ and $z = z_2$.

Evaluating $R_{z=z_1}$, we have

$$R_{z=z_1} = \frac{z^{n+1}}{z - z_2} \Bigg|_{z=z_1}$$

where $z_1 = \gamma e^{j\omega_0 T}$ and $z_2 = \gamma e^{-j\omega_0 T}$

Thus

$$R_{z=z_1} = \gamma^n \frac{e^{j(n+1)\omega_0 T}}{(e^{j\omega_0 T} - e^{-j\omega_0 T})} \qquad (5.3\text{-}13)$$

Using the identity

$$\sin(\omega_0 T) = \frac{1}{2j}(e^{j\omega_0 T} - e^{-j\omega_0 T})$$

(5.3-13) becomes

$$R_{z=z_1} = \frac{\gamma^n}{2j} \frac{e^{j(n+1)\omega_0 T}}{\sin(\omega_0 T)} \qquad (5.3\text{-}14)$$

Similarly,

$$R_{z=z_2} = -\frac{\gamma^n}{2j} \frac{e^{-j(n+1)\omega_0 T}}{\sin(\omega_0 T)} \qquad (5.3\text{-}15)$$

Substitution of (5.3-14) and (5.3-15) in (5.3-12) results in

$$h(n) = \frac{\gamma^n}{\sin(\omega_0 T)}\{\sin[(n+1)\omega_0 T]\}, \qquad n \geq 0 \qquad (5.3\text{-}16)$$

which is the desired impulse response. A plot of the same is depicted in Fig. 5.3-1. We observe that the envelope of the impulse response is in the form of a damped sinusoid whose frequency is ω_0 radians/second. In Example 5.5-4, we will see that ω_0 is actually the *resonant frequency* of the second-order system defined by (5.3-7), when the condition $\{K_1^2/4 + K_2\} < 0$ is satisfied.

From (5.3-16) it is clear that if $|\gamma| < 1$, then $h(n)$ goes to zero as n tends to infinity. Hence the system is stable if $|\gamma| < 1$. We can also come to this conclusion regarding stability by noting that $|\gamma| < 1$ ensures that the complex conjugate poles z_1 and z_2 both lie within the unit circle, since $z_1 = \gamma e^{j\omega_0 T}$ and $z_2 = \gamma e^{-j\omega_0 T}$.

CASE 2

$$\frac{K_1^2}{4} + K_2 > 0$$

In this case z_1 and z_2 are real valued, as apparent from (5.3-9). As such

$$H(z) = \frac{z^2}{(z - z_1)(z - z_2)}$$

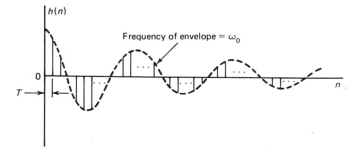

Fig. 5.3-1 Plot of $h(n)$ in (5.3-16).

where z_1 and z_2 are real numbers.

To find the impulse response we obtain the IZT of $H(z)$ via the residue method. Forming

$$G(z) = \frac{z^{n+1}}{(z - z_1)(z - z_2)}$$

we have

$$h(n) = R_{z=z_1} + R_{z=z_2}$$

where

$$R_{z=z_1} = \left. \frac{z^{n+1}}{z - z_2} \right|_{z=z_1} = \frac{z_1^{n+1}}{z_1 - z_2}$$

$$R_{z=z_2} = \left. \frac{z^{n+1}}{z - z_1} \right|_{z=z_2} = \frac{z_2^{n+1}}{z_2 - z_1}$$

Hence we obtain the impulse response to be

$$h(n) = \frac{1}{z_1 - z_2} (z_1^{n+1} - z_2^{n+1}), \qquad n \geq 0 \tag{5.3-17}$$

It also follows that as long as $|z_1| < 1$ and $|z_2| < 1$, the system is stable since $h(n) \to 0$ as $n \to \infty$.

CASE 3

$$\frac{K_1^2}{4} + K_2 = 0$$

Here the roots z_1 and z_2 are equal, as apparent from (5.3-9). Again,

$$z_1 = z_2 = \frac{K_1}{2}$$

where z_1 and z_2 are real numbers. The corresponding transfer function in (5.3-8) becomes

$$H(z) = \frac{z^2}{[z - (K_1/2)]^2} \tag{5.3-18}$$

The impulse response is easily found by resorting to the residue method. Thus we form

$$G(z) = \frac{z^{n+1}}{[z - (K_1/2)]^2}$$

which means that

$$h(n) = R_{z=K_1/2}$$

where $R_{z=K_1/2}$ is evaluated using (3.5-15a) with $m = 2$. This process results in

$$R_{z=K_1/2} = \frac{d}{dz}\{z^{n+1}\}\Big|_{z=K_1/2} = \left(\frac{K_1}{2}\right)^n (n + 1), \qquad n \geq 0 \quad (5.3\text{-}19)$$

Hence

$$h(n) = \left(\frac{K_1}{2}\right)^n (n + 1), \qquad n \geq 0$$

which means that the system is stable if $|K_1| < 2$.

Example 5.3-3: In Fig. 5.3-2, $H_1(z)$ and $H_2(z)$ are the transfer functions of the two first-order IIR systems given by

$$y_1(n) = x_1(n) - 0.5y_1(n - 1) \qquad (5.3\text{-}20)$$

and $\qquad\qquad y_2(n) = x_2(n) + y_2(n - 1), \qquad n \geq 0 \qquad (5.3\text{-}21)$

(a)   Find $H_1(z)$ and $H_2(z)$.

(b)   Evaluate the output $y(n)$, if $x(n) = (-1)^n, n \geq 0$.

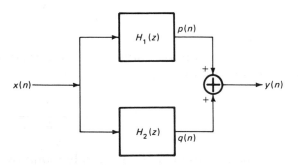

Fig. 5.3-2 Connection of two transfer functions.

Solution:

(a) With zero initial conditions, the ZT of (5.3-20) and (5.3-21) results in

$$Y_1(z) = X(z) - 0.5z^{-1}Y_1(z) \tag{5.3-22}$$

and

$$Y_2(z) = X(z) + z^{-1}Y_2(z)$$

It is apparent that (5.3-22) yields the desired transfer functions as

$$H_1(z) = \frac{z}{z + 0.5} \tag{5.3-23}$$

and

$$H_2(z) = \frac{z}{z - 1}$$

(b) From Fig. 5.3-2 it follows that

$$Y(z) = P(z) + Q(z)$$
$$= H_1(z)X(z) + H_2(z)X(z)$$

which yields

$$Y(z) = H(z)X(z) \tag{5.3-24}$$

where $H(z) = H_1(z) + H_2(z)$
Using (5.3-23), we evaluate $H(z)$ to obtain

$$H(z) = \frac{z(2z - 0.5)}{(z + 0.5)(z - 1)} \tag{5.3-25}$$

Table 3.2-1 yields the ZT of $x(n) = (-1)^n$, $n \geq 0$, to be

$$X(z) = \frac{z}{z + 1} \tag{5.3-26}$$

Substitution of (5.3-25) and (5.3-26) in (5.3-24) leads to

$$Y(z) = \frac{z^2(2z - 0.5)}{(z + 0.5)(z - 1)(z + 1)} \tag{5.3-27}$$

Let us find the IZT of $Y(z)$ in (5.3-27) using the PFE method. Thus we write

$$\frac{Y(z)}{z} = \frac{A}{z + 0.5} + \frac{B}{z - 1} + \frac{C}{z + 1} \tag{5.3-28}$$

where $A = (z + 0.5)\dfrac{Y(z)}{z}\bigg|_{z=-0.5} = -1$

$\quad\; B = (z - 1)\dfrac{Y(z)}{z}\bigg|_{z=1} = \dfrac{1}{2}$

$\quad\; C = (z + 1)\dfrac{Y(z)}{z}\bigg|_{z=-1} = 2.5$

Thus (5.3-28) yields

$$y(n) = Z^{-1}\left\{\frac{-z}{z + 0.5}\right\} + \frac{1}{2}Z^{-1}\left\{\frac{z}{z - 1}\right\} + 2.5Z^{-1}\left\{\frac{z}{z + 1}\right\} \tag{5.3-29}$$

The IZTs indicated in (5.3-29) are obtained by referring to Table 3.2-1. This results in the output

$$y(n) = -(-0.5)^n + 0.5 + 2.5(-1)^n, \qquad n \geq 0$$

5.4 SINUSOIDAL STEADY-STATE CONSIDERATIONS

Consider the following CT system. Here $h(t)$ denotes the impulse response, and $x(t)$ and $y(t)$ are the input and output, respectively. Now, if $x(t) = Xe^{j(\omega t + \psi_x)}$, then from Section 1.11 it is known that the steady-state output $y(t)_{ss}$ is also sinusoidal and is given by

$$y(t)_{ss} = H(\omega)Xe^{j(\omega t + \psi_x)} \tag{5.4-1}$$

$$x(t) \circ\!\!-\!\!\longrightarrow \boxed{h(t)} \longrightarrow\!\!\circ\, y(t)$$

In (5.4-1), $H(\omega)$ is the steady-state transfer function while X and ψ_x are the input magnitude and phase angle, respectively, using phasor notation.

From (5.4-1) it is clear that a fundamental property of the preceding system is that the sinusoidal steady-state response has the same frequency as the input, but its magnitude and phase angle could be different relative to the input. This property is also valid for DT systems, as illustrated in Fig. 5.4-1, where the input and output are also in phasor form. In what follows, our objective will be to relate X and Y to the system transfer function $H(z)$.

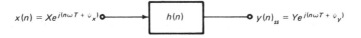

$$x(n) = Xe^{j(n\omega T + \psi_x)} \quad \boxed{h(n)} \quad y(n)_{ss} = Ye^{j(n\omega T + \psi_y)}$$

Fig. 5.4-1 Steady-state analysis of discrete-time systems.

We write $x(n)$ as a product to obtain

$$x(n) = Xe^{j\psi_x}e^{jn\omega T} \tag{5.4-2}$$

The ZT of (5.4-2) yields

$$Z\{x(n)\} = Z\{e^{jn\omega T}\}\, Xe^{j\psi_x} \tag{5.4-3}$$

Table (3.2-1) implies that

$$Z\{e^{jn\omega T}\} = \frac{z}{z - e^{j\omega T}} \tag{5.4-4}$$

Substitution of (5.4-4) in (5.4-3) leads to

$$X(z) = \left(\frac{z}{z - e^{j\omega T}}\right) Xe^{j\psi_x} \tag{5.4-5}$$

Thus the ZT of the corresponding output is given by

$$Y(z) = H(z)X(z) \tag{5.4-6}$$

$$= \left(\frac{zXe^{j\psi_x}}{z - e^{j\omega T}}\right) H(z)$$

Next, the IZT of (5.4-6) consists of two parts as follows:

$y(n) =$ {transient response due to poles of $H(z)$} + {steady-state response due to the pole $z = e^{j\omega T}$ of $X(z)$}

It is the steady-state portion that is of interest to us. In Fig. 5.4-1 this portion is denoted by $y(n)_{ss}$. To evaluate the same, we use the residue method and hence form

$$G(z) = z^{n-1}Y(z)$$

$$= \left(\frac{z^n X e^{j\psi_x}}{z - e^{j\omega T}}\right) H(z)$$

Then
$$y(n)_{ss} = R_{z=e^{j\omega T}} \tag{5.4-7}$$

where $R_{z=e^{j\omega T}}$ is the residue of $G(z)$ at $z = e^{j\omega T}$. Evaluating this residue, we obtain

$$R_{z=e^{j\omega T}} = z^n X e^{j\psi_x} H(z)\Big|_{z=e^{j\omega T}}$$

$$= H(e^{j\omega T}) X e^{j(n\omega T + \psi_x)}$$

Thus (5.4-7) yields

$$y(n)_{ss} = H(e^{j\omega T}) X e^{j(n\omega T + \psi_x)} \tag{5.4-8}$$

Now, the assumed form of $y(n)_{ss}$ is given in Fig. 5.4-1 to be

$$y(n)_{ss} = Y e^{j(n\omega T + \psi_y)} \tag{5.4-9}$$

Hence we equate the right-hand sides of (5.4-8) and (5.4-9) to obtain

$$Y e^{j(n\omega T + \psi_y)} = H(e^{j\omega T}) X e^{j(n\omega T + \psi_x)}$$

which yields

$$Y^\dagger = H(e^{j\omega T}) X^\dagger \tag{5.4-10}$$

where
$$X^\dagger = X e^{j\psi_x}$$
$$Y^\dagger = Y e^{j\psi_y}$$

Equation (5.4-10) implies that

$$H(e^{j\omega T}) = \frac{Y^\dagger}{X^\dagger} = \frac{Y}{X} e^{j(\psi_y - \psi_x)} \tag{5.4-11}$$

which is called the *steady-state transfer function* or *frequency response* of the system.

It is important to note that $H(e^{j\omega T})$ in (5.4-11) is merely the transfer function $H(z)$ evaluated *on* the unit circle at the point $z = e^{j\omega T}$, as illustrated next.

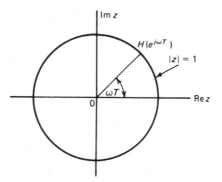

Magnitude and Phase Responses

The steady-state transfer function

$$H(e^{j\omega T}) = H(z) \Big|_{z = e^{j\omega T}} \qquad (5.4\text{-}12)$$

enables us to define magnitude and phase responses for DT systems by noting that $H(e^{j\omega T})$ is a complex number, and hence can be written as

$$H(e^{j\omega T}) = |H(e^{j\omega T})| \, e^{j\psi(\omega T)} \qquad (5.4\text{-}13)$$

where $\psi(\omega T) = \tan^{-1}\left[\dfrac{\text{Im }\{H(e^{j\omega T})\}}{\text{Re }\{H(e^{j\omega T})\}}\right]$

In (5.4-13) $|H(e^{j\omega T})|$ and $\psi(\omega T)$ are the *magnitude (amplitude)* and *phase responses*, respectively. The quantity $|H(e^{j\omega T})|^2$ is referred to as the *squared-magnitude response*. The value of $|H(e^{j\omega T})|$ for a given ω is called the system *gain* at that frequency.

The preceding quantities can also be expressed in terms of the frequency variable f; that is,

$$H(e^{j2\pi fT}) = |H(e^{j2\pi fT})| \, e^{j\psi(2\pi fT)} \qquad (5.4\text{-}14)$$

It is convenient to adopt the following notation in connection with magnitude and phase responses:

- $H(e^{j\omega T}) = \bar{H}(\omega)$
- $H(e^{j2\pi fT}) = \bar{H}(f)$
- $\psi(\omega T) = \bar{\psi}(\omega)$
- $\psi(2\pi fT) = \bar{\psi}(f)$

Then corresponding to (5.4-13) and (5.4-14), we have

$$\tilde{H}(\omega) = |\tilde{H}(\omega)|e^{j\tilde{\psi}(\omega)} \qquad (5.4\text{-}15)$$

and

$$\tilde{H}(f) = |\tilde{H}(f)|e^{j\tilde{\psi}(f)} \qquad (5.4\text{-}16)$$

A fundamental property of the magnitude and phase responses is that they are periodic. This is because they are both functions of the variable $e^{j\omega T}$, which has a period of 2π. In particular, $|\tilde{H}(\omega)|$ and $\tilde{\psi}(\omega)$ are periodic with period $\omega_s = 2\pi/T$ radians/second, while the period of $|\tilde{H}(f)|$ and $\tilde{\psi}(f)$ is $f_s = 1/T$ hertz. In addition, the magnitude and phase responses are even and odd functions, respectively, in each period, as illustrated for $|\tilde{H}(f)|$ and $\tilde{\psi}(f)$ in Fig. 5.4-2. In Fig. 5.4-2 we observe that *all* the information related to $|\tilde{H}(f)|$ and $\tilde{\psi}(f)$ is available in the region $0 \leq f \leq f_s/2$. This is because if $|\tilde{H}(f)|$ and $\tilde{\psi}(f)$ are known in the region $0 \leq f \leq f_s/2$ then they can be extended to any region of the f-axis. We recall that the frequency $f_s/2$ is the Nyquist or folding frequency and is denoted by f_N. In other words, we need to be concerned with $|\tilde{H}(f)|$ and $\tilde{\psi}(f)$ in this region, since

$$|\tilde{H}(f_N + f)| = |\tilde{H}(f_N - f)| \qquad (5.4\text{-}17)$$

and

$$\tilde{\psi}(f_N + f) = -\tilde{\psi}(f_N - f), \qquad 0 \leq f \leq f_N$$

Next, it is convenient to refer to magnitude and phase responses in terms of the following *normalized frequency variable* ν:

$$\nu = \frac{f}{f_N} \qquad (5.4\text{-}18)$$

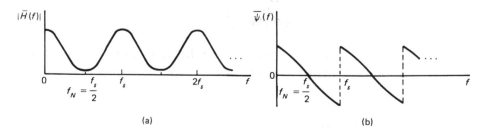

(a)   (b)

Fig. 5.4-2 Magnitude and phase responses to illustrate that they are periodic and have even and odd function properties.

where $0 \leq \nu \leq 1$ since f varies between 0 and f_N. We will denote the magnitude and phase responses that are expressed in terms of ν by $|\bar{H}(\nu)|$ and $\bar{\psi}(\nu)$, respectively.

It is instructive to show how magnitude and phase responses are evaluated in a systematic manner. We do so via examples in the next section.

5.5 ILLUSTRATIVE EXAMPLES

Example 5.5-1: Consider the first-order FIR system

$$y(n) = \frac{1}{2} [x(n) + x(n - 1)], \qquad n \geq 0 \tag{5.5-1}$$

which is sometimes referred to as a *simple moving average filter.*

(a)   Find $|\bar{H}(\nu)|^2$ and $\bar{\psi}(\nu)$.

(b)   What is the bandwidth of this filter if the sampling frequency is 10,000 sps.

Solution:

(a) We first find $H(z)$ and hence take the ZT of (5.5-1) with zero initial conditions to obtain

$$Y(z) = \frac{1}{2} [X(z) + z^{-1}X(z)]$$

which yields

$$H(z) = \frac{Y(z)}{X(z)} = \frac{1}{2}(1 + z^{-1}) \tag{5.5-2}$$

Next, from (5.5-2) we have

$$H(e^{j\omega T}) = H(z) \bigg|_{z = e^{j\omega T}} = \frac{1}{2}(1 + e^{-j\omega T}) \tag{5.5-3}$$

or $$H(e^{j\pi\nu}) = \frac{1}{2}(1 + e^{-j\pi\nu})$$

since $\omega T = 2\pi f/f_s = \pi\nu$. Equation (5.5-3) can be written as

$$\bar{H}(\nu) = \frac{1}{2}\{[1 + \cos(\pi\nu)] - j\sin(\pi\nu)\}, \qquad 0 \le \nu \le 1 \qquad (5.5\text{-}4)$$

which is the steady-state transfer function in terms of the normalized frequency variable ν. Thus

$$|\bar{H}(\nu)|^2 = [\text{Re}\,\{\bar{H}(\nu)\}]^2 + [\text{Im}\,\{\bar{H}(\nu)\}]^2 \qquad (5.5\text{-}5a)$$

and
$$\bar{\psi}(\nu) = \tan^{-1}\left[\frac{\text{Im}\,\{\bar{H}(\nu)\}}{\text{Re}\,\{\bar{H}(\nu)\}}\right] \qquad (5.5\text{-}5b)$$

where $\text{Re}\,\{\bar{H}(\nu)\}$ and $\text{Im}\,\{\bar{H}(\nu)\}$ denote the real and imaginary parts of $\bar{H}(\nu)$, respectively.

From (5.5-4) it follows that

$$\text{Re}\,\{\bar{H}(\nu)\} = \frac{1}{2}[1 + \cos(\pi\nu)]$$
$$(5.5\text{-}6)$$

and
$$\text{Im}\,\{\bar{H}(\nu)\} = -\frac{1}{2}\sin(\pi\nu)$$

Substitution of (5.5-6) in (5.5-5a) leads to

$$|\bar{H}(\nu)|^2 = \frac{1}{2}[1 + \cos(\pi\nu)], \qquad 0 \le \nu \le 1 \qquad (5.5\text{-}7)$$

which is the squared-magnitude response.

Again, substitution of (5.5-6) in (5.5-5b) results in

$$\bar{\psi}(\nu) = \tan^{-1}\left\{\frac{-\sin(\pi\nu)}{1 + \cos(\pi\nu)}\right\} \qquad (5.5\text{-}8)$$

Using the identities

$$\sin(\pi\nu) = 2\sin\left(\frac{\pi\nu}{2}\right)\cos\left(\frac{\pi\nu}{2}\right)$$

and
$$\cos(\pi\nu) = 2\cos^2\left(\frac{\pi\nu}{2}\right) - 1$$

we simplify (5.5-8) to obtain the phase response

$$\bar{\psi}(v) = \tan^{-1}\left\{-\tan\left(\frac{\pi v}{2}\right)\right\} \tag{5.5-9}$$

$$= -\frac{\pi v}{2}, \quad 0 \le v \le 1$$

Figure 5.5-1 shows plots of $|\bar{H}(v)|^2$ and $\bar{\psi}(v)$ in (5.5-7) and (5.5-9), respectively. From these plots it is clear that the difference equation in (5.5-1) represents a linear phase FIR low-pass filter.

(b) Let v_B denote the *normalized bandwidth;* that is,

$$v_B = \frac{B}{f_N} \tag{5.5-10}$$

where B is the actual bandwidth in hertz.

We note that the maximum value of $|\bar{H}(v)|^2$ in (5.5-7) occurs at $v = 0$. Thus v_B represents that value of v at which the squared-magnitude response equals *one-half*[†] its maximum value, which means that

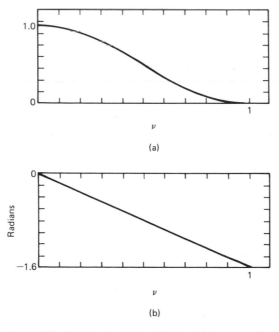

Fig. 5.5-1 (a) Squared-magnitude response and (b) phase response of $H(z)$ in (5.5-2).

[†]That is, the 3-dB point on a logarithmic scale.

$$|\bar{H}(\nu_B)|^2 = \frac{1}{2}|H(0)|^2 = \frac{1}{2} \tag{5.5-11}$$

since $|H(0)|^2 = 1$.

From (5.5-7) and (5.5-11), we have

$$\frac{1}{2}[1 + \cos(\pi\nu_B)] = \frac{1}{2}$$

which means that

$$\pi\nu_B = \frac{\pi}{2}$$

or

$$\nu_B = 0.5$$

Thus (5.5-10) yields the desired bandwidth to be

$$B = \nu_B f_N$$

$$= 0.5 \times 5000 = 2500 \text{ Hz}$$

since a sampling frequency of 10,000 sps yields $f_N = 5000$ Hz.

Example 5.5-2: Consider the first-order IIR system

$$y(n) = \frac{1 - \alpha}{2}\{x(n) + x(n - 1)\} + \alpha y(n - 1), \qquad n \geq 0 \tag{5.5-12}$$

where $0 < \alpha < 1$ is a parameter.

(a) Show that

$$|\bar{H}(\nu)|^2 = \frac{(1 - \alpha)^2}{2} \left\{ \frac{1 + \cos(\pi\nu)}{1 + \alpha^2 - 2\alpha \cos(\pi\nu)} \right\}, \qquad 0 \leq \nu \leq 1 \tag{5.5-13}$$

(b) Assuming that $|\bar{H}(\nu)|^2$ attains its maximum value at $\nu = 0$ in the range $0 \leq \nu \leq 1$, show that the values of α that cause the system to have a normalized bandwidth of ν_B are given by

$$\alpha = \frac{1 - \sin(\pi\nu_B)}{\cos(\pi\nu_B)}$$

Solution:

(a) Taking the ZT of (5.5-12) with zero initial conditions, we obtain

$$Y(z) = \frac{1 - \alpha}{2} \{X(z) + z^{-1}X(z)\} + \alpha z^{-1}Y(z)$$

which results in

$$H(z) = \frac{1 - \alpha}{2} \left(\frac{1 + z^{-1}}{1 - \alpha z^{-1}} \right)$$

Next, we have

$$H(e^{j\omega T}) = \frac{1 - \alpha}{2} \left(\frac{1 + e^{-j\omega T}}{1 - \alpha e^{-j\omega T}} \right) \tag{5.5-14}$$

Since $\omega T = \pi \nu$, it follows that (5.5-14) can be written as

$$\bar{H}(\nu) = \frac{1 - \alpha}{2} \left(\frac{1 + e^{-j\pi\nu}}{1 \quad \alpha e^{-j\pi\nu}} \right)$$

Thus

$$|\bar{H}(\nu)|^2 = \frac{(1 - \alpha)^2}{4} \left(\frac{|1 + e^{-j\pi\nu}|^2}{|1 - \alpha e^{-j\pi\nu}|^2} \right) \tag{5.5-15}$$

Now

$$|1 + e^{-j\pi\nu}|^2 = [1 + \cos(\pi\nu)]^2 + [\sin(\pi\nu)]^2$$
$$= 2[1 + \cos(\pi\nu)]$$

and

$$|1 - \alpha e^{-j\pi\nu}|^2 = [1 - \alpha \cos(\pi\nu)]^2 + \alpha^2[\sin(\pi\nu)]^2$$
$$= [1 + \alpha^2 - 2\alpha \cos(\pi\nu)] \tag{5.5-16}$$

Substituting (5.5-16) in (5.5-15), we obtain $|H(\nu)|^2$ in (5.5-13).

(b) To find the normalized bandwidth ν_B, we note that the maximum value of $|H(\nu)|^2$ is $|H(0)|^2$. Thus ν_B must be such that

$$|H(\nu_B)|^2 = 0.5 |H(0)|^2 \tag{5.5-17}$$
$$= 0.5$$

since $|H(0)|^2 = 1$; see (5.5-13).

Substitution of (5.5-13) with $\nu = \nu_B$ in (5.5-17) leads to

$$\frac{(1-\alpha)^2}{2}\left\{\frac{1+\cos(\pi\nu_B)}{1+\alpha^2-2\alpha\cos(\pi\nu_B)}\right\} = \frac{1}{2}$$

which upon simplification results in

$$\alpha^2\cos(\pi\nu_B) - 2\alpha + \cos(\pi\nu_B) = 0 \qquad (5.5\text{-}18)$$

We note that (5.5-18) is a quadratic equation in terms of α. Thus, solving for α, we obtain

$$\alpha = \frac{2 \pm \sqrt{4 - 4\cos^2(\pi\nu_B)}}{2\cos(\pi\nu_B)} = \frac{1 \pm \sin(\pi\nu_B)}{\cos(\pi\nu_B)}$$

Hence $\qquad \alpha = \dfrac{1 - \sin(\pi\nu_B)}{\cos(\pi\nu_B)}$

yields $0 < \alpha < 1$ for values of ν_B in the range $0 < \nu_B < 0.5$.

Example 5.5-3: Given the FIR system

$$y(n) = \frac{1}{N}\underbrace{\{x(n) + x(n-1) + \cdots + x(n - [N-1])\}}_{N \text{ samples}}, \qquad n \geq 0$$

$$(5.5\text{-}19)$$

which is referred to as a *moving average filter*, a special case (i.e., $N = 2$) of which was considered in Example 5.5-1. Show that

$$|\bar{H}(\nu)| = \frac{1}{N}\left|\frac{\sin(N\pi\nu/2)}{\sin(\pi\nu/2)}\right|$$

and find the phase response $\bar{\psi}(\nu)$, $0 \leq \nu \leq 1$.

Solution: To find $H(z)$, we assume zero initial conditions and take the ZT of (5.5-19). This yields

$$Y(z) = \frac{1}{N}\{1 + z^{-1} + z^{-2} + \cdots + z^{-(N-1)}\}X(z)$$

which means that

$$H(z) = \frac{Y(z)}{X(z)} = \frac{1}{N}\{1 + z^{-1} + z^{-2} + \cdots + z^{-(N-1)}\}$$

Thus

$$H(e^{j\omega T}) = \frac{1}{N} \{1 + e^{-j\omega T} + e^{-2j\omega T} + \cdots + e^{-(N-1)j\omega T}\}$$

(5.5-20)

or

$$\bar{H}(\nu) = \frac{1}{N} \{1 + e^{-j\pi\nu} + e^{-2j\pi\nu} + \cdots + e^{-(N-1)j\pi\nu}\}$$

which is a geometric series with common ratio $e^{-j\pi\nu}$. Hence (5.5-20) can be equivalently written as

$$\bar{H}(\nu) = \frac{1}{N} \left[\frac{1 - e^{-jN\pi\nu}}{1 - e^{-j\pi\nu}} \right]$$

(5.5-21)

Next we manipulate (5.5-21) by using the fact that

$$1 = e^{j\pi\nu/2} \cdot e^{-j\pi\nu/2} = e^{jN\pi\nu/2} \cdot e^{-jN\pi\nu/2}$$

Thus (5.5-21) becomes

$$\bar{H}(\nu) = \frac{1}{N} \left[\frac{e^{jN\pi\nu/2} e^{-jN\pi\nu/2} - e^{-jN\pi\nu}}{e^{j\pi\nu/2} e^{-j\pi\nu/2} - e^{-j\pi\nu}} \right]$$

(5.5-22)

Factoring $e^{-jN\pi\nu/2}$ and $e^{-j\pi\nu/2}$ from the numerator and denominator of (5.5-22), respectively, we get

$$\bar{H}(\nu) = \frac{1}{N} \left[\frac{e^{jN\pi\nu/2} - e^{-jN\pi\nu/2}}{e^{j\pi\nu/2} - e^{-j\pi\nu/2}} \right] \frac{e^{-jN\pi\nu/2}}{e^{-j\pi\nu/2}}$$

(5.5-23)

Use of the identity

$$\sin \theta = \frac{1}{2j} (e^{j\theta} - e^{-j\theta})$$

in (5.5-23) leads to

$$\bar{H}(\nu) = \frac{1}{N} \left[\frac{\sin (N\pi\nu/2)}{\sin (\pi\nu/2)} \right] e^{-j(N-1)\pi\nu/2}$$

(5.5-24)

In (5.5-24), since $\left| e^{-j(N-1)\pi\nu/2} \right| = 1$, we have

$$|\bar{H}(\nu)| = \frac{1}{N} \left| \frac{\sin (N\pi\nu/2)}{\sin (\pi\nu/2)} \right|$$

which is the desired result.

We observe that the FIR system in (5.5-19) has a $\sin (x)/x$ type of magnitude response. A plot of this response for $N = 7$ is shown in Fig. 5.5-2, from which it is apparent that the moving average filter is a low-pass filter.

Next, to find the phase spectrum, $\bar{\psi}(\nu)$, we write (5.5-24) as

$$\bar{H}(\nu) = A(\nu)e^{j\theta(\nu)} \qquad (5.5\text{-}25)$$

where $A(\nu) = \dfrac{1}{N}\left[\dfrac{\sin (N\pi\nu/2)}{\sin (\pi\nu/2)} \right]$

$\theta(\nu) = -\dfrac{(N - 1)\pi\nu}{2}$

From (5.5-25) it is apparent that the phase associated with $\bar{H}(\nu)$ [i.e., $\bar{\psi}(\nu)$] consists of two contributions, one due to $A(\nu)$ and the other due to $\theta(\nu)$. Now $A(\nu)$ is purely a real number, which can be positive or negative. Hence its contribution is 0 or $\pm \pi$, depending upon whether it is positive or negative, respectively. On the other hand, the contribution of $\theta(\nu)$ is $-(N - 1)\pi\nu/2$. Thus

$$\bar{\psi}(\nu) = \begin{cases} -\dfrac{(N - 1)\pi\nu}{2}, & \text{for positive } A(\nu) \\[3mm] -\dfrac{(N - 1)\pi\nu}{2} \pm \pi, & \text{for negative } A(\nu) \end{cases} \qquad (5.5\text{-}26)$$

is the desired phase response.

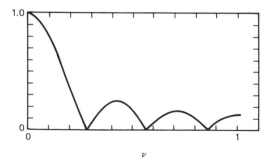

Fig. 5.5-2 Magnitude response of a moving average filter for $N = 7$.

COMMENT. Equation (5.5-26) implies that $\bar{\psi}(\nu)$ is a *piecewise* linear function of ν. However, we observe that

$$T_d(\nu) = -\frac{d\bar{\psi}(\nu)}{d\nu} = \frac{(N-1)\pi}{2} = \text{constant}, \quad 0 \le \nu \le 1 \quad (5.5\text{-}27)$$

is *independent* of ν, where $T_d(\nu)$ is called the *time (group) delay function* of a DT system.

Systems (filters) for which $T_d(\nu)$ is a *constant*, and hence independent of ν, are referred to as *constant time delay* or *linear phase* DT systems.[†] This is because such systems avoid phase distortion by causing all sinusoidal components in the input to the system to be delayed by the *same* amount. The moving average filter defined by (5.5-19) belongs to this class of DT systems. Procedures to design linear phase filters will be introduced in Chapter 7.

Example 5.5-4: Consider the second-order IIR system defined by (5.3-7) with $K_1 = 0.94$ and $K_2 = -0.64$; that is,

$$y(n) = 0.94y(n-1) - 0.64y(n-2) + x(n), \quad n \ge 0 \quad (5.5\text{-}28)$$

(a) Plot the magnitude response, and verify that it peaks at 1500 Hz when the sampling frequency is 10,000 sps.

(b) Use the information available in Example 5.3-2 to explain why the magnitude response peaks at 1500 Hz.

Solution:

(a) With zero initial conditions, the ZT of (5.5-28) leads to

$$Y(z) = 0.94z^{-1}Y(z) - 0.64z^{-2}Y(z) + X(z)$$

which yields

$$H(z) = \frac{Y(z)}{X(z)} = \frac{z^2}{z^2 - 0.94z + 0.64} \quad (5.5\text{-}29)$$

Thus

$$H(e^{j\omega T}) = \frac{e^{2j\omega T}}{e^{2j\omega T} - 0.94e^{j\omega T} + 0.64} \quad (5.5\text{-}30)$$

or

$$\tilde{H}(\nu) = \frac{e^{2j\pi\nu}}{e^{2j\pi\nu} - 0.94e^{j\pi\nu} + 0.64}$$

[†]A thorough discussion of the linear phase concept is given in [1] and [2]. It suffices to restrict our attention to linear phase as it pertains to the time-delay function being a constant (i.e., independent of ν).

From (5.5-30) it is obvious that finding the real and imaginary parts of $\bar{H}(\nu)$ would be a very tedious task. Hence $|\bar{H}(\nu)|$ is evaluated by resorting to a computer. A FORTRAN program for doing so is given in Appendix 5.1. The resulting plot of $|\bar{H}(\nu)|$ is shown in Fig. 5.5-3. It was obtained by computing $|\bar{H}(\nu)|$ at 128 points in the region $0 \le \nu \le 1$. The abscissa in Fig. 5.5-3 is in hertz, where each value of frequency f_i is obtained from ν_i using (5.4-18); that is, $f_i = \nu_i f_N$, where $f_N = 5000$ Hz, which corresponds to a sampling frequency of 10,000 sps. From Fig. 5.5-3 it is apparent that the magnitude response peaks at 1500 Hz.

   (b) With $K_1 = 0.94$ and $K_2 = -0.64$, it follows that

$$\left(\frac{K_1^2}{4} + K_2 \right) < 0$$

which corresponds to case 1 in Example 5.3-2. This implies that $H(z)$ in (5.5-29) has complex-conjugate poles. From (5.3-11) it follows that these poles are given by

$$z_1 = \gamma e^{j\omega_0 T} \quad \text{and} \quad z_2 = \gamma e^{-j\omega_0 T}$$

where $\gamma = 0.8$ and $\omega_0 T = 0.9428$ rad. Clearly, this pair of complex-conjugate poles provide a *resonance effect*, which contributes to a peak in Fig. 5.5-3. Furthermore, as this pole-pair is brought closer to the unit circle, the resonance effect is more pronounced. In this case the resonant frequency f_0 equals $\omega_0/2\pi$, and since $\omega_0 T = 0.9428$, we have

$$f_0 = \frac{0.9428 \times 10^4}{2\pi} = 1500 \text{ Hz}$$

which is the frequency at which the peak in the magnitude response occurs.

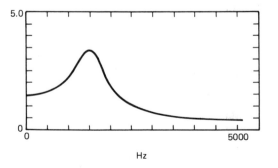

Fig. 5.5-3 Magnitude response of system defined by (5.5-26).

5.6 REALIZATION FORMS

A given $H(z)$ can be realized in several forms or configurations. These are known as (1) direct form 1, (2) direct form 2, (3) cascade, and (4) parallel realizations.

Direct Form 1 Realization

From (5.1-4a) we have

$$H(z) = \frac{a_0 + a_1 z^{-1} + \cdots + a_k z^{-k}}{1 + b_1 z^{-1} + \cdots + b_k z^{-k}} = \frac{N(z)}{D(z)} \qquad (5.6\text{-}1)$$

If $H(z)$ in (5.6-1) is substituted in

$$Y(z) = H(z)X(z) \qquad (5.6\text{-}2)$$

we obtain

$$Y(z)[1 + b_1 z^{-1} + \cdots + b_k z^{-k}]$$
$$= [a_0 + a_1 z^{-1} + \cdots + a_k z^{-k}]X(z) \qquad (5.6\text{-}3)$$

Use of the IZT in connection with (5.6-3) leads to

$$y(n) + b_1 y(n-1) + \cdots + b_k y(n-k)$$
$$= a_0 x(n) + a_1 x(n-1) + \cdots + a_k x(n-k)$$

which can be written in a more compact form as

$$y(n) = \sum_{i=0}^{k} a_i x(n-i) - \sum_{i=1}^{k} b_i y(n-i) \qquad (5.6\text{-}4)$$

Equation (5.6-4) can be realized as shown in Fig. 5.6-1a, where the following notation is used:

1. $f(n) \xrightarrow{\ a_i\ }$ means that the output of this stage is the product $f(n)a_i$.

2. $f(n) \rightarrow \boxed{z^{-1}} \rightarrow$ means that the output of this stage is $f(n-1)$, which implies that $\boxed{z^{-1}}$ can be viewed as being a unit delay element.

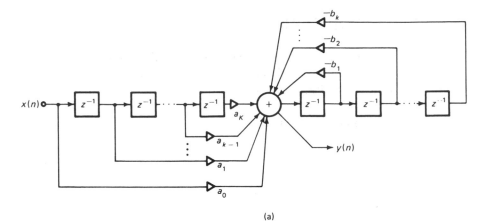

(a)

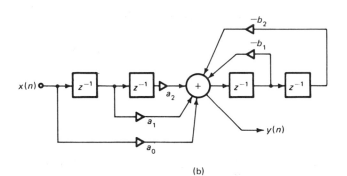

(b)

Fig. 5.6-1 Direct form 1 realization: (a) kth-order $H(z)$; (b) second-order $H(z)$.

From Fig. 5.6-1a it is apparent that $2k$ registers or memory locations are required to store the $x(n - i)$ and $y(n - i)$, $1 \le i \le k$.

It is worthwhile verifying that the direct form 1 realization does indeed represent the fundamental relation $Y(z) = H(z)X(z)$ by tracing through it. To this end, we consider the direct form 1 realization for the second-order case, which is shown in Fig. 5.6-1b. Examining Fig. 5.6-1b at the summing junction, we obtain

$$a_0X(z) + a_1z^{-1}X(z) + a_2z^{-2}X(z) + (-b_1)z^{-1}Y(z) + (-b_2)z^{-2}Y(z) = Y(z)$$

which yields

$$Y(z)(1 + b_1z^{-1} + b_2z^{-2}) = X(z)(a_0 + a_1z^{-1} + a_2z^{-2})$$

That is,

$$Y(z) = \left[\frac{a_0 + a_1 z^{-1} + a_2 z^{-2}}{1 + b_1 z^{-1} + b_2 z^{-2}} \right] X(z)$$

which is the desired result since the quantity in brackets is the transfer function $\Pi(z)$ in (5.6-1) for $k - 2$.

Direct Form 2 Realization

From (5.6-1) and (5.6-2) we have

$$Y(z) = \frac{N(z)}{D(z)} \cdot X(z) \tag{5.6-5}$$

Let us define

$$W(z) = \frac{X(z)}{D(z)} \tag{5.6-6}$$

Then (5.6-5) and (5.6-6) imply that

$$Y(z) = N(z)W(z) \tag{5.6-7}$$

Again, (5.6-6) yields

$$X(z) = W(z)[1 + b_1 z^{-1} + b_2 z^{-2} + \cdots + b_k z^{-k}]$$

which leads to

$$x(n) = w(n) + b_1 w(n - 1) + b_2 w(n - 2) + \cdots + b_k w(n - k), \quad n \geq 0 \tag{5.6-8}$$

Next (5.6-7) can be written as

$$Y(z) = W(z)[a_0 + a_1 z^{-1} + a_2 z^{-2} + \cdots + a_k z^{-k}]$$

which yields

$$y(n) = a_0 w(n) + a_1 w(n - 1) + a_2 w(n - 2) + \cdots + a_k w(n - k), \quad n \geq 0 \tag{5.6-9}$$

We now combine (5.6-8) and (5.6-9) to obtain the direct form realization shown in Fig. 5.6-2 for the general case and for the special case $k = 2$.

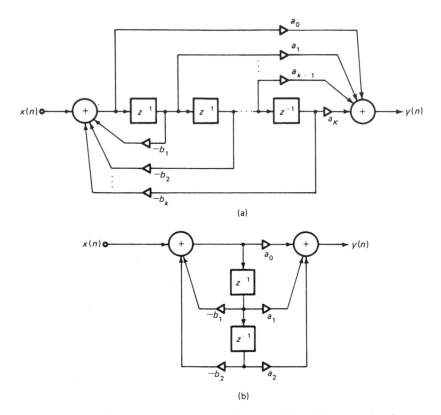

Fig. 5.6-2 Direct form 2 realization: (a) kth-order $H(z)$; (b) second-order $H(z)$.

We observe that this realization requires k *registers* to realize the kth *order* $H(z)$ in (5.6-1), as opposed to *2k registers* required for the direct form 1 realization; see Fig. 5.6-1. Hence the direct form 2 realization is also referred to as a *canonic* form.

Cascade (Series) Form Realization

Using straightforward polynomial factorization techniques, the numerator and denominator polyonimals of $H(z)$ in (5.6-1) can be factored into first- and second-order polynomials. As such, $H(z)$ in (5.6-1) can be expressed as

$$H(z) = a_0 H_1(z) H_2(z) \cdots H_\ell(z) \qquad (5.6\text{-}10)$$

where ℓ is a positive integer, and each $H_i(z)$ is a first- or second-order transfer function; that is,

$$H_i(z) = \frac{1 + a_{i1}z^{-1}}{1 + b_{i1}z^{-1}}$$

$$(5.6\text{-}11)$$

or

$$H_i(z) = \frac{1 + a_{i1}z^{-1} + a_{i2}z^{-2}}{1 + b_{i1}z^{-1} + b_{i2}z^{-2}}$$

Substitution of (5.6-10) in (5.6-2) leads to

$$Y(z) = a_0[H_1(z)\ H_2(z)\ \cdots\ H_\ell(z)]\ X(z)$$

which yields the *cascade* or *series* realization shown in Fig. 5.6-3.

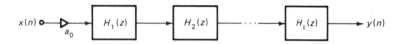

Fig. 5.6-3 Cascade or series realization.

Parallel Form Realization

Here we express $H(z)$ in (5.6-1) as

$$H(z) = c + H_1(z) + H_2(z) + \cdots + H_r(z)$$

$$(5.6\text{-}12)$$

where c is a constant, r is a positive integer, and $H_i(z)$ is a first- or second-order transfer function; that is,

$$H_i(z) = \frac{a_{i0}}{1 + b_{i1}z^{-1}}$$

$$(5.6\text{-}13)$$

or

$$H_i(z) = \frac{a_{i0} + a_{i1}z^{-1}}{1 + b_{i1}z^{-1} + b_{i2}z^{-2}}$$

A given $H(z)$ can be expressed as indicated in (5.6-12) by resorting to a PFE of $H(z)/z$. The desired *parallel realization* is shown in Fig. 5.6-4, and follows directly from (5.6-12).

We now consider an illustrative example.

Example 5.6-1: Given the second-order transfer function

$$H(z) = \frac{0.7(z^2 - 0.36)}{z^2 + 0.1z - 0.72}$$

$$(5.6\text{-}14)$$

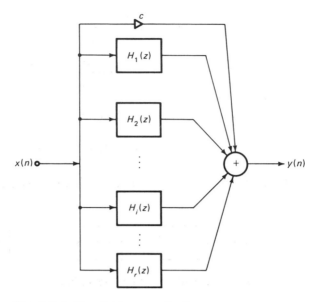

Fig. 5.6-4 Parallel-form realization.

Obtain the following realizations:

(a) The direct form 2 (or canonic).

(b) Series form in terms of first-order sections.

(c) Parallel form in terms of first-order sections.

Solution:

(a) We rewrite $H(z)$ in (5.6-14) as

$$H(z) = \frac{0.7(1 - 0.36z^{-2})}{1 + 0.1z^{-1} - 0.72z^{-2}} = \frac{0.7(a_0 + a_1z^{-1} + a_2z^{-2})}{1 + b_1z^{-1} + b_2z^{-2}} \quad (5.6-15)$$

From (5.6-15) and Fig. 5.6-2b, it is clear that the canonic form realization is as shown in the following sketch.

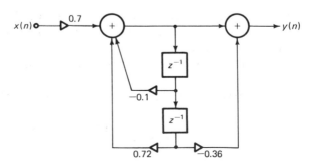

(b) To obtain the series form realization, we factor the numerator and denominator polynomials of $H(z)$ in (5.6-15) to obtain

$$H(z) = \frac{0.7(1 + 0.6z^{-1})(1 - 0.6z^{-1})}{(1 + 0.9z^{-1})(1 - 0.8z^{-1})} \qquad (5.6\text{-}16)$$

There is nothing unique about how one combines the first-order polynomials in (5.6-16) to obtain corresponding first-order transfer functions. For example, one choice is as follows:

$$H_1(z) = \frac{1 + 0.6z^{-1}}{1 - 0.8z^{-1}} \qquad (5.6\text{-}17)$$

and
$$H_2(z) = \frac{1 - 0.6z^{-1}}{1 + 0.9z^{-1}}$$

The transfer functions in (5.6-17) can be realized in terms of either of the two direct forms (i.e., 1 or 2). To illustrate, we use the direct form 2 realization in Fig. 5.6-2b with $a_2 = b_2 = 0$, since $H_1(z)$ and $H_2(z)$ are first-order transfer functions. This results in the following series form realization.

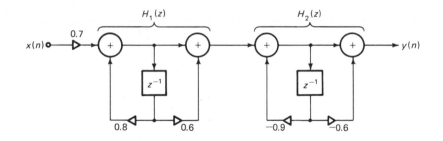

(c) Next we seek a parallel form realization, and hence obtain a PFE of $H(z)/z$, where $H(z)$ is given by (5.6-14); that is,

$$\hat{H}(z) = \frac{H(z)}{z} = \frac{0.7(z + 0.6)(z - 0.6)}{z(z + 0.9)(z - 0.8)}$$

$$= \frac{A}{z} + \frac{B}{z + 0.9} + \frac{C}{z - 0.8}$$

where $A = z\hat{H}(z)\big|_{z=0} = 0.35$

$\qquad B = (z + 0.9)\hat{H}(z)\big|_{z=-0.9} = 0.206$

$\qquad C = (z - 0.8)\hat{H}(z)\big|_{z=0.8} = 0.144$

Thus

$$H(z) = 0.35 + \frac{0.206z}{z + 0.9} + \frac{0.144z}{z - 0.8}$$

which can be written as

$$H(z) = 0.35 + \frac{0.206}{1 + 0.9z^{-1}} + \frac{0.144}{1 - 0.8z^{-1}} \tag{5.6-18}$$

$$= H_1(z) + H_2(z) + H_3(z)$$

One way of realizing $H_2(z)$ and $H_3(z)$ in (5.6-18) is by using the direct form 2 in Fig. 5.6-2b with $a_2 = b_2 = 0$, since $H_2(z)$ and $H_3(z)$ are first-order transfer functions. This approach yields the following parallel realization.

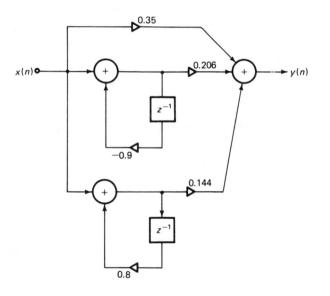

5.7 OUTPUT SPECTRA

We now consider the problem of expressing output spectra of a DT system in terms of its transfer function and the input spectra. To this end, let $x(m)$ be a real input sequence and the corresponding output sequence be $y(m)$, as indicated in Fig. 5.7-1. Then from the DT property in (5.2-4), we have

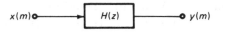

Fig. 5.7-1 Pertaining to input-output spectra.

$$Y(z) = H(z)X(z)$$

With $z = e^{j\omega T}$, we obtain

$$Y(e^{j\omega T}) = H(e^{j\omega T})X(e^{j\omega T}) \qquad (5.7\text{-}1)$$

where $H(e^{j\omega T})$ is the steady-state transfer function; $X(e^{j\omega T})$ and $Y(e^{j\omega T})$ are the FTs of the input and output sequences, respectively, as is apparent from (3.1-8).

Equation (5.7-1) can also be written as

$$\bar{Y}(\omega) = \bar{H}(\omega)\bar{X}(\omega) \qquad (5.7\text{-}2)$$

where $\bar{Y}(\omega) = Y(e^{j\omega T})$, $\bar{H}(\omega) = H(e^{j\omega T})$ and $\bar{X}(\omega) = X(e^{j\omega T})$

Next, (5.4-15) enables us to write $\bar{H}(\omega)$ as

$$\bar{H}(\omega) = |\bar{H}(\omega)|e^{j\bar{\psi}(\omega)} \qquad (5.7\text{-}3)$$

where $|\bar{H}(\omega)|$ and $\bar{\psi}(\omega)$ are the system magnitude and phase responses, respectively. Since $\bar{X}(\omega)$ and $\bar{Y}(\omega)$ are complex numbers, they can be expressed as

$$\bar{X}(\omega) = |\bar{X}(\omega)|\, e^{j\bar{\psi}_x(\omega)} \qquad (5.7\text{-}4)$$

and

$$\bar{Y}(\omega) = |\bar{Y}(\omega)|\, e^{j\bar{\psi}_y(\omega)}$$

In (5.7-4), $|\bar{X}(\omega)|$ and $|\bar{Y}(\omega)|$ are defined as the *FT amplitude spectra* of $x(m)$ and $y(m)$, respectively, while $\bar{\psi}_x(\omega)$ and $\bar{\psi}_y(\omega)$ are the corresponding *phase spectra*.

Substitution of (5.7-3) and (5.7-4) in (5.7-2) leads to

$$|\bar{Y}(\omega)|e^{j\bar{\psi}_y(\omega)} = |\bar{H}(\omega)||\bar{X}(\omega)|e^{j[\bar{\psi}(\omega) + \bar{\psi}_x(\omega)]} \qquad (5.7\text{-}5a)$$

and

$$|\bar{Y}(\omega)|^2 e^{j2\bar{\psi}_y(\omega)} = |\bar{H}(\omega)|^2|\bar{X}(\omega)|^2 e^{j2[\bar{\psi}(\omega) + \bar{\psi}_x(\omega)]} \qquad (5.7\text{-}5b)$$

The quantities $|\bar{X}(\omega)|^2$ and $|\bar{Y}(\omega)|^2$ in (5.7-5b) are defined as the input and output *FT power spectra*, respectively, and $|\bar{H}(\omega)|^2$ is the squared-magnitude response.

Thus from (5.7-5) we obtain the following relations:

1.
$$|\bar{Y}(\omega)| = |\bar{H}(\omega)||\bar{X}(\omega)| \qquad (5.7\text{-}6)$$

which implies that the output amplitude spectrum is the product of the magnitude response and the input amplitude spectrum. This product relationship is also true for the squared-magnitude response and the input power spectrum, since

$$|\bar{Y}(\omega)|^2 = |\bar{H}(\omega)|^2|\bar{X}(\omega)|^2 \qquad (5.7\text{-}7)$$

2.
$$\bar{\psi}_y(\omega) = \bar{\psi}(\omega) + \bar{\psi}_x(\omega) \qquad (5.7\text{-}8)$$

which implies that the output phase spectrum is the sum of the system phase response and the input phase spectrum.

Again, we note that one can obtain relationships corresponding to (5.7-6) to (5.7-8) in terms of the DFT spectra for the input and output, respectively. This can be done via the relation between the FT and DFT in (4.5-27) and the DFT spectra in (4.5-12) and (4.5-13). Thus it can be shown that

$$P_n^y = |\bar{H}(n\omega_0)|^2 \, P_n^x \qquad (5.7\text{-}9a)$$

$$p_n^y = |\bar{H}(n\omega_0)| \, p_n^x \qquad (5.7\text{-}9b)$$

and
$$\psi_n^y = \bar{\psi}(n\omega_0) + \psi_n^x, \qquad 0 \le n \le N/2 \qquad (5.7\text{-}9c)$$

where P_n, p_n, and ψ_n denote power, amplitude, and phase spectra, respectively; the superscripts indicate whether these are input or output spectra; N is the number of points taken to evaluate the DFT coefficients; $\omega_0 = 2\pi/NT$ is the fundamental radian frequency, where T is the sampling interval.

We conclude this discussion with an illustrative example. Suppose $H(z)$ in Fig. 5.7-1 is

$$H(z) = \frac{0.8394 - 1.5511z^{-1} + 0.8394z^{-2}}{1 - 1.5511z^{-1} + 0.6791z^{-2}} \qquad (5.7\text{-}10)$$

and the input is the 64-periodic sequence $x(m)$ displayed in Fig. 5.7-2. Let the sampling frequency be 960 sps; that is, the sampling interval T equals $\frac{1}{960}$ sec.

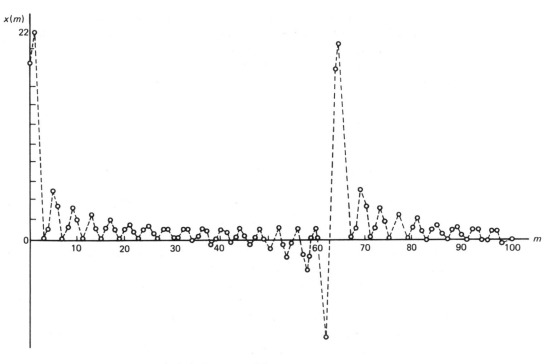

Fig. 5.7-2 Plot of $x(m)$, $0 \leq m \leq 99$.

Using the FFT program in Appendix 4.1, the DFT power spectrum P_n^x in (4.5-12) is computed using 64 points of $x(m)$. The resulting spectrum is as shown in Fig. 5.7-3, from which it is clear that all spectral points beyond $n = 16$ are zero. The fundamental frequency is given by

$$f_0 = \frac{1}{NT} = 15 \text{ Hz} \qquad (5.7\text{-}11)$$

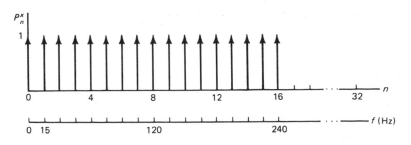

Fig. 5.7-3 Power spectrum of $x(m)$ in Fig. 5.7-2.

since $N = 64$ and $T = \frac{1}{960}$. Hence $n = 16$ corresponds to 240 ($= 16 \times 15$) Hz, which is the highest frequency component in $x(m)$. This means that the bandwidth of $x(m)$ is 240 Hz.

Next, with $H(z)$ in (5.7-10) substituted in $Y(z) = H(z)X(z)$, we obtain the following input–output difference equation:

$$y(m) = \sum_{i=0}^{2} a_i x(m - i) - \sum_{j=1}^{2} b_j y(m - j), \qquad m \geq 0 \qquad (5.7\text{-}12)$$

with $a_0 = 0.8394$, $a_1 = -1.5511$, $a_2 = 0.8394$, $b_1 = -1.5511$, and $b_2 = 0.6791$.

A computer evaluation of (5.7-12) results in the output sequence $y(m)$ displayed in Fig. 5.7-4. To compute the output DFT spectrum P_n^y, we again use a 64-point FFT with the set of output points $36 \leq y(m) \leq 99$. The reason for skipping the first 36 points of $y(m)$ is to avoid the transient portion. The output power spectrum so obtained is plotted in Fig. 5.7-5, from which it is clear that there is no 60-Hz component, and spectral components in the vicinity of 60 Hz are attenuated. Comparing P_n^x and P_n^y in Figs. 5.7-3 and 5.7-5, respectively, we conclude that $H(z)$ in (5.7-10) is the transfer function of a 60-Hz notch filter. One can also come to this conclusion by examining $|\bar{H}(\nu)|$, $0 \leq \nu \leq 1$; see Problem 5-12.

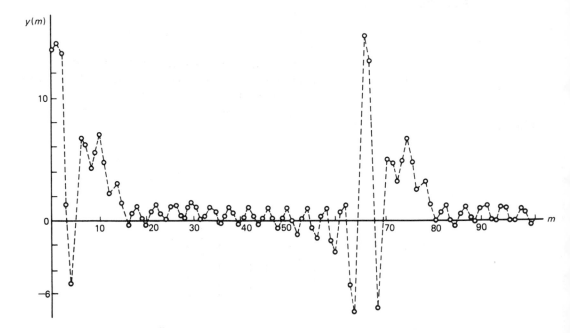

Fig. 5.7-4 Output sequence $y(m)$, $0 \leq m \leq 99$.

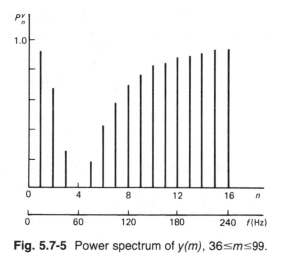

Fig. 5.7-5 Power spectrum of $y(m)$, $36 \le m \le 99$.

5.8 FAST CONVOLUTION

From the material discussed in this chapter, it is apparent the DT convolution plays a fundamental role in characterizing DT systems. Thus it is important that we be able to compute DT convolution in an efficient manner. Methods for doing so employ the FFT and have been studied in great detail. It is convenient to collectively refer to such methods as *fast convolution*. This section is devoted to studying the elements of fast convolution with respect to two cases. These concern the convolution of periodic and aperiodic sequences, respectively.

Circular[†] or Periodic Convolution

Given two real N-periodic sequences $x(m)$ and $y(m)$, their *circular* or *periodic convolution* sequence $z(m)$ is also an N-periodic sequence, and given by (4.5-17) as

$$z(m) = \sum_{h=0}^{N-1} x(h)y(m - h), \qquad 0 \le m \le N - 1 \qquad (5.8\text{-}1)$$

To illustrate, let us consider the case $N = 4$. Then (5.8-1) yields the following set of equations:

[†]Also known as *cyclic convolution*.

$$m = 0: \quad z(0) = x(0)y(0) + x(1)y(-1) + x(2)y(-2) + x(3)y(-3)$$
$$m = 1: \quad z(1) = x(0)y(1) + x(1)y(0) + x(2)y(-1) + y(3)y(-2)$$
$$m = 2: \quad z(2) = x(0)y(2) + x(1)y(1) + x(2)y(0) + x(3)y(-1) \quad (5.8\text{-}2)$$
$$m = 3: \quad z(3) = x(0)y(3) + x(1)y(2) + x(2)y(1) + x(3)y(0)$$

Since $y(m)$ is a 4-periodic sequence, it follows that $y(-\ell) = y(4 - \ell)$, $0 \le \ell \le 3$. Hence (5.8-2) can equivalently be written as follows:

$$z(0) = x(0)y(0) + x(1)y(3) + x(2)y(2) + x(3)y(1)$$
$$z(1) = x(0)y(1) + x(1)y(0) + x(2)y(3) + x(3)y(2) \quad (5.8\text{-}3)$$
$$z(2) = x(0)y(2) + x(1)y(1) + x(2)y(0) + x(3)y(3)$$
$$z(3) = x(0)y(3) + x(1)y(2) + x(2)y(1) + x(3)y(0)$$

The information in (5.8-3) can conveniently be represented in graphical form, as shown in Fig. 5.8-1. It is evident that $z(m)$ can be computed via successive sums and shifts of $x(m)$ with four values of the "reversed sequence" $y(0)$, $y(3)$, $y(2)$, and $y(1)$. It follows that the general form of the reversed sequence is $y(0)$, $y(N - 1)$, $y(N - 2)$, ..., $y(1)$, and is also N-periodic.

Examination of Fig. 5.8-1 shows that to compute each $z(m)$ we need four real multiplications and three real additions. Thus the number of real arithmetic operations (multiplications and additions), Σ_4, needed to compute $z(m)$, $0 \le m \le 3$, is

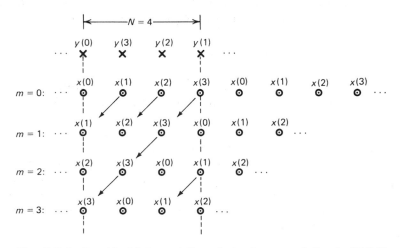

Fig. 5.8-1 Graphical interpretation of circular convolution in (5.8-1) for $N = 4$.

$$\Sigma_4 = 4[4 + (4 - 1)]$$

It is apparent that the general form of Σ_4 is given by

$$\Sigma_N = N[N + (N - 1)] = 2N^2 - N \tag{5.8-4}$$

where Σ_N is the total number of real arithmetic operations needed to compute $z(m)$, $0 \le m \le N - 1$. For convenience, let us call this brute-force approach for computing circular convolution the "direct method." From (5.8-4) we see that as N increases, $\Sigma_N \approx 2N^2$. Thus the number of arithmetic operations needed to compute circular convolution via the direct method is *proportional to N^2*.

We now consider an alternate approach which uses the property that the convolution of the periodic sequences $x(m)$ and $y(m)$ is equivalent to multiplying their DFTs $X(n)$ and $Y(n)$, respectively; see (4.5-18). Thus

$$Z(n) = X(n)Y(n), \quad 0 \le n \le N - 1 \tag{5.8-5}$$

From (5.8-5) it follows that the desired circular convolution coefficients $z(m)$ can be obtained as summarized in Fig. 5.8-2. Here N is assumed to be an integer power of 2. From Sections 4.6 and 4.7 it is known that approximately $N \log_2 N$ complex arithmetic operations (multiplications and additions) are required to compute an N-point FFT or IFFT. Thus the number of arithmetic operations required to compute circular convolution via the FFT is *proportional to $N \log_2 N$* rather than N^2, as is the case with the direct method. As such, the FFT approach yields a substantial savings in computations as N increases; for example, $N = 1024$ is not uncommon in practice. Hence the scheme in Fig. 5.8-2 is referred to as a fast convolution algorithm for circular convolution.

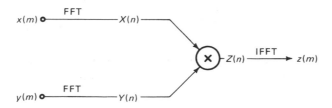

Fig. 5.8-2 Fast convolution algorithm for circular convolution.

Linear or Aperiodic Convolution

Linear or *aperiodic convolution* concerns the case where the sequences to be convolved are aperiodic; that is, they have finite durations. An example of linear convolution is (5.1-12), which defines the input–output relation of an FIR system. It can be expressed as

$$y(m) = \sum_{i=0}^{P-1} h(i)x(m - i) \tag{5.8-6}$$

where $y(m)$ is the output at time m; $h(i)$ is the ith impulse response coefficient; $x(m)$ is the input at time m; and P is the number of impulse response values and is finite.

Suppose the input is the finite duration sequence $x(m)$, $0 \le m \le M - 1$. Then from (5.8-6) it is clear that the output is also a finite duration sequence and consists of $M + P - 1$ samples (points); that is, $y(m)$, $0 \le m \le M + P - 2$.

From the discussion on circular convolution, we know that if two $(M + P - 1)$-periodic sequences are convolved then the resulting circular sequence is also $(M + P - 1)$-periodic. Thus one way of computing *linear convolution* in (5.8-6) is to simply append the sequences $h(m)$ and $x(m)$ with zeros, until each has $M + P - 1$ samples, and then compute their circular convolution. A single period of the resulting $(M + P - 1)$-periodic *circular convolution* sequence will be identical to the desired *linear convolution* sequence, which is of finite duration and consists of $M + P - 1$ samples [4]. This important concept is best illustrated by a simple example. To this end, suppose we wish to compute the linear convolution sequence $y(m)$ in (5.8-6) for $P = M = 4$ via circular convolution. First, we form the following appended sequences $h'(i)$ and $x'(m)$, which are periodic with period 7 ($= M + P - 1$):

$$h'(i) = \begin{cases} h(i), & 0 \le i \le 3 \ (= P - 1) \\ 0, & 3 < i \le 6 \ (= M + P - 2) \end{cases} \tag{5.8-7a}$$

and

$$x'(m) = \begin{cases} x(m), & 0 \le m \le 3 \ (= M - 1) \\ 0, & 3 < m \le 6 \ (= M + P - 2) \end{cases} \tag{5.8-7b}$$

Next we compute the circular convolution of $h'(i)$ and $x'(m)$, using (5.8-6) with $N = 7 \ (= M + P - 1)$, to obtain

$$y'(m) = \sum_{\ell=0}^{6} h'(\ell)x'(m - \ell), \qquad 0 \le m \le 6 \tag{5.8-8}$$

where $y'(m)$ is the circular convolution sequence which is 7-periodic. Using the graphical interpretation of circular convolution (see Fig. 5.8-1), we evaluate (5.8-8) and summarize the corresponding results in Fig. 5.8-3. It is observed that the first row in Fig. 5.8-3 is the reversed sequence corresponding to $h'(m)$, $0 \le m \le 6$; that is, $h'(0)$, $h'(6)$, $h'(5)$, ..., $h'(2)$, $h'(1)$. Point-by-point multiplication of this reversed sequence with each of the shifted input sequences shown in the remaining rows of Fig. 5.8-3 leads to the following values for $y'(m)$ in (5.8-8):

$$y'(0) = h(0)x(0)$$

$$y'(1) = h(0)x(1) + h(1)x(0)$$

$$y'(2) = h(0)x(2) + h(2)x(0) + h(1)x(1)$$

$$y'(3) = h(0)x(3) + h(3)x(0) + h(2)x(1) + h(1)x(2) \qquad (5.8-9)$$

$$y'(4) = h(3)x(1) + h(2)x(2) + h(1)x(3)$$

$$y'(5) = h(3)x(2) + h(2)x(3)$$

$$y'(6) = h(3)x(3)$$

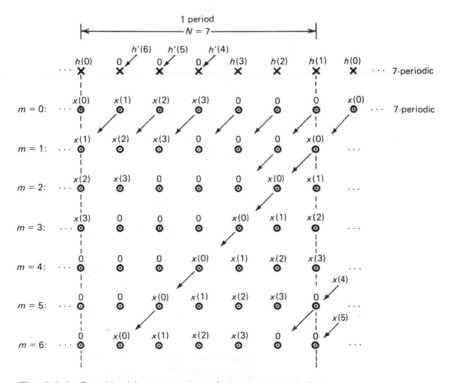

Fig. 5.8-3 Graphical interpretation of circular convolution.

Let us now verify that $y'(m)$ in (5.8-9) is indeed equivalent to the linear convolution sequence $y(m)$ obtained by direct evaluation of (5.8-6) for $P = M = 4$. The steps involved to compute $y(m)$ are summarized in Fig. 5.8-4. These steps follow from the graphical interpretation of linear convolution discussed earlier in Section 5.2; see Fig. 5.2-2a.

An inspection of Fig. 5.8-4 leads to the following linear convolution sequence $y(m)$:

$$y(0) = h(0)x(0)$$

$$y(1) = h(0)x(1) + h(1)x(0)$$

$$y(2) = h(0)x(2) + h(1)x(1) + h(2)x(0)$$

$$y(3) = h(0)x(3) + h(1)x(2) + h(2)x(1) + h(3)x(0) \qquad (5.8\text{-}10)$$

$$y(4) = h(1)x(3) + h(2)x(2) + h(3)x(1)$$

$$y(5) = h(2)x(3) + h(3)x(2)$$

$$y(6) = h(3)x(3)$$

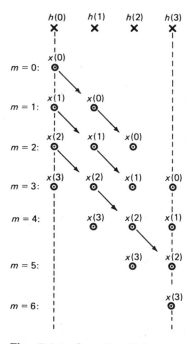

Fig. 5.8-4 Graphical interpretation of linear convolution in (5.8-6), for $M = P = 4$.

Comparing (5.8-9) and (5.8-10), we see that

$$y(m) \equiv y'(m), \qquad 0 \le m \le 6 \qquad (5.8\text{-}11)$$

which is the desired result.

The fact that linear convolution can be computed via circular convolution is a significant result, since it enables us to compute the same using the FFT. To do so, we need only to make the length of each of the sequences $h'(i)$ and $x'(m)$ in (5.8-7) an integer power of 2, as follows:

$$h'(i) = \begin{cases} h(i), & 0 \le i \le P - 1 \\ 0, & P - 1 < i \le L - 1 \end{cases} \qquad (5.8\text{-}12a)$$

and
$$x'(m) = \begin{cases} x(m), & 0 \le m \le M - 1 \\ 0, & M - 1 < m \le L - 1 \end{cases} \qquad (5.8\text{-}12b)$$

where $L \ge M + P - 1$ is an *integer power of 2*. Then $y'(m)$ can be obtained using the fast convolution algorithm in Fig. 5.8-5, which readily follows from Fig. 5.8-2. The desired $y(m)$ in (5.8-6) is then obtained as

$$y(m) = y'(m), \qquad 0 \le m \le M + P - 2 \qquad (5.8\text{-}13)$$

and the remaining $y'(m)$, $M + P - 2 < m \le L - 1$, will each be zero.

The fast convolution scheme we have just described enables one to compute linear convolution more rapidly and efficiently, compared to the direct evaluation of (5.8-6). This is so even for moderate values of $M + P - 1$, such as those on the order of 30 [2]. Since (5.8-6) yields the output of an FIR system with impulse response $h(i)$, fast convolution methods provide an effective means for implementing such systems. Implementation aspects of FIR systems will be addressed in Chapter 7.

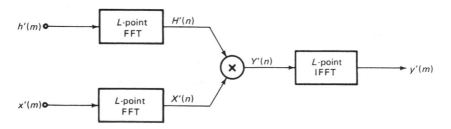

Fig. 5.8-5 Fast convolution algorithm to compute linear convolution using the fast Fourier transform.
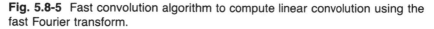

Other Considerations

In certain applications where FIR systems are used, the input sequence $x(m)$ is much longer than the impulse response sequence $h(i)$ (e.g., in the processing of radar and speech signals). In such cases, $M \gg P$ in (5.8-6), which in turn causes $M + P - 1$ to be large. Consequently, the corresponding L-point FFTs could be difficult to compute, where $L \geq M + P - 1$ is an integer power of 2. This problem can be alleviated by computing smaller-sized convolutions, and then piecing them together to form the desired output sequence in (5.8-6).

Two fast convolution techniques that enable us to do so are referred to as the *overlap-add* and *overlap-save* methods. In what follows, we discuss the basic concepts of the overlap-add method.

For convenience, let us assume that the length of $x(m)$ is indefinite, and $h(i)$ consists of P samples. The sequence $x(m)$ is partitioned into subsequences, each consisting of Q samples; see Fig. 5.8-6. A good rule of thumb is to choose Q on the order of P, where P is the duration of the impulse response sequence $h(i)$. Thus we see that the input sequence $x(m)$ can be represented as

$$x(m) = \sum_{k=1}^{\infty} x_k(m) \tag{5.8-14}$$

where $\quad x_k(m) = \begin{cases} x(m), & (k-1)Q \leq m \leq kQ - 1 \\ 0, & \text{elsewhere} \end{cases}$

Substitution of (5.8-14) into (5.8-6) leads to

$$y(m) = \sum_{i=0}^{P-1} h(i) \sum_{k=1}^{\infty} x_k(m - i)$$

$$= \sum_{k=1}^{\infty} \left[\sum_{i=0}^{P-1} h(i) x_k(m - i) \right]$$

That is,

$$y(m) = \sum_{k=1}^{\infty} y_k(m) \tag{5.8-15}$$

where $\quad y_k(m) = \sum_{i=0}^{P-1} h(i) x_k(m - i)$

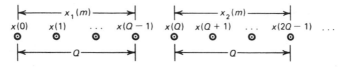

Fig. 5.8-6 Partitioning of x(m) related to the overlap-add method.

Equation (5.8-15) implies that the desired linear convolution sequence $y(m)$ in (5.8-6) can be computed as the sum of smaller linear convolutions $y_k(m)$. Since each subsequent $x_k(m)$ has Q samples, it follows that each $y_k(m)$ in (5.8-15) has $(P + Q - 1)$ samples. However, there is a region of $(P - 1)$ samples over which the kth convolution *overlaps* the $(k + 1)$th convolution, and thus the sequences $y_k(m)$ must be added in an appropriate manner. Hence the name *overlap-add* method. The mechanics of first forming the $y_k(m)$ and then adding them is best illustrated by a simple numerical example.

Suppose $h(i)$, $0 \le i \le 2$, and $x(m)$, $0 \le m \le 8$, as shown in Fig. 5.8-7, which implies that $P = 3$ and $M = 9$. It can be easily shown (Problem 5-13) that the corresponding linear convolution sequence $y(m)$ obtained via (5.8-6) is as indicated in Fig. 5.8-7. We observe that $y(m)$ consists of 11 $(= M + P - 1)$ samples.

Let us now verify that $y(m)$ in Fig. 5.8-7 can also be obtained via smaller linear convolutions using the overlap-add method. To this end, suppose we choose $Q = 3$, which results in 3 subsequences $x_1(m)$, $x_2(m)$, and $x_3(m)$; see Fig. 5.8-8. Next we compute the linear convolutions of each of these subsequences and $h(i)$, $0 \le i \le 2$. It can be shown that (Problem 5-13) these convolutions result in the sequences $y_1(m)$, $y_2(m)$, and $y_3(m)$ shown in Fig. 5.8-8. We note that the starting points of $x_i(m)$ and $y_i(m)$ coincide for $1 \le i \le 3$. The desired convolution sequence is then the sum of $y_i(m)$, $1 \le i \le 3$, which results in $y(m)$, $0 \le m \le 10$, shown in Fig. 5.8-8. Comparing Figs. 5.8-7 and 5.8-8 we see that the sequences $y(m)$ are identical, which is the desired result. In practice, the sequences $y_k(m)$ are computed using L-point FFTs, where $L \ge P + Q - 1$ is an integer power of 2, as discussed earlier in this section; see Fig. 5.8-5.

i:	0	1	2
$h(i)$:	1	2	1

m:	0	1	2	3	4	5	6	7	8
$x(m)$:	1	−1	2	1	2	−1	1	3	1

m:	0	1	2	3	4	5	6	7	8	9	10
$y(m)$:	1	1	1	4	6	4	1	4	8	5	1

Fig. 5.8-7 Linear convolution of h(i) and x(m).

i:	0	1	2
$h(i)$:	1	2	1

m:	0	1	2	3	4	5	6	7	8	9	10
$x(m)$:	1	−1	2	1	2	−1	1	3	1		

$\overset{|\!\leftarrow Q=3\rightarrow\!|}{}\quad\overset{|\!\leftarrow Q=3\rightarrow\!|}{}\quad\overset{|\!\leftarrow Q=3\rightarrow\!|}{}$

$x_1(m)$:	1	−1	2								
$x_2(m)$:				1	2	−1					
$x_3(m)$:							1	3	1		

$y_1(m)$:	1	1	1	3	2						
$y_2(m)$:				1	4	4	0	−1			
$y_3(m)$:							1	5	8	5	1

$\left.\begin{array}{c}\\\\\\\end{array}\right\} +$

$y(m)$:	1	1	1	4	6	4	1	4	8	5	1

Fig. 5.8-8 Linear convolution of *h(i)* and *x(m)* using the overlap-add method.

As mentioned during the beginning of this discussion related to computing smaller-sized convolutions, a second technique for doing so is known as the overlap-save method. Here the input subsequences are overlapped rather than output subsequences, as was the case with the overlap-add method, which we just discussed. The interested reader may refer to [1, 2] for the pertinent details.

5.9 SUMMARY

By means of the ZT we have shown that the notion of transfer functions can be associated with DT systems. It was also shown that convolution in the DT domain is equivalent to multiplication in the ZT domain; that is, the ZT of the output equals the transfer function, times the ZT of the input. This background was then used to study some important sinusoidal steady-state aspects of DT systems. Realization forms for DT system transfer functions were introduced. These realization forms will be used in Chapter 8, when we discuss digital hardware implementation aspects of DT systems. Some aspects of the output spectra of DT systems were discussed, and their relation to corresponding input spectra and transfer functions was explored. In conclusion, the notion of fast convolution was introduced, and some related algorithms were developed. These algorithms will be used in Chapter 7 in connection with implementing FIR systems.

PROBLEMS

5–1   Consider the IIR system

$$y(n) = \beta x(n) + \alpha y(n - 1), \qquad n \geq 0$$

with $y(-1) = 0$.
(a) Find $H(z)$.
(b) Find the response $y_1(n)$ to the input

$$x_1(n) = (-1)^n, \qquad n \geq 0$$

5–2   The magnitude response $|\bar{H}(v)|$ of a certain system is shown in the following sketch. Use this information to sketch $|\bar{H}(v)|$ in the interval $1 \leq v \leq 2$.

5–3   The unit-step response $y_s(n)$ of a low-pass IIR system is experimentally determined to be

$$y_s(n) = 1 - 0.7^{n+1}, \qquad n \geq 0$$

(a) Use this information to show that the system transfer function is given by

$$H(z) = \frac{0.3z}{z - 0.7}$$

(b) Evaluate $|\bar{H}(v)|^2$, $0 \le v \le 1$.

(c) Express the system bandwidth in terms of f_N.

5-4   The transfer function of a certain IIR system is given by

$$H(z) = \frac{z(z - 1)}{(z^2 - z + 1)(z + 0.8)}$$

Determine whether this system is stable or marginally stable.

5-5   An IIR system is described by the difference equation

$$3y(n) = 3.7y(n - 1) - 0.7y(n - 2) + x(n - 1), \qquad n \ge 0$$

Determine whether this system is stable, marginally stable, or unstable.

5-6   Given the FIR system

$$y(n) = x(n) + 2x(n - 1) + 4x(n - 2) + 2x(n - 3) + x(n - 4),$$
for $n \ge 0$

(a) Find $H(z)$.

(b) Show that $|\bar{H}(v)| = |2 \cos (2\pi v) + 4 \cos (\pi v) + 4|$.

5-7   A second-order IIR system is defined by the difference equation

$$y(n) = K^2 y(n - 2) + x(n), \qquad n \ge 0$$

where $|K| < 1$. Show that the unit-step response $y_s(n)$ is given by

$$y_s(n) = \frac{K^{n+1}}{2} \left[\frac{1}{K - 1} + \frac{(-1)^n}{1 + K} \right] + \frac{1}{1 - K^2}$$

What is the steady-state value of $y_s(n)$?

5-8   The unit-step response of an IIR system is given by

$$y_s(n) = \frac{\beta}{1 - \alpha} \left[1 - \alpha^{n+1} \right], \qquad n \ge 0$$

where $|\alpha| < 1$.

(a) Find $H(z)$.

(b) Show that the response of this system to the sequence γ^n, $n \ge 0$, can be expressed as

$$y(n) = \frac{\beta}{\gamma - \alpha} \left[\gamma^{n+1} - \alpha^{n+1} \right], \qquad n \geq 0$$

where $|\gamma| < 1$.

5–9   Find the bandwidth of an IIR system that is represented by the difference equation

$$y(n) = 0.634\{x(n) + x(n - 1)\} - 0.268y(n - 1), \qquad n \geq 0$$

if the sampling frequency is 300 sps.

5–10   Given:

$$H(z) = \frac{0.5(z^2 - 1.1z + 0.3)}{z^3 - 2.4z^2 + 1.91z - 0.504}$$

Obtain the following realizations:
(a) The direct form 2 (or canonic).
(b) Series form in terms of first-order sections.
(c) Parallel form in terms of first-order sections.

5–11   Given the IIR system

$$y(n) = Ky(n - 1) + (1 - K)x(n), \qquad n \geq 0$$

where $|K| < 1$.
(a) Evaluate $|\bar{H}(\nu)|^2$, $0 \leq \nu \leq 1$.
(b) Assuming that this system is a low-pass filter for $0 < K < 1$, find K such that it has a bandwidth of 200 Hz when the sampling frequency is 1000 sps.

5–12   Use $H(z)$ in (5.7-10) and the computer program in Appendix 5.1 to compute $|\bar{H}(\nu)|$, $0 \leq \nu \leq 1$, and hence verify that it represents a 60-Hz notch filter when the sampling frequency is 960 sps.

5–13 (a)   Using the graphical interpretation (Fig. 5.2-2a) of linear convolution in (5.8-6), verify that the linear convolution of $h(i)$ and $x(m)$ in Fig. 5.8-7 is as given in the same figure by $y(m)$.

(b)   Given: $h(0) = 1$, $h(1) = 2$, $h(2) = 1$; verify that the linear convolutions of $h(i)$, $0 \leq i \leq 2$, and the sequences

$$x_1(0) = 1, \qquad x_1(1) = -1, \qquad x_1(2) = 2$$

$$x_2(3) = 1, \qquad x_2(4) = 2, \qquad x_2(5) = -1$$

$$x_3(6) = 1, \qquad x_3(7) = 3, \qquad x_3(8) = 1$$

are given by

$$y_1(0) = 1, \quad y_1(1) = 1, \quad y_1(2) = 1, \quad y_1(3) = 3, \quad y_1(4) = 2$$

$$y_2(3) = 1, \quad y_2(4) = 4, \quad y_2(5) = 4, \quad y_2(6) = 0, \quad y_2(7) = -1$$

$$y_3(6) = 1, \quad y_3(7) = 5, \quad y_3(8) = 8, \quad y_3(9) = 5, \quad y_3(10) = 1$$

respectively.

APPENDIX 5.1[†] COMPUTER PROGRAM TO EVALUATE THE MAGNITUDE AND PHASE RESPONSES OF A DISCRETE-TIME SYSTEM

The program presented in this appendix yields the magnitude and phase responses of a given DT transfer function $H(z)$ in the form

$$H(z) = \frac{a_0 + a_1 z^{-1} + a_2 z^{-2} + \cdots + a_k z^{-k}}{1 + b_1 z^{-1} + b_2 z^{-2} + \cdots + b_m z^{-m}}$$

Thus the numbers of numerator and denominator coefficients in $H(z)$ are $(k + 1)$ and $(m + 1)$, respectively. To illustrate, let

$$H(z) = \frac{0.951 z^{-1}}{1 - 0.618 z^{-1} + z^{-2}} \tag{A.5-1}$$

Suppose the sampling frequency is 500 sps, and the magnitude and phase responses are desired at 50 points between 0 Hz and 250 Hz. Then the queries and related responses are as follows:

```
DISCRETE TRANSFER FUNCTION RESPONSE

NUMBER OF NUMERATOR COEFFICIENTS (1-128) ? 2
        A( 0)(Z**- 0)= 0
        A( 1)(Z**- 1)=  .951

NUMBER OF DENOMINATOR COEFFICIENTS (1-128) ? 3
        B( 0)(Z**- 0)= 1
        B( 1)(Z**- 1)= -.618
        B( 2)(Z**- 2)= 1

NUMBER OF STEPS (1-512) ? 50

SAMPLING FREQUENCY (HERTZ) ? 500

NORMALIZE RESPONSE (YES/NO) ? NO

POWER RATHER THAN AMPLITUDE RESPONSE (YES/NO) ? NO
```

[†]This program was developed by Steven Steps, Fred Ratcliffe, Myron Flickner, and David Martz.

The resulting output is as given next, from which it is apparent that $H(z)$ in (A.5-1) represents a DT system that is resonant at about 102 Hz.

FREQ	MAGNITUDE	PHASE
.00	.6881	.0000
5.10	.6902	.0000
10.20	.6964	.0000
15.31	.7070	.0000
20.41	.7223	.0000
25.51	.7429	.0000
30.61	.7695	.0000
35.71	.8032	.0000
40.82	.8456	.0000
45.92	.8987	-.0000
51.02	.9656	-.0000
56.12	1.0509	-.0000
61.22	1.1616	.0000
66.33	1.3088	-.0000
71.43	1.5119	-.0000
76.53	1.8070	.0000
81.63	2.2705	-.0000
86.73	3.0963	.0000
91.84	4.9625	.0000
96.94	13.0619	.0000
102.04	19.4631	3.1414
107.14	5.5013	3.1416
112.24	3.1837	3.1416
117.35	2.2332	3.1414
122.45	1.7173	3.1414
127.55	1.3944	3.1416
132.65	1.1742	3.1414
137.76	1.0149	3.1414
142.86	.8947	3.1414
147.96	.8012	3.1416
153.06	.7267	3.1416
158.16	.6662	3.1416
163.27	.6164	3.1414
168.37	.5747	3.1414
173.47	.5397	9.4246
178.57	.5100	3.1414
183.67	.4846	3.1416
188.78	.4629	3.1416
193.88	.4442	3.1416
198.98	.4282	3.1416
204.08	.4145	-3.1416
209.18	.4029	-3.1414
214.29	.3930	-3.1416
219.39	.3847	-3.1416
224.49	.3780	-3.1416
229.59	.3726	-3.1414
234.69	.3685	-3.1414

FREQ	MAGNITUDE	PHASE
239.80	.3656	-3.1414
244.90	.3638	-3.1416
250.00	.3633	-3.1416

```
C***********************************************************************
C
        COMPILER FREE
        DIMENSION A(128),B(128)
        DIMENSION AMAG(512),PHASE(512)
        INTEGER IANS(1)
C
3       FORMAT (10X,"A(",I3,")(Z**-",I3,")= ",Z)
5       FORMAT (10X,"B(",I3,")(Z**-",I3,")= ",Z)
8       FORMAT ("<12>NORMALIZE RESPONSE (YES/NO) ?",Z)
9       FORMAT ("<12>POWER RATHER THAN AMPLITUDE RESPONSE (YES/NO)?",Z)
12      FORMAT (S1)
13      FORMAT ("<12>",F8.2,5X,F10.4,5X,F10.4)
14      FORMAT ("<12>")
15      FORMAT (3X," FREQ",9X,"MAGNITUDE",7X,"PHASE")
16      FORMAT ("<33><14>")
C
C       CLEAR PAGE TO START
C
        TYPE"<33><14>"
C
C       SET UP PARAMETERS
C
        ANGLE=0.0
        ARG=0.0
        MCOEFF=128
        MSTEP=512
        PI=3.141493
        PIT2=PI*2.0
        SUM=0.0
        TEMP=0.0
        YLAST=0.0
        TYPE"<12>DISCRETE TRANSFER FUNCTION RESPONSE"
100     CONTINUE
C
C       READ IN NUMERATOR COEFFICIENTS
C
        ACCEPT"<12><12>NUMBER OF NUMERATOR COEFFICIENTS (1-128) ?",NN
        IF(NN.LT.1.OR.NN.GT.MCOEFF) GO TO 100
                DO 200 I=1,NN
                J=I-1
                WRITE (10,3) J,J
                READ (11) A(I)
200             CONTINUE
250     CONTINUE
C
C       READ IN DENOMINATOR COEFFICIENTS
C
        ACCEPT("<12>NUMBER OF DENOMINATOR COEFFICIENTS (1-128) ? "),ND
        IF(ND.LT.1.OR.ND.GT.MCOEFF) GO TO 250
                DO 300 I=1,ND
                J=I-1
                WRITE (10,5) J,J
```

```
                    READ (11) B(I)
300                 CONTINUE
350         CONTINUE
C
C       READ IN STEP SIZE
C
        ACCEPT'<12>NUMBER OF STEPS (1-512) ? ',NSTEP
        IF(NSTEP.LT.1.OR.NSTEP.GT.MSTEP) GO TO 350
C
C       READ IN SAMPLING FREQUENCY
C
        ACCEPT'<12>SAMPLING FREQUENCY (HERTZ) ? ',SF
        SFD2=SF/2
C
C       ASK IF NORMALIZED RESPONSE IS DESIRED
C
        WRITE(10,8)
        READ(11,12) IANS(1)
        IANS0=0
        IF(IANS(1).EQ.'Y<0>') IANS0=1
C
C       ASK IF POWER RESPONSE IS DESIRED
C
        WRITE(10,9)
        READ(11,12) IANS(1)
        IANS1=0
        IF(IANS(1).EQ.'Y<0>') IANS1=1
C
C       CALCULATE THE STEP SIZE
C
        SSIZE=PI/FLOAT(NSTEP-1)
C
C       DETERMINE POWER AND PHASE RESPONSE
C
                DO 500 I=1,NSTEP
C
C       EVALUATE NUMERATOR POLYNOMIAL FOR THIS STEP
C
                XN=A(1)
                YN=0.0
                IF(NN.LT.2) GO TO 425
                        DO 400 J=2,NN
                        ANG=ARG*FLOAT(J-1)
                        XN=XN+A(J)*COS(ANG)
                        YN=YN-A(J)*SIN(ANG)
400                     CONTINUE
425             CONTINUE
C
C       EVALUATE DENOMINATOR POLYNOMIAL FOR THIS STEP
C
                XD=B(1)
                YD=0.0
                IF(ND.LT.2)GO TO 475
                        DO 450 J=2,ND
                        ANG=ARG*FLOAT(J-1)
                        XD=XD+B(J)*COS(ANG)
                        YD=YD-B(J)*SIN(ANG)
450                     CONTINUE
475             CONTINUE
C
C       CALCULATE POWER RESPONSE FOR THIS STEP
C
```

```
               AMAG(I)=(XN*XN+YN*YN)/(XD*XD+YD*YD)
               SUM=SUM+AMAG(I)
C
C       CALCULATE PHASE RESPONSE FOR THIS STEP
C
               X=XN*XD+YN*YD
               Y=XD*YN-XN*YD
               IF(X.LT.0.0.AND.Y.GT.0.0.AND.YLAST.LT.0.0)ANGLE=ANGLE-PIT2
               IF(X.LT.0.0.AND.Y.LT.0.0.AND.YLAST.GT.0.0)ANGLE=ANGLE+PIT2
               PHASE(I)=ATAN2(Y,X)+ANGLE
               YLAST=Y
C
C       INCREMENT ARG
C
               ARG=ARG+SSIZE
500     CONTINUE
C
C       NEED TO CALCULATE THE AMPLITUDE RESPONSE ?
C
        IF(IANS1.NE.1) GO TO 600
C
C       NEED TO NORMALIZE THE POWER RESPONSE ?
C
        IF(IANS0.NE.1) GO TO 750
C
C       CALCULATE NORMALIZED POWER RESPONSE
C
               DO 550 I=1,NSTEP
               AMAG(I)=AMAG(I)/SUM
550            CONTINUE
        GO TO 750
600     CONTINUE
C
C       CALCULATE THE AMPLITUDE RESPONSE
C
               DO 650 I=1,NSTEP
               AMAG(I)=SQRT(AMAG(I))
               IF(TEMP.LT.AMAG(I)) TEMP=AMAG(I)
650            CONTINUE
C
C       NEED TO NORMALIZE THE AMPLITUDE RESPONSE ?
C
        IF(IANS0.NE.1) GO TO 750
C
C       NORMALIZE AMPLITUDE RESPONSE
C
               DO 700 I=1,NSTEP
               AMAG(I)=AMAG(I)/TEMP
700            CONTINUE
750     CONTINUE
C
C       PRINT OUT MAGNITUDE AND PHASE ARRAY
C
        STEPF=SFD2/(NSTEP-1)
        FREQ=0.0
        WRITE(12,15)
        WRITE(12,14)
        DO 1000 K=1,NSTEP
        WRITE(12,13) FREQ,AMAG(K),PHASE(K)
        FREQ=FREQ+STEPF
1000    CONTINUE
        WRITE(12,16)
        STOP***NORMAL TERMINATION***
        END
```

REFERENCES

1. A. V. OPPENHEIM and R. SCHAFER, *Digital-Signal Processing*, Prentice-Hall, Englewood Cliffs, N.J., 1975.
2. L. R. RABINER and B. GOLD, *Theory and Application of Digital Signal Processing*, Prentice-Hall, Englewood Cliffs, N.J., 1975.
3. W. D. STANLEY, *Digital Signal Processing*, Reston, Reston, Va., 1975.
4. T. G. STOCKHAM, JR., "High Speed Convolution and Correlation," 1966 Spring Joint Computer Conference, AFIPS Proc., 1966, pp. 229–233.

Infinite Impulse Response Discrete-Time Filters

Up to this point we have considered DT systems. In this chapter our attention is restricted to *filters,* that are a special class of such systems. By the term *DT filter* we mean a DT system that is designed to alter the spectral content of an input DT signal (or sequence) in a specified manner. The term *digital filter* means that the input, output, and coefficient values of the DT filter are quantized and coded in binary form.

The filter design process concerns procedures for determining filter transfer functions that satisfy certain specifications in the frequency domain. Examples of such specifications are illustrated in the sketches on page 227, where $|\bar{H}(f)|$ denotes the magnitude response of the desired DT filter.

Thus, filter design is the process of determining the filter coefficients a_i, b_j in the input–output relation

$$y(n) = \sum_{i=0}^{k} a_i x(n - i) - \sum_{j=1}^{m} b_j y(n - j), \qquad n \geq 0$$

so that the related transfer function has the desired frequency-domain properties. In this chapter we assume that at least one $b_j \neq 0$; that is, we shall be concerned with IIR filters. The design of finite impulse response (FIR) filters will be undertaken in Chapter 7.

Some advantages of DT filters relative to their analog counterparts are as follows:

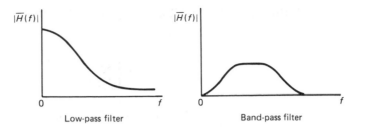

1. *High reliability.* Temperature changes and aging do not affect DT filter components, as is the case with analog filters. As such, the filter parameters are stable resulting in reliable performance.

2. *Better performance.* Stringent frequency-domain specifications can be realized relatively easily. In addition, exact linear phase response can be attained, as is the case with FIR filters, whose design will be considered in Chapter 7.

3. *Flexibility.* Filter characteristics such as bandwidth and resonant frequency can be readily modified by appropriately changing the values of filter coefficients.

     A difficulty related to DT filters arises when we attempt to implement them in hardware. This is because the accuracy with which filter coefficients can be represented depends upon the word length used to encode them in binary form. Word lengths will have to be sufficiently long to ensure enough accuracy for encoding filter coefficients, and hence achieve the desired performance. Some aspects of implementing DT filters will be addressed in Chapter 8.
     Two methods for designing IIR filters will be studied, and Section 6.1 is devoted to the first of these methods.

6.1 IMPULSE INVARIANT METHOD

Let $Y(s)$ denote the Laplace transform of the output $y(t)$ of a CT filter, when subjected to an input $x(t)$. For convenience, we shall assume that $Y(s)$ consists only of simple poles. Then a PFE of $Y(s)$ can be expressed as

$$Y(s) = \sum_{i=1}^{m} \frac{A_i X(s)}{s + s_i} = H(s)X(s) \qquad (6.1\text{-}1)$$

where s_i is the ith pole of the transfer function $H(s)$, and $X(s)$ is the LT of $x(t)$.

With $X(s) = 1$, the ILT of $H(s)$ in (6.1-1) yields the impulse response $h(t)$; that is,

$$h(t) = L^{-1}\left\{\sum_{i=1}^{m} \frac{A_i}{s + s_i}\right\} \tag{6.1-2}$$

where L^{-1} denotes the ILT. Thus (6.1-2) yields

$$h(t) = \sum_{i=1}^{m} A_i e^{-s_i t} \tag{6.1-3}$$

The essence of the impulse invariant method is to obtain a DT filter whose impulse response $h(n)$ is exactly equal to sampled values of $h(t)$ that are T seconds apart, where T is the sampling interval. Thus (6.1-3) yields

$$h(n) = \sum_{i=1}^{m} A_i e^{-s_i n T}, \qquad n \geq 0 \tag{6.1-4}$$

From Table 3.2-1 (line 6), it is known that

$$Z\{e^{-s_i n T}\} = \frac{z}{z - e^{-s_i T}} \tag{6.1-5}$$

Equations (6.1-4) and (6.1-5) imply that

$$H(z) = \sum_{i=1}^{m} \frac{A_i z}{z - e^{-s_i T}} \tag{6.1-6}$$

is the transfer function of the DT filter corresponding to the CT filter whose transfer function is

$$H(s) = \sum_{i=1}^{m} \frac{A_i}{s + s_i}$$

Similarly, the impulse invariant method can be used to determine the $H(z)$ that corresponds to any given $H(s)$. For example, if

$$H(s) = \frac{b}{(s + a)^2 + b^2} \tag{6.1-7}$$

then it can be shown that (Problem 6-1)

$$H(z) = \frac{e^{-aT} \sin (bT)z}{z^2 - 2e^{-aT} \cos (bT)z + e^{-2aT}} \tag{6.1-8}$$

We conclude this discussion with an illustrative example.

Example 6.1-1: Let

$$H(s) = \frac{1}{(s + 1)(s + 2)} \tag{6.1-9}$$

(a) Find the corresponding $H(z)$ using the impulse invariant method.
(b) Let $f_s = 5$ sps. Plot $|H(f)|$ and $\psi(f)$ corresponding to $H(s)$ and $T |\bar{H}(f)|$ and $\bar{\psi}(f)$ corresponding to $H(z)$.

Solution: A PFE of $H(s)$ in (6.1-9) yields

$$H(s) = \frac{A_1}{s + 1} + \frac{A_2}{s + 2} \tag{6.1-10}$$

where

$$A_1 = (s + 1)H(s)\Big|_{s = -1} = 1$$

$$A_2 = (s + 2)H(s)\Big|_{s = -2} = -1$$

which yields

$$H(s) = \frac{1}{s + 1} - \frac{1}{s + 2}$$

The impulse response is

$$h(t) = e^{-t} - e^{-2t}, \quad t \geq 0 \tag{6.1-11}$$

and hence from (6.1-5) and (6.1-11) we obtain

$$H(z) = \frac{z}{z - e^{-T}} - \frac{z}{z - e^{-2T}} \tag{6.1-12}$$

or

$$H(z) = \frac{(e^{-T} - e^{-2T})z}{z^2 - (e^{-T} + e^{-2T})z + e^{-3T}}$$

Since $f_s = 5$ sps, $T = \frac{1}{5}$ sec, and (6.1-12) yields

$$H(z) = \frac{0.148z}{z^2 - 1.489z + 0.5488} \tag{6.1-13}$$

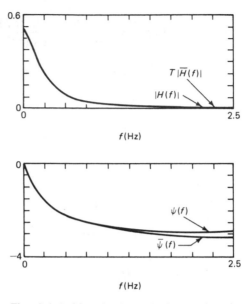

Fig. 6.1-1 Magnitude and phase plots for Example 6.1-1

The programs in Appendices 1.1 and 5.1 are used to obtain the desired magnitude and phase responses of $H(s)$ and $H(z)$, respectively. The results are displayed in Fig. 6.1-1, from which it is apparent that they compare closely. We note that $T|\bar{H}(f)|$ is compared with $|H(f)|$ rather than $|\bar{H}(f)|$, because of the scale factor $1/T$ that appears in relation between the FT of a CT function and its DT counterpart; see (4.1-12).

Perhaps the most commonly used technique for designing IIR filters is that which involves a mapping known as the bilinear transform (BLT). This mapping relates points on the s- and z-planes. We shall refer to the related design technique as the BLT method and discuss it in Section 6.4. The BLT will be introduced in the next section, and a frequency scaling procedure that plays a role in the BLT method will be discussed in Section 6.3.

6.2 BILINEAR TRANSFORM

As mentioned earlier, the BLT is a mapping or transformation that relates points on the s- and z-planes, respectively. It is defined as

$$s = \frac{z - 1}{z + 1} \tag{6.2-1}$$

or *equivalently* as

$$z = \frac{1 + s}{1 - s} \qquad (6.2\text{-}2)$$

For example, (6.2-1) implies that the point $z = 1 + j1$ on the z-plane is mapped into the point $s = (1 + 2j)/5$ in the s-plane, while (6.2-2) implies that the point $s = j2$ on the s-plane is mapped into the point $z = (-3 + j4)/5$ on the z-plane, and so on.

It can be shown that the general mapping properties of the BLT are as illustrated in Fig. 6.2-1, and can be summarized as follows:

1. The left-half s-plane is mapped *into* the unit circle in the z-plane.

2. The right-half s-plane is mapped *outside* the unit circle in the z-plane.

3. The $(0 \to \infty)$ portion of the $j\omega$-axis in the s-plane is mapped *onto* the $(0 \to \pi)$ portion of the unit circle in the z-plane.

4. The $(0 \to -\infty)$ portion of the $j\omega$-axis in the s-plane is mapped *onto* the $(0 \to -\pi)$ portion of the unit circle in the z-plane.

It is worthwhile exploring the mappings cited in properties 3 and 4 in more detail. To this end, we substitute $s = j\omega_A$ and $z = e^{j\omega_D T}$ in (6.2-1). These substitutions for s and z imply that we are concerned with points on the $j\omega$-axis in the s-plane and corresponding points on the unit circle of the z-plane, as illustrated in Fig. 6.2-2. The quantity ω_A can be viewed as an analog frequency variable, corresponding to a frequency ω_D of the desired DT filter. Specified values of ω_D are called *critical frequencies* and will be denoted by ω_{D_i}. Now (6.2-1) yields

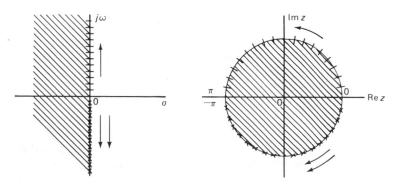

Fig. 6.2-1 Mapping properties of the bilinear transform.

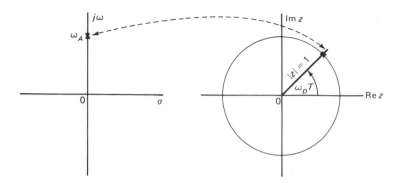

Fig. 6.2-2 On relating points on the $j\omega$-axis to those on the unit circle.

$$j\omega_A = \frac{e^{j\omega_D T} - 1}{e^{j\omega_D T} + 1} \tag{6.2-3}$$

Since $e^{j\omega_D T/2} e^{-j\omega_D T/2} = 1$, (6.2-3) can be written as

$$j\omega_A = \frac{e^{j\omega_D T} - e^{j\omega_D T/2} e^{-j\omega_D T/2}}{e^{j\omega_D T} + e^{j\omega_D T/2} e^{-j\omega_D T/2}} \tag{6.2-4}$$

which leads to

$$j\omega_A = \frac{e^{j\omega_D T/2} - e^{-j\omega_D T/2}}{e^{j\omega_D T/2} + e^{-j\omega_D T/2}} \tag{6.2-5}$$

From (6.2-5) it follows that

$$\omega_A = \tan\left(\frac{\omega_D T}{2}\right) \tag{6.2-6}$$

Equation (6.2-6) is a fundamental relation that enables us to locate a point ω_A on the $j\omega$-axis for a given point on the unit circle, where the latter is defined by ω_D and T; see Fig. 6.2-2. The frequencies ω_A and ω_D are such that

$$H(s)\Big|_{s=j\omega_A} = H(z)\Big|_{z=e^{j\omega_D T}} \tag{6.2-7}$$

where $H(z)$ is the transfer function of the DT filter, and $H(s)$ is the transfer function of an analog filter with desired frequency characteristics (i.e., low pass, high pass, etc.).

It is important to note that the BLT provides a *one-to-one* mapping

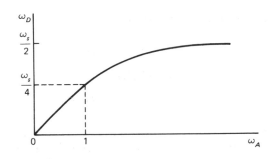

Fig. 6.2-3 Plot of transformation in (6.2-6).

of the points of the $j\omega$-axis onto the unit circle. This is *not* the case, however, with the transformation $z = e^{sT}$ in (3.1-3), which was used to define the ZT. The BLT causes the entire $j\omega_A$ axis to be mapped uniquely onto the unit circle, or, equivalently, onto the Nyquist band $|\omega| < \omega_s/2$. This follows from (6.2-6) and is illustrated for positive values of ω_A and ω_D in Fig. 6.2-3. We note that the frequency band $0 < \omega_A < 1$ is mapped onto the interval $0 < \omega_D < \omega_s/4$, while the infinite band $\omega_A \geq 1$ is compressed onto $\omega_s/4 < \omega_D < \omega_s/2$. This compression effect associated with the BLT is known as *frequency warping*, and its nonlinear nature is apparent from Fig. 6.2-3. We also note that the point $\omega_A = 0$ is mapped to $\omega_D = 0$, and the point $\omega_A = \infty$ is mapped to $\omega_D = \omega_s/2$.

The frequency warping phenomenon must be taken into effect when designing DT filters using the BLT method. This can be done by prewarping the critical frequencies and using frequency scaling. We shall briefly discuss frequency scaling in the next section.

6.3 FREQUENCY SCALING

The notion of frequency scaling is best introduced by considering the transfer function $H(s)$ of a CT system. Next we define the transfer function $\hat{H}(s)$ such that

$$\hat{H}(s) = H(s)\bigg|_{s=s/\alpha} = H\left(\frac{s}{\alpha}\right) \tag{6.3-1}$$

where the factor α is a real positive number.
With $s = j\omega$, (6.3-1) yields

$$\hat{H}(\omega) = H\left(\frac{\omega}{\alpha}\right) \tag{6.3-2}$$

From (6.3-2) it follows that

$$\hat{H}(\alpha) = H(1)$$
$$\hat{H}(2\alpha) = H(2) \qquad\qquad (6.3\text{-}3)$$
$$\hat{H}(0.5\alpha) = H(0.5), \qquad \text{etc.}$$

which means that the magnitude and phase responses of $H(s)$ and $\hat{H}(s)$ have *identical shapes*. However, points on these responses of $\hat{H}(s)$ are *scaled* with respect to their counterparts of $H(s)$ by a factor of α. Thus we shall refer to $\hat{H}(s)$ as the scaled transfer function corresponding to a given $H(s)$.

To illustrate, consider the transfer function

$$H(s) = \frac{100}{s + 100} \qquad\qquad (6.3\text{-}4)$$

whose magnitude and phase responses are displayed in Figure 6.3-1. From $|H(\omega)|$ it is apparent that $H(s)$ represents a low-pass filter with a bandwidth of 100 radians/second. Now, suppose we desire a corresponding low-pass filter with a bandwidth of 1 radian/second. Equations (6.3-1) and (6.3-3) imply that this can be accomplished by choosing $\alpha = 0.01$. Thus we have

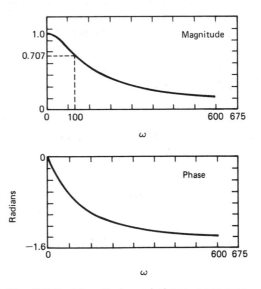

Fig. 6.3-1. Magnitude and phase responses of $H(s)$ in (6.3-4)

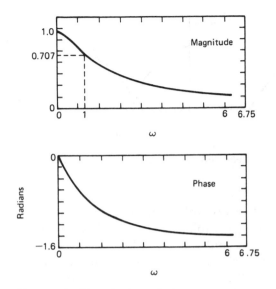

Fig. 6.3-2. Magnitude and phase responses of $\hat{H}(s)$ in (6.3-5).

$$\hat{H}(s) = H(s)\Big|_{s = s/0.01}$$

(6.3-5)

$$= \frac{1}{s + 1}$$

where $\hat{H}(s)$ is the desired transfer function, as is apparent from its magnitude and phase responses in Fig. 6.3-2. We also observe that the magnitude and phase responses of $H(s)$ and $\hat{H}(s)$ shown in Figs. 6.3-1 and 6.3-2 have identical shapes.

6.4 BILINEAR TRANSFORM DESIGN METHOD

We are now in a position to design DT filters using the BLT. One starts with an $H(s)$ that has the desired frequency characteristics. For example, suppose we seek a low-pass DT filter whose bandwidth is ω_{D_1} as illustrated in Fig. 6.4-1a. Then ω_{D_1} is identified as the critical frequency, and we choose an $H(s)$ that is the transfer function of a low-pass filter with bandwidth 1 radian/second. Similarly, if a notch filter with notch frequency ω_{D_1} is desired, we commence with an $H(s)$ that represents a notch filter whose notch frequency occurs at 1 radian/second; see Fig. 6.4-1b.

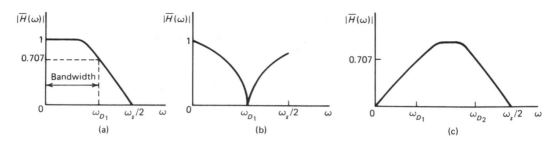

Fig. 6.4-1 Critical frequencies: (a) low-pass filter; (b) notch filter; (c) band-pass filter.

It is possible to have more than one critical frequency, as is the case, for example, with the DT band-pass filter shown in Fig. 6.4-1c, where ω_{D_1} and ω_{D_2} are called the lower and upper cutoff frequencies, respectively. We shall discuss this case in Section 6.5, and restrict the current discussion to cases where only one critical frequency is involved.

There are basically three steps involved in the BLT design procedure, once ω_{D_1} is identified and an appropriate $H(s)$ has been selected. They are as follows:

1. Prewarp the critical frequency ω_{D_1} using (6.2-6) to obtain the analog frequency ω_{A_1}; that is,

$$\omega_{A_1} = \tan\left(\frac{\omega_{D_1}T}{2}\right) \qquad (6.4\text{-}1)$$

2. Frequency scale the chosen $H(s)$ with $\alpha = \omega_{A_1}$, as indicated in (6.3-1) to obtain

$$\hat{H}(s) = H(s)\Big|_{s=s/\omega_{A_1}} = H\left(\frac{s}{\omega_{A_1}}\right) \qquad (6.4\text{-}2)$$

where $\hat{H}(s)$ is the scaled transfer function corresponding to $H(s)$.
3. Replace s in $\hat{H}(s)$ by $(z-1)/(z+1)$ to obtain $H(z)$; that is,

$$H(z) = \hat{H}(s)\Big|_{s=(z-1)/(z+1)} \qquad (6.4\text{-}3)$$

From the mapping properties of the BLT discussed in Section 6.3, it is apparent that (6.4-3) causes the poles and zeros of $\hat{H}(s)$ in the s-plane to be mapped into the z-plane, to become the poles and zeros, respectively of $H(z)$. Furthermore, the magnitude response of $H(z)$ will be similar to that of $\hat{H}(s)$, since it is simply the response of $\hat{H}(s)$ compressed into the interval $|\omega| < \omega_s/2$. Thus $H(z)$ in (6.4-3) is the desired DT system transfer function and satisfies the property in (6.2-7); that is,

$$H(z)\Big|_{z=e^{j\omega_D T}} = H(s)\Big|_{s=j\omega_A}$$

where $\omega_A = \tan(\omega_D T/2)$.

We now illustrate these steps by means of examples.

Example 6.4-1: The transfer function of the simple RC low-pass filter illustrated is given by

$$H(s) = \frac{O(s)}{I(s)} = \frac{1}{s+1} \qquad (6.4\text{-}4)$$

and its bandwidth is known from (6.3-5) to be 1 radian/second.

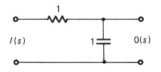

Use $H(s)$ in (6.4-4) and the BLT method to design a corresponding DT low-pass filter whose bandwidth is 20 Hz at a sampling frequency of 60 sps. Plot the magnitude and phase responses of $H(z)$.

Solution: The critical frequency here is the filter bandwidth. Thus

$$\omega_{D_1} = 2\pi(20) \text{ radians/second}$$

Next we follow the three design steps associated with the BLT method.

STEP 1.

$$\omega_{A_1} = \tan\left(\frac{\omega_{D_1}T}{2}\right) \qquad (6.4\text{-}5)$$

Since $T = 1/60$ sec, (6.4-5) yields

$$\omega_{A_1} = \tan\frac{\pi}{3} = \sqrt{3}$$

STEP 2. $H(s)$ is known to have a bandwidth of 1 radian/second.

We thus use frequency scaling to obtain $\hat{H}(s)$, which has a bandwidth of $\omega_{A_1} = \sqrt{3}$; that is,

$$\hat{H}(s) = H(s)\bigg|_{s = s/\sqrt{3}} \qquad (6.4\text{-}6)$$

From (6.4-4) and (6.4-6) we obtain

$$\hat{H}(s) = \frac{\sqrt{3}}{s + \sqrt{3}}$$

STEP 3. Thus (6.4-3) yields the desired transfer function to be

$$H(z) = \left[\frac{\sqrt{3}}{s + \sqrt{3}}\right]_{s = (z-1)/(z+1)}$$

which yields

$$H(z) = \frac{\sqrt{3}z + \sqrt{3}}{(1 + \sqrt{3})z + (\sqrt{3} - 1)}$$

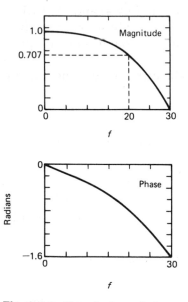

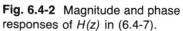

Fig. 6.4-2 Magnitude and phase responses of $H(z)$ in (6.4-7).

That is,

$$H(z) = 0.634 \frac{1 + z^{-1}}{1 + 0.268z^{-1}} \qquad (6.4\text{-}7)$$

The magnitude and phase responses are computed using the computer program in Appendix 5.1. The plots so obtained are displayed in Fig. 6.4-2, from which it is clear that $H(z)$ in (6.4-7) is indeed the transfer function of a low-pass filter whose cutoff frequency is 20 Hz.

Example 6.4-2: The transfer function of the circuit shown is

$$H(s) = \frac{O(s)}{I(s)} = \frac{s^2 + 1}{s^2 + s + 1} \qquad (6.4\text{-}8)$$

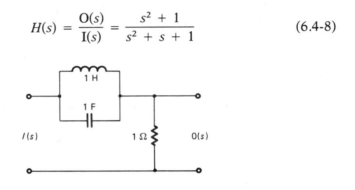

It can be shown that (Problem 6-3) this circuit is a notch filter with a frequency of 1 radian/second.

Design a DT notch filter with the following specifications:

1. Notch frequency = 60 Hz.

2. Sampling frequency = 960 sps.

Plot the corresponding magnitude and phase responses.

Solution: The critical frequency is 60 Hz, which corresponds to $\omega_{D_1} = 2\pi(60)$ radians/second. Next, we proceed with the three design steps.

STEP 1.

$$\omega_{A_1} = \tan\left\{\frac{\omega_{D_1}T}{2}\right\}$$

$$= \tan\left\{\frac{2\pi \times 60}{2 \times 960}\right\} \qquad (6.4\text{-}9)$$

since $T = \frac{1}{960}$ sec.

Evaluation of ω_{A_1} in (6.4-9) results in

$$\omega_{A_1} = \tan\frac{\pi}{16} = 0.1989 \qquad (6.4\text{-}10)$$

STEP 2. We obtain the scaled transfer function

$$\hat{H}(s) = H(s)\Big|_{s = s/\omega_{A_1}}$$

where $H(s)$ and ω_{A_1} are given by (6.4-8) and (6.4-10), respectively. This computation leads to

$$\hat{H}(s) = \frac{s^2 + 0.0396}{s^2 + 0.1989s + 0.0396} \qquad (6.4\text{-}11)$$

STEP 3. Using $\hat{H}(s)$ in (6.4-11), we evaluate

$$H(z) = \hat{H}(s)\Big|_{s = (z-1)/(z+1)}$$

to obtain

$$H(z) = \frac{1.0396 - 1.9208z^{-1} + 1.0396z^{-2}}{1.2385 - 1.9208z^{-1} + 0.8407z^{-2}} \qquad (6.4\text{-}12)$$

whose magnitude and phase responses are plotted in Fig. 6.4-3. It is evident that the notch frequency does occur at 60 Hz, as specified.

Example 6.4-3: The following $H(s)$ can be shown (Problem 6-4) to be the transfer function of an analog circuit that is resonant at 1 radian/second:

$$H(s) = \frac{5s + 1}{s^2 + 0.4s + 1} \qquad (6.4\text{-}13)$$

Design a corresponding DT resonant filter that resonates at 10 Hz when the sampling frequency is 60 Hz. Also plot its magnitude and phase responses.

Solution: In this case the critical frequency is the 10 Hz (i.e., the resonant frequency). Thus $\omega_{D_1} = 2\pi(10)$ radians/second.

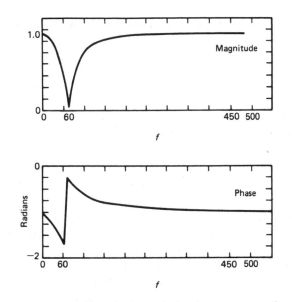

Fig. 6.4-3 Magnitude and phase responses of $H(z)$ in (6.4-12).

STEP 1. With $\omega_{D_1} = 2\pi(10)$ and $T = \frac{1}{60}$ sec, the relation

$$\omega_{A_1} = \tan\left\{\frac{\omega_{D_1}T}{2}\right\}$$

yields

$$\omega_{A_1} = \tan\frac{\pi}{6} = \frac{1}{\sqrt{3}}$$

STEP 2. Frequency scaling $H(s)$ in (6.4-13) by the factor $1/\sqrt{3}$, we obtain

$$\hat{H}(s) = H(s)\bigg|_{s = \sqrt{3}s}$$

$$= \frac{5(\sqrt{3})s + 1}{3s^2 + (\sqrt{3})(0.4)s + 1}$$

That is,

$$\hat{H}(s) = \frac{8.6603s + 1}{3s^2 + 0.6928s + 1} \tag{6.4-14}$$

STEP 3. The transfer function of the resonant filter we seek is given by

$$H(z) = \hat{H}(s) \Big|_{s = (z-1)/(z+1)} \qquad (6.4\text{-}15)$$

where $\hat{H}(s)$ is defined in (6.4-14).
Evaluating (6.4-15), we get

$$H(z) = \frac{9.6603 + 2.0z^{-1} - 7.6603z^{-2}}{4.6928 - 4.0z^{-1} + 3.3072z^{-2}} \qquad (6.4\text{-}16)$$

Evaluation of the magnitude and phase responses associated with $H(z)$ leads to the plots in Fig. 6.4-4, from which it is apparent that $H(z)$ represents a DT resonator.

From the preceding examples it is apparent that the most tedious part of the design process is that of evaluating $H(z)$ from $\hat{H}(s)$ via the BLT $s = (z - 1)/(z + 1)$. By exploiting some interesting properties of the BLT, it has been shown that the process of evaluating $H(z)$ from $\hat{H}(s)$ can be carried out by means of an efficient algorithm. The interested reader may refer to [1, 2] for a detailed discussion of the same. For our

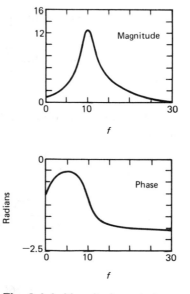

Fig. 6.4-4 Magnitude and phase responses of $H(z)$ in (6.4-16).

purposes it suffices to use a computer program that implements this algorithm, a listing of which is given in Appendix 6.1. This program accepts the coefficients of a given $\hat{H}(s)$ and yields the corresponding $H(z)$.

6.5 BUTTERWORTH DISCRETE-TIME FILTERS

From the discussion in the previous section, it is apparent that the BLT is a very convenient and powerful design method that utilizes known filtering characteristics of analog transfer functions. A great deal of knowledge of such transfer functions is readily available since analog filters have already been investigated in detail. Several textbooks in this area are available (e.g., see [3 to 5]).

One type of analog filter that is very well known and useful is the *Butterworth filter*. Our objective in this section is to use a class of analog Butterworth filter transfer functions and design corresponding DT filters via the BLT method. We shall consider low-pass, high-pass, band-pass, and band-stop Butterworth filters.

Low-pass Filters

The magnitude response

$$|P_n(\omega)| = \frac{1}{\sqrt{1 + \omega^{2n}}}, \qquad n \geq 1 \qquad (6.5\text{-}1)$$

is known as the nth order analog Butterworth response. In (6.5-1) we observe that the maximum value of $|P_n(\omega)|$ is 1, and it occurs at $\omega = 0$. Again, $|P_n(1)| = 0.707$, which means that the *bandwidths* of the low-pass filters represented by $P_n(s)$ equal 1 radian/second for *all n*, as illustrated in Fig. 6.5-1 for $n = 1$, 2, and 4. We also observe that $n \to \infty$ yields an ideal magnitude response in that frequencies beyond 1 radian/second are eliminated, while those below 1 radian/second are passed without attenuation. Hence the regions $0 \leq \omega \leq 1$ and $\omega \geq 1$ are referred to as the *passband* and *stopband*, respectively. The frequency between the passband and stopband is called the *cutoff frequency*, which also equals 1 radian/second for *all n*; see Fig. 6.5-1.

Again, from (6.5-1) it follows that

$$|P_n(\omega)| = (1 + \omega^{2n})^{-1/2} = 1 - \frac{1}{2}\omega^{2n} + \frac{3}{8}\omega^{4n} - \cdots, \qquad \omega^2 < 1$$

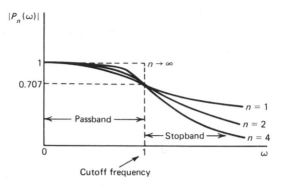

Fig. 6.5-1 Magnitude response of low-pass analog Butterworth filters.

is the Maclaurin series representation for the Butterworth response. From this expansion it is clear that the first $(2n - 1)$ derivatives of $|P_n(\omega)|$, evaluated at $\omega = 0$, are zero. Because of this property, the Butterworth response in (6.5-1) is said to be *maximally flat* in the passband region $0 \leq \omega \leq 1$.

It is apparent that the poles associated with the Butterworth response in (6.5-1) are defined by the equation

$$1 + (-s^2)^n = 0$$

which means that the poles s_k are given by

$$s_k = \exp\left\{\frac{j(2k + n - 1)\pi}{2n}\right\}, \qquad 1 \leq k \leq 2n \qquad (6.5\text{-}2a)$$

Inspection of (6.5-2a) reveals that the poles s_k are located on a unit circle in the s-plane, as illustrated in Fig. 6.5-2 for $n = 4$ and $n = 5$. We note the symmetry present in the pole locations, with respect to the real and imaginary axes. Also, the poles are separated by π/n radians.

To form the nth-order transfer function $P_n(s)$ in (6.5-1), we choose only the poles in the left-half s-plane (e.g., see Fig. 6.5-2). These poles are given by (6.5-2a) to be

$$s_k = \exp\left\{\frac{j(2k + n - 1)\pi}{2n}\right\}, \qquad 1 \leq k \leq n \qquad (6.5\text{-}2b)$$

which in turn yields

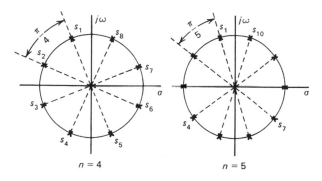

Fig. 6.5-2 Pole locations of the Butterworth response function for $n = 4$ and $n = 5$.

$$P_n(s) = \frac{1}{(s - s_1)(s - s_2) \cdots (s - s_n)}$$

$$= \frac{1}{s^n + a_{n-1}s^{n-1} + \cdots + a_1 s + a_0} \qquad (6.5\text{-}3)$$

In (6.5-3), the coefficients a_i are real numbers; for example, see $P_n(s)$, $1 \le n \le 4$, in Table 6.5-1.

Table 6.5-1   Some Low-pass Analog Butterworth Filter Transfer Functions

n	$P_n(s)$
1	$\dfrac{1}{s + 1}$
2	$\dfrac{1}{s^2 + \sqrt{2}s + 1}$
3	$\dfrac{1}{s^3 + 2s^2 + 2s + 1}$
4	$\dfrac{1}{s^4 + 2.6131s^3 + 3.4142s^2 + 2.6131s + 1}$

By means of an example, we now demonstrate that the BLT design method can be applied to the analog transfer function $P_n(s)$ to obtain corresponding DT low-pass Butterworth filters.

Example 6.5-1: Design a second-order DT Butterworth filter whose cut-off frequency is 1 kHz at a sampling frequency of 10^4 sps. Plot the corresponding magnitude and phase responses.

Solution: The critical frequency is the cutoff frequency; that is, $\omega_{D_1} = (2\pi)(1000)$ radians/second. Next we apply the three steps related to the BLT design method.

STEP 1. Using (6.4-1), we find the point on the $j\omega$-axis that corresponds to $\omega_{D_1}T$ on the unit circle; that is,

$$\omega_{A_1} = \tan\left(\frac{(2\pi)(1000)(10^{-4})}{2}\right)$$

$$= \tan\left(\frac{\pi}{10}\right)$$

which yields

$$\omega_{A_1} = 0.325$$

STEP 2. From Table 6.5-1 we choose a second-order transfer function, which is

$$P_2(s) = \frac{1}{s^2 + \sqrt{2}s + 1}$$

This transfer function is known to represent a low-pass filter whose cutoff frequency is 1 radian/second. We therefore use frequency scaling to obtain

$$\hat{P}_2(s) = P_2\left(\frac{s}{0.325}\right)$$

$$= \frac{0.1058}{s^2 + 0.461s + 0.1058}$$

(6.5-4)

which represents a low-pass filter with a cutoff frequency of 0.325 radians/second.

STEP 3. Equation (6.4-3) is now used to obtain the desired $H(z)$. Hence

$$H(z) = \hat{P}_2(s)\Big|_{s=(z-1)/(z+1)}$$

Substitution of (6.5-4) for $\hat{P}_2(s)$ leads to

$$H(z) = \frac{0.0676(z^2 + 2z + 1)}{z^2 - 1.1422z + 0.4124}$$

(6.5-5)

or
$$H(z) = \frac{0.0676(1 + 2z^{-1} + z^{-2})}{1 - 1.1422z^{-1} + 0.4124z^{-2}}$$

The corresponding magnitude and phase responses are plotted in Fig. 6.5-3, from which it is clear that $H(z)$ represents a DT low-pass filter whose cutoff frequency is 1 kHz.

High-pass Filters

High-pass Butterworth transfer functions $Q_n(s)$ are obtained from low-pass filters via the relation

$$Q_n(s) = P_n(s) \bigg|_{s = 1/s} = P_n\left(\frac{1}{s}\right)$$

(6.5-6)

where the $P_n(s)$ are low-pass Butterworth transfer functions, some of which were listed in Table 6.5-1.

For example, $n = 1$ in (6.5-6) yields

$$Q_1(s) = \frac{1}{s + 1} \bigg|_{s = 1/s} = \frac{s}{s + 1}$$

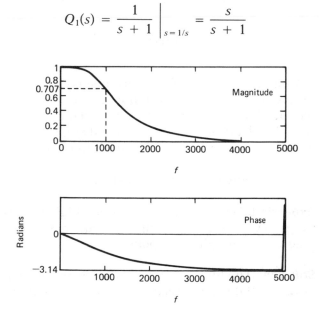

Fig. 6.5-3 Magnitude and phase responses of $H(z)$ in (6.5-5).

since $P_1(s) = 1/(s + 1)$; see Table 6.5-1.

Similarly, we can evaluate $Q_n(s)$ for $n = 2, 3, 4$, and so on. We list $Q_n(s)$, $1 \leq n \leq 4$, in Table 6.5-2. From Tables 6.5-1 and 6.5-2 it is apparent that the denominator polynomials of $P_n(s)$ and $Q_n(s)$ are the same. This means that low- and high-pass Butterworth transfer functions have identical poles s_k, which are given by (6.5-2b). We note that $Q_n(s)$ has an additional nth-order zero at the origin.

Table 6.5-2   Some High-pass Analog Filter Transfer Functions

n	$Q_n(s)$
1	$\dfrac{s}{s + 1}$
2	$\dfrac{s^2}{s^2 + \sqrt{2}s + 1}$
3	$\dfrac{s^3}{s^3 + 2s^2 + 2s + 1}$
4	$\dfrac{s^4}{s^4 + 2.6131s^3 + 3.4142s^2 + 2.6131s + 1}$

It can be shown that the magnitude response $|Q_n(j\omega)|$ is given by

$$|Q_n(\omega)| = \frac{\omega^n}{\sqrt{\omega^{2n} + 1}}, \qquad n \geq 1 \tag{6.5-7}$$

Examples of plots of $|Q_n(\omega)|$ are displayed in Fig. 6.5-4, from which it follows that as $n \to \infty$ all frequencies below 1 radian/second are eliminated, while those above 1 radian/second are passed without attenuation. This leads to the passband and stopband regions identified in Fig. 6.5-4, which implies that the cutoff frequency is 1 radian/second for *all* n.

We now consider an illustrative example.

Example 6.5-2: Design a first-order high-pass DT filter whose cutoff frequency is 1 kHz at a sampling frequency of 10^4 sps. Plot the resulting magnitude and phase responses.

Solution: The filter cutoff frequency is the critical frequency. Hence $\omega_{D_1} = (2\pi)(1000)$ radians/second, and we now follow the BLT design steps.

STEP 1. Use of (6.4-1) leads to

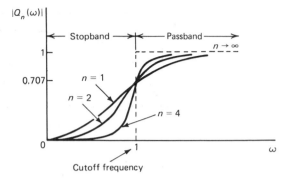

Fig. 6.5-4. Magnitude response of high-pass analog Butterworth filters.

$$\omega_{A_1} = \tan\left(\frac{(2\pi)(1000)(10^{-4})}{2}\right)$$

$$= \tan\left(\frac{\pi}{10}\right)$$

or $\qquad\qquad \omega_{A_1} = 0.325$

Thus ω_{A_1} is the point on the $j\omega$-axis that corresponds to the critical frequency.

STEP 2. Table 6.5-2 yields

$$Q_1(s) = \frac{s}{s + 1} \qquad (6.5\text{-}8)$$

which is known to have a cutoff frequency of 1 radian/second. If $\hat{Q}_1(s)$ denotes the transfer function of a corresponding high-pass filter with a cutoff of $\omega_{A_1} = 0.325$ radians/second, then we use frequency scaling to obtain

$$\hat{Q}_1(s) = Q_1\left(\frac{s}{0.325}\right) \qquad (6.5\text{-}9)$$

Substituting (6.5-8) in (6.5-9), we obtain

$$\hat{Q}_1(s) = \frac{s}{s + 0.325} \qquad (6.5\text{-}10)$$

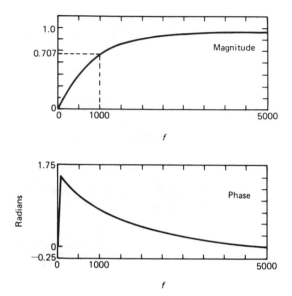

Fig. 6.5-5 Magnitude and phase responses of $H(z)$ in (6.5-12).

STEP 3. The transfer function $H(z)$ of the desired high-pass filter is obtained as

$$H(z) = \hat{Q}_1(s) \Big|_{s=(z-1)/(z+1)} \tag{6.5-11}$$

From (6.5-10) and (6.5-11) we have

$$H(z) = \frac{z - 1}{1.325z - 0.675}$$
$$\tag{6.5-12}$$
$$H(z) = \frac{1 - z^{-1}}{1.325 - 0.675z^{-1}}$$

The magnitude and phase responses of $H(z)$ in (6.5-12) are shown in Fig. 6.5-5. The magnitude response verifies that the cutoff frequency of the DT high-pass filter is 1 kHz, as desired.

Band-pass Filters

Transfer functions of analog *band-pass* filters are obtained from their *low-pass* counterparts by means of a simple transformation. It has been shown that this transformation is equivalent to replacing s by $(s^2 + \omega_0^2)/$

Ws in the low-pass filter transfer function, where ω_0 is called the *center frequency* of the band-pass filter, and W is its *bandwidth*. Thus we have

$$R_{2n}(s) = P_n(s) \Big|_{s=(s^2+\,\omega_0^2\,)/Ws} \tag{6.5-13a}$$

where $R_{2n}(s)$ denotes a $2n$th-order Butterworth band-pass filter transfer function.

For example, since $P_1(s) = 1/(s + 1)$ (see Table 6.5-1), (6.5-13a) yields

$$R_2(s) = \frac{1}{s + 1} \Big|_{s=(s^2+\,\omega_0^2\,)/Ws}$$

or

$$R_2(s) = \frac{Ws}{s^2 + Ws + \omega_0^2} \tag{6.5-13b}$$

We note that while $P_1(s)$ has a pole at $s = -1$, $R_2(s)$ has one zero at the origin and a pole-pair at

$$s = 0.5 \left[-W \pm j\sqrt{4\omega_0^2 - W^2} \right]$$

as summarized in the following illustration.

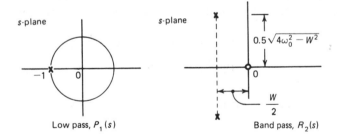

Low pass, $P_1(s)$ Band pass, $R_2(s)$

This property holds in general, and hence $R_{2n}(s)$ has n zeros at the origin, and n pole-pairs corresponding to the n poles of $P_n(s)$.

A concise expression is available for the squared-magnitude response $|R_{2n}(\omega)|^2$, and is given by

$$|R_{2n}(\omega)|^2 = \frac{W^{2n}\omega^{2n}}{W^{2n}\omega^{2n} + (\omega^2 - \omega_0^2)^{2n}}, \quad n \geq 1 \tag{6.5-14}$$

Plots of $|R_{2n}(\omega)|^2$ are illustrated in Fig. 6.5-6, where ω_ℓ and ω_h are called

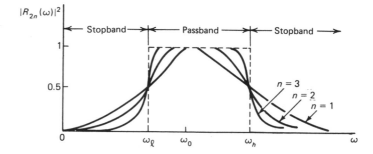

Fig. 6.5-6 Squared-magnitude response of Butterworth band-pass filters.

the lower and upper cutoff frequencies, and the center frequency ω_0 is defined as

$$\omega_0 = \sqrt{\omega_\ell \omega_h} \qquad (6.5\text{-}15)$$

The reason why ω_0 is called the "center frequency" is apparent from (6.5-15) by noting that ω_0 is the midpoint of ω_ℓ and ω_h on a *logarithmic scale*.

From Fig. 6.5-6 it is also clear that as $n \to \infty$ all frequencies in the regions $0 \le \omega < \omega_\ell$ and $\omega > \omega_h$ are eliminated, while those in the region $\omega_\ell < \omega < \omega_h$ are passed without attenuation. As such, band-pass filters possess two stopbands and one passband.

Next we examine (6.5-14) and make the following observations:

$$|R_{2n}(\omega_0)|^2 = 1, \qquad \text{for all } n \qquad (6.5\text{-}16a)$$

$$|R_{2n}(\omega_\ell)|^2 = |R_{2n}(\omega_h)|^2 = 0.5, \qquad \text{for all } n \qquad (6.5\text{-}16b)$$

Thus (6.5-16) verifies that the filter bandwidth is indeed W, and is given by

$$W = \omega_h - \omega_\ell \qquad (6.5\text{-}17)$$

While designing DT band-pass filters, we need to consider *two* critical frequencies, say ω_{D_1} and ω_{D_2}, which are the specified lower and upper cutoff frequencies. We recall that up to this point only one critical frequency has been addressed for using the BLT approach for design purposes. It is therefore necessary to slightly modify steps 1 and 2 of the BLT design method to accommodate the two critical frequencies. The pertinent changes are best conveyed by means of the example that follows.

Example 6.5-3: Design a second-order DT Butterworth band-pass filter that meets the following specifications:

Lower cutoff frequency = 210 Hz

Upper cutoff frequency = 330 Hz

Sampling frequency = 960 sps

Plot the corresponding magnitude and phase responses.

Solution: We treat the lower and upper cutoff frequencies as critical frequencies. Hence

$$\omega_{D_1} = 2\pi(210) \text{ radians/second} \tag{6.5-18}$$

and

$$\omega_{D_2} = 2\pi(330) \text{ radians/second}$$

STEP 1. Let ω_{A_1} and ω_{A_2} be the analog frequency variables corresponding to the critical frequencies ω_{D_1} and ω_{D_2} in (6.5-18). Then ω_{A_1} and ω_{A_2} are given by (6.4-1) to be

$$\omega_{A_1} = \tan\frac{\omega_{D_1}T}{2} \tag{6.5-19}$$

and

$$\omega_{A_2} = \tan\frac{\omega_{D_2}T}{2}$$

Substitution of (6.5-18) in (6.5-19) and using the fact that $T = \frac{1}{960}$, we obtain

$$\omega_{A_1} = 0.8207$$

and

$$\omega_{A_2} = 1.8709$$

STEP 2. Now ω_{A_1} and ω_{A_2} become the lower and upper cutoff frequencies, ω_ℓ and ω_h, respectively, of a second-order analog band-pass filter whose transfer function is $R_2(s)$ in (6.5-13b); that is,

$$R_2(s) = \frac{Ws}{s^2 + Ws + \omega_0^2}$$

where $W = \omega_h - \omega_\ell = 1.8709 - 0.8207 = 1.0502$ and $w_0^2 = \omega_\ell \omega_h = (0.8207)(1.8709) = 1.5354$

Thus we have

$$R_2(s) = \frac{1.0502s}{s^2 + 1.0502s + 1.5354} \qquad (6.5\text{-}20)$$

STEP 3. The BLT is now used to map the poles and zeros of $R_2(s)$ in (6.5-20) into the z-plane to obtain the desired $H(z)$. Thus

$$H(z) = R_2(s)\bigg|_{s=(z-1)/(z+1)} \qquad (6.5\text{-}21)$$

Use of the BLT program in Appendix 6.1 to evaluate (6.5-21) yields

$$H(z) = \frac{1.0502(z^2 - 1)}{3.5856z^2 + 1.0708z + 1.48542} \qquad (6.5\text{-}22)$$

The corresponding magnitude and phase responses are plotted in Fig. 6.5-7. From the magnitude response it is clear that the lower and

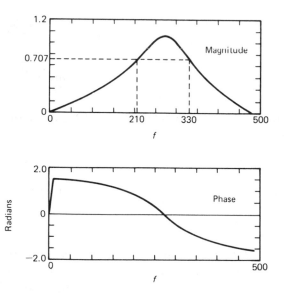

Fig. 6.5-7. Magnitude and phase responses of $H(z)$ in (6.5-22).

upper cutoff frequencies realized equal those that are specified, 210 and 330 Hz, respectively.

Band-stop Filters

The design of band-stop or band-rejection filters is very similar to that of band-pass filters. *Band-stop* filter transfer functions can be obtained from corresponding *high-pass* transfer functions via a transformation. This transformation is equivalent to replacing s in the high-pass filter transfer function by $(s^2 + \omega_0^2)/Ws$, where ω_0 is the center frequency of the band-stop filter and

$$W = \omega_h - \omega_\ell \tag{6.5-23}$$

where ω_ℓ and ω_h are the lower and upper cutoff frequencies. Thus, if $T_{2n}(s)$ denotes a $2n$th-order band-stop Butterworth filter, we have

$$T_{2n}(s) = Q_n(s)\bigg|_{s = (s^2 + \omega_0^2)/Ws} \tag{6.5-24}$$

where $Q_n(s)$ is the transfer function of an nth-order high-pass Butterworth filter; see Table 6.5-2.

To illustrate, consider $n = 1$. Then Table 6.5-2 yields $Q_1(s) = s/(s + 1)$, and hence

$$T_2(s) = \left[\frac{s}{s + 1}\right]\bigg|_{s = (s^2 + \omega_0^2)/Ws} \tag{6.5-25}$$

which yields

$$T_2(s) = \frac{s^2 + \omega_0^2}{s^2 + Ws + \omega_0^2{}_o} \tag{6.5-26}$$

From (6.5-26) it is apparent that while $Q_1(s)$ has a zero at the origin and a pole at $s = -1$, $T_2(s)$ has a pair of zeros at $s = \pm j\omega_0$, and a pole-pair at

$$s = 0.5[-W \pm j\sqrt{4\omega_0^2 - W^2}]$$

as illustrated in the following sketch.

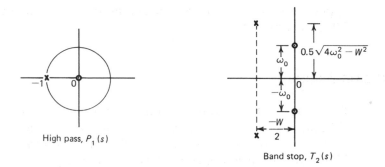

High pass, $P_1(s)$

Band stop, $T_2(s)$

This property holds in general, and thus $T_{2n}(s)$ has n pairs of zeros at $s = \pm j\omega_0$, and n pole-pairs corresponding to the n poles of $Q_n(s)$.

As in the case of band-pass filters, a concise expression is available for the squared-magnitude response $|T_{2n}(\omega)|^2$. It is given by

$$|T_{2n}(\omega)|^2 = \frac{(\omega^2 - \omega_0^2)^{2n}}{W^{2n}\omega^{2n} + (\omega^2 - \omega_0^2)^{2n}}, \qquad n \geq 1 \qquad (6.5\text{-}27)$$

where $\omega_0^2 = \omega_\ell \omega_h$; see (6.5-15).

Figure 6.5-8 shows some plots of $|T_{2n}(\omega)|^2$, from which it is observed that as $n \rightarrow \infty$ all frequencies in the region $0 \leq \omega \leq \omega_\ell$ and $\omega \geq \omega_h$ are passed without attenuation, while those in the region $\omega_\ell \leq \omega < \omega_h$ are eliminated. Thus band-stop filters consist of two passbands and one stopband, as illustrated in Fig. 6.5-8. Since frequencies in the $\omega_\ell \leq \omega < \omega_h$ region are eliminated (rejected), such filters are also called *band-rejection* filters.

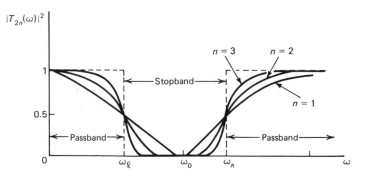

Fig. 6.5-8 Squared-magnitude response of analog Butterworth band-stop filters.

The procedure for designing DT band-stop Butterworth filters using the BLT parallels that of designing band-pass filters as outlined in Example 6.5-3. We illustrate the same by an additional example.

Example 6.5-4: Design a second-order DT Butterworth band-stop filter that meets the following specifications:

Lower cutoff frequency = 210 Hz

Upper cutoff frequency = 330 Hz

Sampling frequency = 960 sps

Plot the magnitude and phase responses of the resulting $H(z)$.

Solution: As in the case of band-pass filters, there are two critical frequencies, $\omega_\ell = 2\pi(210)$ and $\omega_h = 2\pi(330)$ radians/second.

STEP 1. As in Example 6.5-3, we obtain $\omega_{A_1} = 0.8207$ and $\omega_{A_2} = 1.8709$.

STEP 2. The ω_{A_1} and ω_{A_2} become the lower and upper cutoff frequencies ω_ℓ and ω_h, respectively, of a second-order analog band-stop filter whose transfer function is $T_2(s)$ in (6.5-26); that is,

$$T_2(s) = \frac{s^2 + \omega_0^2}{s^2 + Ws + \omega_0^2}$$

where $W = 1.0502$ and $\omega_0^2 = 1.5354$ as in Example 6.5-3.
Hence we obtain

$$T_2(s) = \frac{s^2 + 1.5354}{s^2 + 1.0502s + 1.5354} \tag{6.5-28}$$

STEP 3. The desired $H(z)$ is obtained by replacing s by $(z - 1)/(z + 1)$ in (6.5-28). This process yields

$$H_2(z) = \frac{2.5354z^2 + 1.0708z + 2.5354}{3.5856z^2 + 1.0708z + 1.4852} \tag{6.5-29}$$

the magnitude and phase responses of which are displayed in Fig. 6.5-9. From this magnitude response we conclude that $H(z)$ is indeed the transfer function of a band-stop filter with lower and upper cutoff frequencies of 210 and 330 Hz, respectively. We also note the presence of

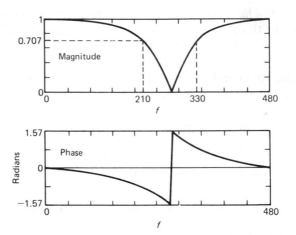

Fig. 6.5-9 Magnitude and phase responses of $H(z)$ in (6.5-29).

a notch in the magnitude response. This occurs at the center frequency $f_0 = \sqrt{\omega_\ell \omega_h}/2\pi$, which is approximately 263 Hz.

6.6 DESIGN OF HIGH-ORDER BUTTERWORTH FILTERS

The notion of cascade realizations was introduced in Section 5.6. The basic idea is to realize a desired $H(z)$ as a product (cascade) of lower-order transfer functions $H_i(z)$; see Fig. 5.6-3. Stearns [7] has shown that by exploiting the fact that the poles of even-ordered analog Butterworth transfer functions occur in complex-conjugate pairs, it is possible to design DT Butterworth filters in cascade form. The procedure is especially useful when high-order filters are required (e.g., fourth order or higher). Cascade realizations are also preferred for hardware implementation since they require shorter word lengths and are less prone to overflow problems. Such hardware implementation aspects will be discussed in Chapter 8.

Let us consider even-ordered low-pass Butterworth filters. We observe that the poles s_k in (6.5-2b) of the corresponding transfer functions occur in complex-conjugate pairs for even values of n; for example, see Fig. 6.5-2 for $n = 4$, where (s_1, s_4) and (s_2, s_3) form complex conjugate pairs. In general, the complex conjugate of s_k is s_{n+1-k}, and

$$s_k = e^{j\theta_k}, \qquad 1 \le k \le \frac{n}{2} \qquad (6.6\text{-}1a)$$

$$\text{where} \qquad \theta_k = \frac{(2k + n - 1)\pi}{2n} \qquad\qquad (6.6\text{-}1b)$$

Thus an even-ordered $P_n(s)$ in (6.5-3) can be expressed as

$$P_n(s) = P_1'(s) \, P_2'(s) \cdots P_{n/2}'(s) \qquad\qquad (6.6\text{-}2)$$

$$\text{where} \quad P_k'(s) = \frac{1}{(s - s_k)(s - \bar{s}_k)}, \qquad 1 \le k \le \frac{n}{2}$$

and $\bar{s}_k$ denotes the complex conjugate of s_k. Substitution of $s_k = e^{j\theta_k}$ in the preceding expression for $P_k'(s)$ leads to

$$P_k'(s) = \frac{1}{s^2 - 2 \cos \theta_k \, s + 1}, \qquad 0 \le k \le \frac{n}{2} \qquad (6.6\text{-}3)$$

For a specified critical frequency ω_{D_1}, we obtain the corresponding analog frequency variable ω_{A_1} via (6.4-1); that is,

$$\omega_{A_1} = \tan \frac{\omega_{D_1} T}{2} \qquad\qquad (6.6\text{-}4)$$

Thus frequency scaling $P_k'(s)$ by ω_{A_1} results in

$$\hat{P}_k(s) = \frac{\omega_{A_1}^2}{s^2 - 2\omega_{A_1} \cos \theta_k s + \omega_{A_1}^2} \qquad\qquad (6.6\text{-}5)$$

We now find $H_k(z)$ corresponding to $\hat{P}_k(s)$ via the bilinear transform; that is,

$$H_k(z) = \hat{P}_k(s) \Big|_{s = (z-1)/(z+1)}$$

which leads to

$$H_k(z) = \frac{(1 + 2z^{-1} + z^{-2}) \, \omega_{A_1}^2}{A_k' + B_k' z^{-1} + C_k' z^{-2}} \qquad\qquad (6.6\text{-}6a)$$

$$\text{where} \quad A_k' = 1 - 2\omega_{A_1} \cos \theta_k + \omega_{A_1}^2$$

$$B_k' = 2(\omega_{A_1}^2 - 1) \qquad\qquad (6.6\text{-}6b)$$

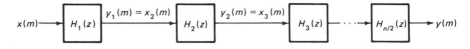

Fig.6.6-1 Cascade realization of low-pass discrete-time Butterworth filters for n even; the output of each stage is computed using (6.6-8).

$$C'_k = 1 + 2\omega_{A_1} \cos \theta_k + \omega_{A_1}^2$$

and the θ_k are given by (6.6-1b).
Hence the output of the kth stage is obtained by writing

$$Y_k(z) = H_k(z)X_k(z) \qquad (6.6\text{-}7)$$

which yields the input–output equation

$$y_k(m) = \frac{1}{A'_k}\{-B'_k y_k(m-1) - C'_k y_k(m-2) + \omega_{A_1}^2[x_k(m) + 2x_k(m-1)$$
$$+ x_k(m-2)]\} \qquad (6.6\text{-}8)$$

where $x_k(m)$ is the input to the kth stage, and $m \ge 0$.

In (6.6-8) we note that $1 \le k \le n/2$, which means that the processing can be carried out in stages; that is, the output of the $(k-1)$th stage $[= y_{k-1}(m)]$ becomes the input to the kth stage $[= x_k(m)]$, as illustrated in Fig. 6.6-1. The desired filtered output is the sequence $y(m)$, which is the output of the last stage; see Fig. 6.6-1.

The preceding design procedures can readily be extended to other DT Butterworth filters discussed in Section 6.5. The results so obtained are summarized next. The related derivations are left as an exercise to the reader; see Problem 6-9. It follows that such cascade realizations can readily be implemented in software (e.g., see [7] or Appendix 6.2).

High-pass Filters

$$\hat{Q}_k(s) = \frac{s^2}{s^2 - 2\omega_{A_1} \cos \theta_k s + \omega_{A_1}^2}$$

where θ_k and ω_{A_1} are given by (6.6-1b) and (6.6-4), respectively.

$$H_k(z) = \hat{Q}_k(s)\Big|_{s=(z-1)/(z+1)} \qquad (6.6\text{-}9)$$

$$= \frac{1 - 2z + z^{-2}}{A'_k + B'_k z^{-1} + C'_k z^{-2}}, \qquad 1 \le k \le n/2$$

where A'_k, B'_k, and C'_k are as given by (6.6-6b).

Band-pass Filters

$$R'_{2k}(s) = \frac{W^2 s^2}{s^4 - 2W \cos \theta_k s^3 + (2\omega_0^2 + W^2)s^2 - 2W\omega_0^2 \cos \theta_k s + \omega_0^4}$$

where $\omega_{A_i} = \tan (\omega_{D_i} T/2)$, $i = 1, 2$

$\qquad W = \omega_{A_2} - \omega_{A_1}$

$\qquad \omega_0^2 = \omega_{A_1}\omega_{A_2}$

and the θ_k is given by (6.6-1b).

$$H_k(z) = R'_{2k}(s)\Big|_{s=(z-1)/(z+1)} \tag{6.6-10}$$

$$= \frac{(1 - 2z^{-2} + z^{-4})W^2}{D_k + E_k z^{-1} + F_k z^{-2} + G_k z^{-3} + J_k z^{-4}}$$

where $D_k = (1 + \omega_0^2)[1 - 2W \cos \theta_k + \omega_0^2] + W^2$

$\qquad E_k = 4(\omega_0^2 - 1)[1 - W \cos \theta_k + \omega_0^2]$

$\qquad F_k = 6\omega_0^4 - 4\omega_0^2 + (6 - 2W^2)$

$\qquad G_k = 4(\omega_0^2 - 1)[1 + W \cos \theta_k + \omega_0^2]$

$\qquad J_k = (1 + \omega_0^2)[1 + 2W \cos \theta_k + \omega_0^2] + W^2$

Band-stop Filters

$$T'_{2k}(s) = \frac{s^4 + 2\omega_0^2 s^2 + \omega_0^4}{s^4 - 2W \cos \theta_k s^3 + (2\omega_0^2 + W^2)s^2 - 2W\omega_0^2 \cos \theta_k s + \omega_0^4}$$

where θ_k, W, and ω_0^2 are evaluated as indicated for band-pass filters.

$$H_k(z) = T'_{2k}(s)\Big|_{s=(z-1)/(z+1)}$$

$$= \frac{D'_k + E'_k z^{-1} + F'_k z^{-2} + E'_k z^{-3} + D'_k z^{-4}}{D_k + E_k z^{-1} + F_k z^{-2} + G_k z^{-3} + J_k z^{-4}} \tag{6.6-11}$$

where D_k, E_k, F_k, G_k, and J_k are defined in (6.6-10), and

$$D_k' = (\omega_0^2 + 1)^2$$

$$E_k' = 4(\omega_0^4 - 1)$$

$$F_k' = 2(\omega_0^2 - 1)^2 + 4(\omega_0^4 + 1).$$

Illustrative Example

We conclude this discussion on Butterworth filters in cascade form with an example. Suppose we wish to extract the predominant component in the helicopter noise data that was displayed in Fig. 4.9-4a, using a band-pass filter. From its PDS in Fig. 4.9-4b it is apparent that the predominant component is around 20 Hz. Let us process the noise data through a five-section filter, where each section has a fourth-order transfer function given by $H_k(z)$ in (6.6-10). Thus the overall transfer function $H(z)$ will be of order 20, which means that $2n = 20$ in (6.6-1), or $n = 10$. Hence (6.6-1b) yields

$$\theta_k = \frac{(2k + 9)\pi}{20}, \qquad 1 \le k \le 5 \tag{6.6-12}$$

which pertains to the five sections.

Next the sampling frequency is given to be 256 sps, which means that $T = \frac{1}{256}$ sec. Suppose we choose the lower and upper cutoff frequencies to be 16 and 24 Hz, respectively. Then we have $f_{D_1} = 16$ and $f_{D_2} = 24$, which result in

$$\omega_{A_1} = \tan(\pi f_{D_1} T) = 0.1989 \tag{6.6-13}$$

and $$\omega_{A_2} = \tan(\pi f_{D_2} T) = 0.3033$$

Thus we have

$$W = \omega_{A_2} - \omega_{A_1} = 0.1044 \tag{6.6-14}$$

and $$\omega_0^2 = \omega_{A_1}\omega_{A_2} = 0.0603$$

The information in (6.6-12) to (6.6-14) is sufficient to evaluate the $H_k(z)$ in (6.6-10) for $1 \le k \le 5$. Then, writing $Y_k(z) = H_k(z)X(z)$, we obtain a fourth-order input–output difference equation for each section of the cascade realization. The output of the fifth section is the desired filtered output.

The results so obtained are displayed in Figs. 6.6-2 and 6.6-3. The

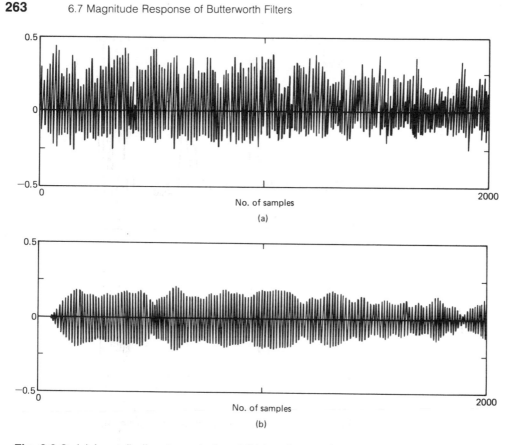

Fig. 6.6-2 (a) Input (helicopter noise) and (b) band-passed output.

input to the five-section band-pass filter is shown in Fig. 6.6-2a, while Fig. 6.6-2b shows the corresponding output. The input and output consist of 2000 samples each and appear as CT signals because of interpolation introduced by the plotting device. From the related input–output power density spectra shown in Fig. 6.6-3, it follows that the output consists mainly of the 20-Hz component in the input, which is the desired result. Details pertaining to computing the output PDS parallel those for computing the input PDS; see Section 4.9.

6.7 MAGNITUDE RESPONSE OF BUTTERWORTH FILTERS

The computer program in Appendix 5.1 was used to evaluate the magnitude response pertaining to Examples 6.5-1 through 6.5-4. However, since magnitude responses of analog Butterworth transfer functions

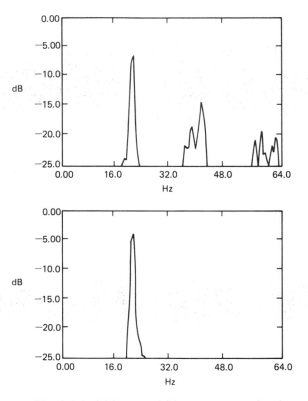

Fig. 6.6-3 (a) Input and (b) output power density spectra.

have closed-form expressions, corresponding closed-form expressions can also be derived for DT Butterworth transfer functions [7, 8]. Such expressions may be used to evaluate magnitude responses of DT Butterworth filters, in lieu of the program in Appendix 5.1.

The key to deriving the desired closed-form expression is the property of the BLT that is given in (6.2-7), which is

$$H(z)\Big|_{z=e^{j\omega_D T}} = H(s)\Big|_{s=j\omega_A}$$

where $\omega_A = \tan(\omega_D T/2)$; $H(z)$ and $H(s)$ denote DT and analog transfer functions, respectively. That is,

$$H(e^{j\omega_D T}) = H(\omega_A) \tag{6.7-1}$$

where $\omega_A = \tan(\omega_D T/2)$ and $H(\omega_A) = H(j\omega_A)$

Let us now consider the relevance of (6.7-1) to the low-pass case.

We have

$$H(z) = \hat{P}_n(s)\Big|_{s=(z-1)/(z+1)} \tag{6.7-2}$$

where $\hat{P}_n(s)$ is the frequency scaled Butterworth transfer function. Equations (6.7-1) and (6.7-2) imply that

$$H(e^{j\omega_D T}) = \hat{P}_n(\omega_A) \tag{6.7-3}$$

Again, $\hat{P}_n(s)$ is related to the Butterworth transfer function $P_n(s)$ via the relation

$$\hat{P}_n(s) = P_n\left(\frac{s}{\omega_{A_1}}\right) \tag{6.7-4}$$

where $\omega_{A_1} = \tan(\omega_{D_1}T/2)$, and ω_{D_1} is the critical frequency of the DT filter.

Combining (6.7-3) and (6.7-4), we get

$$H(e^{j\omega_D T}) = P_n\left(\frac{\omega_A}{\omega_{A_1}}\right) \tag{6.7-5}$$

which means that

$$|H(e^{j\omega_D T})|^2 = \left|P_n\left(\frac{\omega_A}{\omega_{A_1}}\right)\right|^2 \tag{6.7-6}$$

The right-hand side of (6.7-6) is the squared-magnitude response of an nth-order analog Butterworth transfer function, which is given by (6.5-1). Thus, from (6.5-1) and (6.7-6) we have

$$|H(e^{j\omega_D T})|^2 = \frac{1}{1 + (\omega_A/\omega_{A_1})^{2n}} \tag{6.7-7}$$

With $\omega_D T = 2\pi f T$, (6.7-7) can also be written as

$$|\bar{H}(f)|^2 = \frac{1}{1 + \left[\dfrac{\tan(\pi f T)}{\tan(\pi f_{D_1} T)}\right]^{2n}}, \qquad 0 \le f \le f_N \tag{6.7-8}$$

where f_{D_1} is the critical frequency in hertz.

From (6.7-8) it is clear that either the magnitude or squared-magnitude response of a low-pass DT Butterworth can readily be computed when n, f_D, and T are specified.

The preceding analysis can be extended to other DT Butterworth filters (i.e., high pass, band pass, and band stop). The results so obtained are summarized in what follows. The derivations pertaining to the same are left to the reader to pursue; see Problem 6-10.

High-pass Filters

$$H(e^{j\omega_D T}) = Q_n \left(\frac{\omega_A}{\omega_{A_1}} \right)$$

where $\omega_A = \tan \left(\dfrac{\omega_D T}{2} \right)$

$\omega_{A_1} = \tan \left(\dfrac{\omega_{D_1} T}{2} \right)$

$$|H(e^{j\omega_D T})|^2 = \frac{(\omega_A/\omega_{A_1})^{2n}}{1 + (\omega_A/\omega_{A_1})^{2n}}$$

$$= \frac{1}{1 + (\omega_{A_1}/\omega_A)^{2n}}$$

$$|\tilde{H}(f)|^2 = \frac{1}{1 + \left[\dfrac{\tan (\pi f_{D_1} T)}{\tan (\pi f T)} \right]^{2n}}, \qquad 0 \le f \le f_N \qquad (6.7\text{-}9)$$

Band-pass Filters

$$H(e^{j\omega_D T}) = R_{2n}(\omega_A)$$

where $\omega_A = \tan \left(\dfrac{\omega_D T}{2} \right)$

$$|R_{2n}(\omega_A)|^2 = \frac{1}{1 + \left[\dfrac{(\omega^2 - \omega_0^2)}{W \omega_A} \right]^{2n}}, \qquad [\text{see } (6.5\text{-}14)]$$

where $\quad \omega_{A_i} = \tan \left(\dfrac{\omega_{D_i} T}{2} \right), \qquad i = 1, 2$

$W = \omega_{A_2} - \omega_{A_1}$

$\omega_0^2 = \omega_{A_1} \omega_{A_2}$

$$|\bar{H}(f)|^2 = \frac{1}{1 + \left[\dfrac{\tan^2 (\pi f T) - \omega_0^2}{W \tan (\pi f T)} \right]^{2n}}, \quad 0 \le f \le f_N \qquad (6.7\text{-}10)$$

Band-stop Filters

$$H(e^{j \omega_D T}) = T_{2n}(\omega_A)$$

where $\quad \omega_A = \tan \left(\dfrac{\omega_D T}{2} \right)$

$$|T_{2n}(\omega)|^2 = \frac{1}{1 + \left[\dfrac{W \omega_A}{(\omega_A^2 - \omega_0^2)} \right]^{2n}} \qquad [\text{see } (6.5\text{-}27)]$$

where $\quad \omega_{A_i} = \tan \left(\dfrac{\omega_{D_i} T}{2} \right), \qquad i = 1, 2$

$W = \omega_{A_2} - \omega_{A_1}$

$\omega_0^2 = \omega_{A_1} \omega_{A_2}$

$$|\bar{H}(f)|^2 = \frac{1}{1 + \left[\dfrac{W \tan (\pi f T)}{\tan^2 (\pi f T) - \omega_0^2} \right]^{2n}}, \quad 0 \le f \le f_N \qquad (6.7\text{-}11)$$

6.8 SUMMARY

This chapter was devoted to introducing the design of IIR filters. The impulse invariant method and the BLT method for filter design were discussed. Much more emphasis was placed on the BLT method, since it is a powerful technique that is simple to use. The BLT design procedure was illustrated by means of a variety of examples, including a class of DT Butterworth filters. A convenient design procedure to obtain cascade realizations for DT Butterworth filters was developed. This design procedure is very useful when high-order filters are required. Computer programs for designing low-pass, high-pass, band-pass, and band-stop cascaded Butterworth filters are presented in Appendix 6.2.

It is emphasized that the power in the BLT design method lies in the fact that known properties of analog filters can be exploited by designing corresponding DT filters. Although we restricted our attention to Butterworth filters, the related design procedure can also be used to design a variety of other types of DT filters (e.g., Chebyshev and Bessel filters). To do so, one would use transfer functions pertaining to the desired analog filters in place of analog Butterworth filter transfer functions (e.g., see Problem 6-11).

Computer-aided design techniques for IIR filters have been investigated for attaining solutions that are optimum with respect to some criterion. However, such methods require a significant amount of computation time. The interested reader may refer to [9, 10] for a detailed discussion of such methods.

PROBLEMS

6–1   Starting with $H(s)$ in (6.1-7), find $h(t)$, $t \geq 0$, and use it to show that $H(z)$ is as in (6.1-8).

6–2   Given:

$$H(s) = \frac{s + 2}{(s + 1)(s + 3)}$$

(a) Find the corresponding $H(z)$ using the impulse invariant method.

(b) With $f_s = 10$ sps, plot $T|\tilde{H}(f)|$ and $\tilde{\psi}(f)$, and compare them with $|H(f)|$ and $\psi(f)$, respectively.

6–3   (a) For the circuit shown in connection with Example 6.4-2, show that $H(s)$ is as given by (6.4-8).

(b) Evaluate the magnitude response of $H(s)$ in (6.4-8), and hence verify that it is the transfer function of a notch filter whose frequency is 1 radian/second.

6–4   Evaluate the magnitude response of $H(s)$ in (6.4-13). Use this information to verify that $H(s)$ is the transfer function of a resonator with resonant frequency of 1 radian/second.

6–5   Design a third-order DT Butterworth low-pass filter whose bandwidth is 100 Hz at a sampling frequency of 1000 sps. Plot the corresponding magnitude and phase responses.

6–6 Starting with $H(s)$ in (6.4-13), design a DT resonator whose resonant frequency is 1000 Hz when the sampling frequency is 8000 sps. Use the BLT method, and plot the resulting magnitude and phase responses.

6–7 Use the BLT method to design a second-order DT Butterworth high-pass filter whose cutoff frequency is 70 Hz at a sampling frequency of 1000 sps. Plot the corresponding magnitude and phase responses.

6–8 A second-order DT band-pass Butterworth filter with the following specifications is desired:

> Lower cutoff frequency = 10 Hz
> Upper cutoff frequency = 20 Hz
> Sampling frequency = 100 sps

Find $H(z)$ and plot the resulting magnitude and phase responses.

6–9 (a) Derive $H_k(z)$ in (6.6-9). Find the corresponding input–output difference equation for the kth stage of a cascade realization for DT Butterworth high-pass filters.

(b) Derive $H_k(z)$ in (6.6-10). Find the related input–output difference equation for the kth stage of a cascade realization for DT Butterworth band-pass filters.

(c) Derive $H_k(z)$ in (6.6-11). Hence evaluate the corresponding difference equation for the kth stage of a cascade realization for DT Butterworth band-stop filters.

6–10 (a) Derive $|\bar{H}(f)|^2$ in (6.7-9), which is the squared-magnitude response of an nth-order DT Butterworth high-pass filter.

(b) Derive $|\bar{H}(f)|^2$ in (6.7-10) and (6.7-11), which are the squared-magnitude responses of nth-order DT Butterworth band-pass and band-stop filters, respectively.

6–11 The transfer function associated with an nth order analog low-pass Bessel filter is given by

$$G(s) = \frac{1}{b_0 + b_1 s + \cdots + b_n s^n}$$

where the coefficients b_j are as listed in the following table.

n	b_0	b_1	b_2	b_3	b_4
0	1				
1	1	1			
2	3	3	1		
3	15	15	6	1	
4	105	105	45	10	1

Use this information to design the following DT Bessel filters.

(a) Second order; low pass; bandwidth = 100 Hz; sampling frequency = 1000 sps.

(b) Third order; high pass; cutoff frequency = 10 Hz; sampling frequency = 50 sps.

(c) Second order; band pass; lower cutoff frequency = 30 Hz; upper cutoff frequency = 100 Hz; sampling frequency = 500 sps.

(d) Second order; band stop; lower cutoff frequency = 100 Hz; upper cutoff frequency = 200 Hz; sampling frequency = 1000 sps.

In each case, plot the magnitude and phase responses of the filter that results.

6–12 Let $\tau(\omega_A)$ and $\tau'(\omega_D)$ denote the time-delay functions of analog and DT transfer functions $H(s)$ and $H(z)$, respectively. Then, by definition,

$$\tau(\omega_A) = -\frac{d\beta(\omega_A)}{d\omega_A}$$

and
$$\tau'(\omega_D) = -\frac{d\beta'(\omega_D)}{d\omega_D}$$

where $\beta(\omega_A)$ and $\beta'(\omega_D)$ are the phase functions of $H(s)$ and $H(z)$, respectively. If $H(s)$ and $H(z)$ are related via the BLT, show that

$$\tau'(\omega_D) = \frac{T}{2}(1 + \omega_A^2)\tau(\omega_A) \qquad \text{(P6-12-1)}$$

where $\omega_A = \tan\left(\frac{\omega_D T}{2}\right)$

Hint: $\beta'(\omega_D) = \beta(\omega_A)$, and hence

$$\frac{d\beta'(\omega_D)}{d\omega_D} = \frac{d\beta(\omega_A)}{d\omega_A} \cdot \frac{d\omega_A}{d\omega_D}$$

6–13 Use (P6-12-1) to show that the time-delay function of the DT low-pass filter designed in Example 6.4-1 is given by

$$\tau'(\omega_D) = \frac{1 + \omega_A^2}{40(3 + \omega_A^2)}, \qquad 0 \le \omega_D T \le \pi$$

where $\omega_A = \tan\left(\frac{\omega_D T}{2}\right)$

APPENDIX 6.1[†] COMPUTER PROGRAM TO EVALUATE THE BILINEAR TRANSFORM

The computer program listed in this appendix evaluates the BLT. That is, given the numerator and denominator coefficients of $H(s)$, it computes and prints the corresponding numerator and denominator coefficients of $H(z)$ via the relation

$$H(z) = H(s)\Big|_{s = (z-1)/(z+1)}$$

For example, if

$$H(s) = \frac{s^2 + 1}{s^2 + s + 1}$$

then the queries and related responses would be as follows:

```
NUMBER OF NUMERATOR COEFFICIENTS (1-21) ? 3
        A( 2)*(S** 2) = ? 1
        A( 1)*(S** 1) = ? 0
        A( 0)*(S** 0) = ? 1
NUMBER OF DENOMINATOR COEFFICIENTS (1-21) ? 3
        B( 2)*(S** 2) = ? 1
        B( 1)*(S** 1) = ? 1
        B( 0)*(S** 0) = ? 1
```

The following corresponding output is obtained:

```
        A(I)'S                          B(I)'S
    2.0000      *Z**  0          3.0000      *Z**  0
     .00000     *Z** -1           .00000     *Z** -1
    2.0000      *Z** -2          1.0000      *Z** -2
```

Thus the desired DT transfer function is given by

$$H(z) = \frac{2 + 2z^{-2}}{3 + z^{-2}}$$

```
C**************************************************************************
C
        COMPILER FREE
        PARAMETER ITTO=10,ITTI=11
        REAL AS(21),BS(21),AZ(21),BZ(21),COEF(21,21)
100     CONTINUE
        TYPE ' BILINEAR TRANSFORM REV 0.0 FORTRAN 5 '
             DO 1000 I=1,21
             AS(I)=0.0
             BS(I)=0.0
```

[†]Program developed by David Hein and Myron Flickner.

```
1000            CONTINUE
        ACCEPT ' NUMBER OF NUMERATOR COEFFICIENTS (1-21) ? ',NN
                DO 1001 I=NN,1,-1
                II=I-1
                WRITE(ITTO,1) II,II
1               FORMAT('     A(',I2,')*(S**',I2,') = ? ',Z)
                READ(ITTI) AS(I)
1001            CONTINUE
        ACCEPT ' NUMBER OF DENOMINATOR COEFFICIENTS (1-21) ? ',ND
                DO 1002 I=ND,1,-1
                II=I-1
                WRITE(ITTO,2) II,II
2               FORMAT('     B(',I2,')*(S**',I2,') = ? ',Z)
                READ(ITTI) BS(I)

1002            CONTINUE
        IF(ND.LT.NN) STOP ' TRANSFER FUNCTION IS NOT REALISTIC '
        N=ND-1
                DO 1003 K=1,ND
                COEF(1,K)=1.0
1003            CONTINUE
        FACT=1.0
                DO 1004 K=2,ND
                FACT=FACT*FLOAT(N-K+2)/FLOAT(K-1)
                COEF(K,1)=FACT
1004            CONTINUE
                DO 1005 K=2,ND
                DO 1005 J=2,ND
                COEF(K,J)=COEF(K,J-1)-COEF(K-1,J)-COEF(K-1,J-1)
1005            CONTINUE
                DO 1006 K=1,ND
                ASUM=0.0
                BSUM=0.0
                    DO 1007 J=1,ND
                    ASUM=ASUM+COEF(K,J)*AS(J)
                    BSUM=BSUM+COEF(K,J)*BS(J)
1007                CONTINUE
                AZ(K)=ASUM
                BZ(K)=BSUM
1006            CONTINUE
        CALL QUERY(' OUTPUT COEFFICIENTS TO PRINTER (Y/N) ? ',IQ)
        IO=10
        IF(IQ.EQ.1) IO=12
        WRITE(IO,3)
3       FORMAT(/,' ',10X,'A(I)''S',27X,'B(I)''S')
                DO 1008 I=1,ND
                II=-(I-1)
                WRITE(IO,4) AZ(I),II,BZ(I),II
4               FORMAT(' ',G15.5,' *Z**',I3,10X,G15.5,' *Z**',I3)
1008            CONTINUE
        CALL QUERY(' RE-EXECUTE BLT (Y/N) ? ',IQ)
        IF(IQ.EQ.1) GO TO 100
        STOP 'NORMAL TERMINATION OF BLT '
        END
```

APPENDIX 6.2[†] SUBROUTINES FOR DESIGNING BUTTERWORTH FILTERS

Subroutines for designing high-order DT Butterworth filters as described in Section 6.6 are given in this appendix. Subroutines LPDES

[†]Programs written by Myron Flickner, David Martz, and Wayne Blasi.

and HPDES can be used to design nth-order low-pass and high-pass filters, respectively, where n is even. Again, $2n$th-order band-pass and band-stop filters can be designed using subroutines BPDES and BSDES, respectively, for n even. Appendix C of [7] was used as a guide for writing these subroutines.[†]

```
C*********************************************************************************
C
C       CALLING SEQUENCE
C
C               CALL LPDES(FC,T,NSEC,A,B,C)
C
C       PURPOSE
C
C               THE SUBROUTINE WILL RETURN THE COEFFICIENTS TO A CASCADE
C               REALIZATION OF A MULTIPLE SECTION LOWPASS FILTER.  THE
C               KTH SECTION HAS THE FOLLOWING TRANFER FUNCTION:
C
C                       A(K)*(1 + 2*Z**-1 + Z**-2)
C               H(Z)=----------------------------
C                       1 + B(K)*Z**-1 + C(K)*Z**-2
C
C               THUS, IF F(M) AND G(M) ARE THE INPUT AND OUTPUT OF THE
C               KTH SECTION THE FOLLOWING DIFFERENCE EQUATION IS SATISFIED.
C
C               G(M)=A(K)*(F(M)+2*F(M-1)+F(M-2)) - B(K)*G(M-1) - C(K)*G(M-2)
C
C
C       ROUTINE(S) CALLED BY THIS ROUTINE
C
C               NONE
C
C       ARGUMENT(S) REQUIRED FROM THE CALLING ROUTINE
C
C               FC      -       3-DB CUTOFF FREQUENCY IN HERTZ
C               T       -       SAMPLING INTERVAL IN SECONDS
C               NSEC    -       NUMBER OF SECTIONS TO BE IMPLEMENTED
C
C       ARGUMENT(S) SUPPLIED  TO  THE CALLING ROUTINE
C
C               A       -       FILTER COEFFICIENT (VECTOR OF LENGTH NSEC)
C               B       -       FILTER COEFFICIENT (VECTOR OF LENGTH NSEC)
C               C       -       FILTER COEFFICIENT (VECTOR OF LENGTH NSEC)
C
C*********************************************************************************
-
C
        SUBROUTINE LPDES(FC,T,NSEC,A,B,C)
        DIMENSION A(1),B(1),C(1)
        PI=4.0*ATAN(1.0)
        Q1=FC*PI*T
        WCP=SIN(Q1)/COS(Q1)
        WCP2=WCP*WCP
C
```

[†]Subroutines LPDES, HPDES, and BPDES reprinted with permission of Hayden Book Company from *Digital Signal Analysis*, by S.D. Stearns. Copyright 1975.

```
                      DO 1000 K=1,NSEC
                      CS=COS(FLOAT(2*(K+NSEC)-1)*PI/FLOAT(4*NSEC))
                      X=1.0/(1.0+WCP2-2.0*WCP*CS)
                      A(K)=WCP2*X
                      B(K)=2.0*(WCP2-1)*X
                      C(K)=(1.0+WCP2+2.0*WCP*CS)*X
         1000         CONTINUE


         C
               RETURN
               END


C*********************************************************************************
C
C        CALLING SEQUENCE
C
C                 CALL HPDES(FC,T,NSEC,A,B,C)
C
C        PURPOSE
C
C                 THE SUBROUTINE WILL RETURN THE COEFFICIENTS TO A CASCADE
C                 REALIZATION OF A MULTIPLE SECTION HIGHPASS FILTER.  THE
C                 KTH SECTION HAS THE FOLLOWING TRANFER FUNCTION:
C
C                       A(K)*(1 - 2*Z**-1 + Z**-2)
C                 H(Z)=-----------------------------
C                       1 + B(K)*Z**-1 + C(K)*Z**-2
C
C                 THUS, IF F(M) AND G(M) ARE THE INPUT AND OUTPUT OF THE
C                 KTH SECTION THE FOLLOWING DIFFERENCE EQUATION IS SATISFIED.
C
C                 G(M)=A(K)*(F(M)-2*F(M-1)+F(M-2)) - B(K)*G(M-1) - C(K)*G(M-2)
C
C
C        ROUTINE(S) CALLED BY THIS ROUTINE
C
C                 NONE
C
C        ARGUMENT(S) REQUIRED FROM THE CALLING ROUTINE
C
C                 FC      -        3-DB CUTOFF FREQUENCY IN HERTZ
C                 T       -        SAMPLING INTERVAL IN SECONDS
C                 NSEC    -        NUMBER OF SECTIONS TO BE IMPLEMENTED
C
C        ARGUMENT(S) SUPPLIED  TO  THE CALLING ROUTINE
C
C                 A       -        FILTER COEFFICIENT (VECTOR OF LENGTH NSEC)
C                 B       -        FILTER COEFFICIENT (VECTOR OF LENGTH NSEC)
C                 C       -        FILTER COEFFICIENT (VECTOR OF LENGTH NSEC)
C
C*********************************************************************************
C
               SUBROUTINE HPDES(FC,T,NSEC,A,B,C)
               DIMENSION A(1),B(1),C(1)
               PI=4.0*ATAN(1.0)
               Q1=FC*PI*T
               WCP=SIN(Q1)/COS(Q1)
               WCP2=WCP*WCP
         C
```

```
                  DO 1000 K=1,NSEC
                  CS=COS(FLOAT(2*(K+NSEC)-1)*PI/FLOAT(4*NSEC))
                  A(K)=1.0/(1.0+WCP2-2.0*WCP*CS)
                  B(K)=2.0*(WCP2-1)*A(K)
                  C(K)=(1.0+WCP2+2.0*WCP*CS)*A(K)
1000              CONTINUE
C
          RETURN
          END




C**********************************************************************************
C
C         CALLING SEQUENCE
C
C                 CALL BPDES(FCL,FCU,T,NSEC,A,B,C,D,E,F,G)
C
C         PURPOSE
C
C                 THE SUBROUTINE WILL RETURN THE COEFFICIENTS TO A CASCADE
C                 REALIZATION OF A MULTIPLE SECTION BANDPASS FILTER.  THE
C                 KTH SECTION HAS THE FOLLOWING TRANFER FUNCTION:
C
C                         A(K)*(1 + Z**-4)+C(K)*Z**-2
C                 H(Z)=--------------------------------------------------------
C                         1 + D(K)*Z**-1 + E(K)*Z**-2 + F(K)*Z**-3 + G(K)*Z**-4
C
C                 THUS, IF F(M) AND G(M) ARE THE INPUT AND OUTPUT OF THE
C                 KTH SECTION THE FOLLOWING DIFFERENCE EQUATION IS SATISFIED.
C
C                 G(M)=A(K)*(F(M)+F(M-4))+C(K)*F(M-2)
C                     - D(K)*G(M-1) - E(K)*G(M-2) - F(K)*G(M-3) - G(K)*G(M-4)
C
C                 NOTE:
C                         DO NOT USE THIS ROUTINE FOR LOWPASS DESIGN - USE LPDES
C                         DO NOT USE THIS ROUTINE FOR HIGHPASS DESIGN - USE HPDES
C                         DO NOT USE THIS ROUTINE FOR BANDSTOP DESIGN - USE BSDES
C
C         ROUTINE(S) CALLED BY THIS ROUTINE
C
C                 NONE
C
C         ARGUMENT(S) REQUIRED FROM THE CALLING ROUTINE
C
C                 FCL     -       LOWER 3-DB CUTOFF FREQUENCY IN HERTZ
C                 FCU     -       UPPER 3-DB CUTOFF FREQUENCY IN HERTZ
C                 T       -       SAMPLING INTERVAL IN SECONDS
C                 NSEC    -       NUMBER OF SECTIONS TO BE IMPLEMENTED
C
C         ARGUMENT(S) SUPPLIED  TO  THE CALLING ROUTINE
C
C                 A       -       FILTER COEFFICIENT (VECTOR OF LENGTH NSEC)
C                 B       -       FILTER COEFFICIENT (VECTOR OF LENGTH NSEC)
C                 C       -       FILTER COEFFICIENT (VECTOR OF LENGTH NSEC)

C                 D       -       FILTER COEFFICIENT (VECTOR OF LENGTH NSEC)
C                 E       -       FILTER COEFFICIENT (VECTOR OF LENGTH NSEC)
C                 F       -       FILTER COEFFICIENT (VECTOR OF LENGTH NSEC)
C                 G       -       FILTER COEFFICIENT (VECTOR OF LENGTH NSEC)
C
C**********************************************************************************
```

```
C
         SUBROUTINE BPDES(FCL,FCU,T,NSEC,A,B,C,D,E,F,G)
         DIMENSION A(1),B(1),C(1),D(1),E(1),F(1),G(1)
         PI=4.0*ATAN(1.0)
         Q1=FCL*PI*T
         Q2=FCU*PI*T
         W1=SIN(Q1)/COS(Q1)
         W2=SIN(Q2)/COS(Q2)
         WC=W2-W1
         Q=WC*WC+2.0*W1*W2
         S=W1*W1*W2*W2
C
             DO 1000 K=1,NSEC
             CS=COS(FLOAT(2*(K+NSEC)-1)*PI/FLOAT(4*NSEC))
             P=-2.0*WC*CS
             R=P*W1*W2
             X=1.0/(1.0+P+Q+R+S)
             A(K)=WC*WC*X
             B(K)=0.0
             C(K)=-2.0*A(K)
             D(K)=(-4.0-2.0*P+2.0*R+4.0*S)*X
             E(K)=(6.0-2.0*Q+6.0*S)*X
             F(K)=(-4.0+2.0*P-2.0*R+4.0*S)*X
             G(K)=(1.0-P+Q-R+S)*X
1000         CONTINUE
C
     RETURN
     END



C******************************************************************************
C
C       CALLING SEQUENCE
C
C             CALL BSDES(FCL,FCU,T,NSEC,A,B,C,D,E,F,G))
C
C       PURPOSE
C
C             THE SUBROUTINE WILL RETURN THE COEFFICIENTS TO A CASCADE
C             REALIZATION OF A MULTIPLE SECTION BANDSTOP FILTER.   THE
C             KTH SECTION HAS THE FOLLOWING TRANFER FUNCTION:
C
C                     A(K)*(1 + Z**-4)+B(K)*(Z**-1+Z**-3)+C(K)*Z**-2
C       H(Z)=------------------------------------------------------------
C                     1 + D(K)*Z**-1 + E(K)*Z**-2 + F(K)*Z**-3 + G(K)*Z**-4
C
C             THUS, IF F(M) AND G(M) ARE THE INPUT AND OUTPUT OF THE
C             KTH SECTION THE FOLLOWING DIFFERENCE EQUATION IS SATISFIED.
C
C             G(M)=A(K)*(F(M)+F(M-4))+B(K)*(F(M-1)+F(M-3))+C(K)*F(M-2)
C                 - D(K)*G(M-1) - E(K)*G(M-2) - F(K)*G(M-3) - G(K)*G(M-4)
C
C             NOTE:
C                     DO NOT USE THIS ROUTINE FOR LOWPASS DESIGN - USE LPDES
C                     DO NOT USE THIS ROUTINE FOR HIGHPASS DESIGN - USE HPDES
C                     DO NOT USE THIS ROUTINE FOR BANDPASS DESIGN - USE BPDES
C
C       ROUTINE(S) CALLED BY THIS ROUTINE
C
C             NONE
C
C       ARGUMENT(S) REQUIRED FROM THE CALLING ROUTINE
```

```
C
C                    FCL    -       LOWER 3-DB CUTOFF FREQUENCY IN HERTZ
C                    FCU    -       UPPER 3-DB CUTOFF FREQUENCY IN HERTZ
C                    T      -       SAMPLING INTERVAL IN SECONDS
C                    NSEC   -       NUMBER OF SECTIONS TO BE IMPLEMENTED
C
C        ARGUMENT(S) SUPPLIED  TO  THE CALLING ROUTINE
C
C                    A      -       FILTER COEFFICIENT (VECTOR OF LENGTH NSEC)
C                    B      -       FILTER COEFFICIENT (VECTOR OF LENGTH NSEC)
C                    C      -       FILTER COEFFICIENT (VECTOR OF LENGTH NSEC)
C
C                    D      -       FILTER COEFFICIENT (VECTOR OF LENGTH NSEC)
C                    E      -       FILTER COEFFICIENT (VECTOR OF LENGTH NSEC)
C                    F      -       FILTER COEFFICIENT (VECTOR OF LENGTH NSEC)
C                    G      -       FILTER COEFFICIENT (VECTOR OF LENGTH NSEC)
C
C***********************************************************************************
C
         SUBROUTINE BSDES(FCL,FCU,T,NSEC,A,B,C,D,E,F,G)
         DIMENSION A(1),B(1),C(1),D(1),E(1),F(1),G(1)
         PI=4.0*ATAN(1.0)
         Q1=FCL*PI*T
         Q2=FCU*PI*T
         W1=SIN(Q1)/COS(Q1)
         W2=SIN(Q2)/COS(Q2)
         WC=W2-W1
         Q=WC*WC+2.0*W1*W2
         S=W1*W1*W2*W2
         DP=(W1*W2+1.0)**2
         EP=4.0*(S-1.0)
         FP=6.0*S-4.0*W1*W2+6.0
C
             DO 1000 K=1,NSEC
             CS=COS(FLOAT(2*(K+NSEC)-1)*PI/FLOAT(4*NSEC))
             P=-2.0*WC*CS
             R=P*W1*W2
             X=1.0/(1.0+P+Q+R+S)
             A(K)=DP*X
             B(K)=EP*X
             C(K)=FP*X
             D(K)=(-4.0-2.0*P+2.0*R+4.0*S)*X
             E(K)=(6.0-2.0*Q+6.0*S)*X
             F(K)=(-4.0+2.0*P-2.0*R+4.0*S)*X
             G(K)=(1.0-P+Q-R+S)*X
1000         CONTINUE
C
         RETURN
         END
```

REFERENCES

1. E. I. Jury, *Inners and Stability of Dynamic Systems*, Wiley, New York, 1974.

2. D. N. Hein, "Algorithmic Properties of the Bilinear Transform," M.S. thesis, Kansas State University (Elect. Eng. Dept.), Manhattan, 1976.

3. M. E. Van Valkenburg, *Introduction to Modern Network Synthesis*, Wiley, New York, 1967, Chap. 13 (fifth printing).

4.   W. L. CASSELL, *Linear Circuit Analysis*, Wiley, New York, 1964, Chaps. 19, 20.

5.   L. WEINBERG, *Network Analysis and Synthesis*, McGraw-Hill, New York, 1962, Chapter 11.

6.   B. GOLD and C. RADER, *Digital Processing of Signals*, McGraw-Hill, New York, 1969.

7.   S. D. STEARNS, *Digital Signal Analysis*, Hayden, Rochelle Park, N.J., 1975.

8.   R. M. GOLDEN and J. F. KAISER, "Design of Wideband Sampled-Data Filters," *Bell System Tech. J*, Vol. 43, July 1964, pp. 1533–1546.

9.   K. STEIGLITZ, "Computer-Aided Design for Recursive Digital Filters," *IEEE Trans. Audio and Electroacoust.*, Vol. AU-18, 1970, pp. 123–129.

10. A. G. DECZKY, "Synthesis of Recursive Digital Filters Using p-Error Criterion," *IEEE Trans. Audio and Electroacoust.*, Vol. AU-20, 1972, pp. 257–263.

7

Finite Impulse Response

Discrete-Time

Filters

As in Chapter 6 we restrict our attention to DT filters, which is a special class of DT systems. In particular we shall consider finite impulse response (FIR) filters whose input–output difference equation is obtained with all $b_j = 0$ in (5.1-1); that is,

$$y(n) = \sum_{i=0}^{k} a_i x(n - i)$$

The main objective of this chapter is to introduce simple but effective methods for designing FIR filters; that is, procedures for obtaining the coefficients a_i so that the resulting transfer function

$$H(z) = a_0 + a_1 z^{-1} + a_2 z^{-2} + \cdots + a_k z^{-k}$$

approximates a desired magnitude response and has the *linear phase* or *constant time delay* property given in (5.5-27).

Some advantages and disadvantages of FIR filters compared to their IIR counterparts are as follows:

1. They can be realized without "feeding back" *past outputs,* as is the case with IIR filters due to the presence of the term $\sum_{j=1}^{m} b_j y(n - j)$ in (5.1-1); see Figs. 5.6-1 and 5.6-2. Hence FIR filter realizations are stable.

2. The design of linear phase filters is quite straightforward. We recall that

the methods used in the last chapter to design IIR filters did not assure a linear phase filter. It will be seen in this chapter that a desired magnitude response can be approximated, and at the same time linear phase can be *guaranteed*. In applications such as speech processing, linear phase filters are preferred since they avoid phase distortion by causing all the sinusoidal components in the filter input to be delayed by the *same* amount. This is due to the basic relation [see (5.5-27)]

$$T_d(\omega) = -\frac{d\bar{\psi}(\omega)}{d\omega} = \text{constant}$$

where $\bar{\psi}(\omega)$ and $T_d(\omega)$ are the phase and time-delay functions, respectively. The value of $T_d(\omega)$ at a specified ω is the time taken by a sinusoid of frequency ω to pass through the filter.

3. An appreciably higher order FIR filter is normally required to obtain the same "sharpness" in a specified magnitude response, as compared with an IIR filter. Consequently, more computations are required in the case of FIR filters, and long time delays may be involved.

The overall approximation problem of FIR filters is, in general, appreciably more difficult than that of IIR filters. As such, we shall emphasize the most straightforward design technique, which is referred to as the Fourier series (FS) method. Of the other available methods, the frequency sampling method will be discussed in Section 7.5.

7.1 FOURIER SERIES METHOD

This is a straightforward method which utilizes the fact that the steady-state transfer function (or frequency response) of a DT filter is a periodic function with period f_s, where f_s is the sampling frequency. In terms of the normalized frequency variable ν, this is equivalent to a periodicity of 2. It is convenient to proceed with the derivation in terms of the variable ν, and let $\bar{H}_d(\nu) = H_d(e^{j\pi\nu})$ denote the *desired* steady-state transfer function, which is assumed to be *real*.

Since $\bar{H}_d(\nu)$ is real, it can take negative and positive values. Two representatives cases of $\bar{H}_d(\nu)$ are depicted in Fig. 7.1-1. It is important to note that if $\bar{H}_d(\nu)$ is written as

$$\bar{H}_d(\nu) = |\bar{H}_d(\nu)|e^{j\bar{\psi}_d(\nu)} \tag{7.1-1a}$$

where $|\bar{H}_d(\nu)|$ and $\bar{\psi}_d(\nu)$ denote the magnitude and phase responses, respectively, then

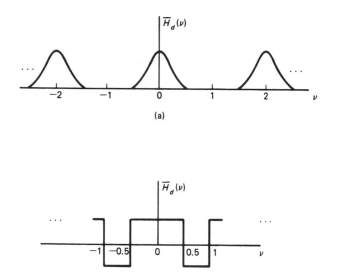

Fig. 7.1-1 Desired steady-state transfer functions.

$$\bar{\psi}_d(\nu) = \begin{cases} \pm 2k\pi, & \text{for positive } \bar{H}_d(\nu) \\ \pm(2k+1)\pi, & \text{for negative } \bar{H}_d(\nu) \end{cases} \qquad (7.1\text{-}1b)$$

for $k \geq 0$.

To illustrate, the magnitude and phase responses of $\bar{H}_d(\nu)$ in Fig. 7.1-1b are depicted in Fig. 7.1-2 for values of ν between ± 1. We observe that $\bar{\psi}_d(\nu)$ is zero in regions where $\bar{H}_d(\nu)$ in Fig. 7.1-1b is positive, and equals $\pm \pi$ in those regions where $\bar{H}_d(\nu)$ is negative. Thus, in summary the phase function $\bar{\psi}_d(\nu)$ of a *real* steady-state transfer function is piecewise constant, and consequently its time-delay function $T_d(\nu) = -d\bar{\psi}_d(\nu)/d\nu$ is zero.

Our objective is to design an FIR filter whose transfer function $H(z)$ is such that its magnitude response approximates the desired magnitude response $|\bar{H}_d(\nu)|$ and has the linear phase property. For convenience, let us assume that $\bar{H}_d(\nu)$ is an *even function* in the interval $|\nu| < 1$.

Since $\bar{H}_d(\nu)$ is a periodic function with period 2, it has the following FS representation, which is given by form III in Table 1.4-1:

$$\bar{H}_d(\nu) = \sum_{n=-\infty}^{\infty} c_n e^{jn\pi\nu}, \qquad |\nu| < 1 \qquad (7.1\text{-}2)$$

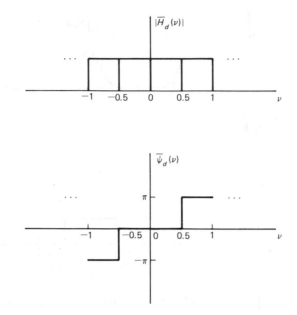

Fig. 7.1-2 Magnitude and phase responses of $\bar{H}_d(\nu)$ in Fig. 7.1-1b.

where the FS coefficient c_n is obtained using the formula

$$c_n = \frac{1}{2} \int_{-1}^{1} \bar{H}_d(\nu) e^{-jn\pi\nu} \, d\nu, \qquad |n| < \infty$$

Noting that $\bar{H}_d(\nu)$ is assumed to be an even function in the interval $|\nu| < 1$, it can be shown that (see Problem 7-1)

$$c_n = \int_{0}^{1} \bar{H}_d(\nu) \cos(n\pi\nu) \, d\nu, \qquad n \geq 0 \qquad (7.1\text{-}3)$$

and $c_{-n} = c_n$.

The coefficients of the filter we seek will be obtained in terms of the FS coefficients c_n in (7.1-3). Clearly, if we wish to realize (implement) such a filter, then it must have a finite number of coefficients. To this end, we use a finite number of c_n in (7.1-3), which is equivalent to truncating the infinite expansion in (7.1-2). This truncation leads to an approximation of $\bar{H}_d(\nu)$, which we denote by $\bar{H}_1(\nu)$; that is,

$$\bar{H}_1(\nu) \simeq \bar{H}_d(\nu) \qquad (7.1\text{-}4)$$

Thus

$$\bar{H}_1(\nu) = \sum_{n=-Q}^{Q} c_n e^{jn\pi\nu}, \qquad |\nu| < 1 \qquad (7.1\text{-}5a)$$

where Q is a finite positive integer, and

$$c_n = \int_0^1 \bar{H}_d(\nu) \cos(n\pi\nu) \, d\nu, \qquad 0 \le n \le Q \qquad (7.1\text{-}5b)$$

with $c_{-n} = c_n$.

In (7.1-5a) we note that the term $e^{jn\pi\nu}$ corresponds to points on the unit circle for each value of n. Thus, if $e^{j\pi\nu}$ is replaced by z, then (7.1-5a) yields

$$H_1(z) = \sum_{n=-Q}^{Q} c_n z^n \qquad (7.1\text{-}6)$$

From (7.1-6) it is clear that $H_1(z)$ is the transfer function of an FIR filter whose impulse response is given by the coefficients $c_{-Q}, c_{-Q+1}, \dots,$ c_{Q-1}, c_Q. However, this filter is *not* physically realizable, since the presence of positive powers of z means that the filter must produce an output that is advanced in time with respect to the input. This difficulty can be overcome by introducing a delay of Q samples. To this end, we define the transfer function

$$H(z) = z^{-Q} H_1(z) \qquad (7.1\text{-}7)$$

where the term z^{-Q} introduces a delay of Q samples at the output.

Substitution of (7.1-6) in (7.1-7) leads to

$$H(z) = z^{-Q} \sum_{n=-Q}^{Q} c_n z^n \qquad (7.1\text{-}8)$$

which yields

$$H(z) = c_{-Q} z^{-2Q} + c_{-Q+1} z^{-2Q+1}$$
$$+ \cdots + c_0 z^{-Q} + \cdots + c_{Q-1} z^{-1} + c_Q \qquad (7.1\text{-}9)$$

In (7.1-9), let $a_0 = c_Q$, $a_1 = c_{Q-1}$, $a_2 = c_{Q-2}$, and so on; that is,

$$a_i = c_{Q-i}, \qquad 0 \le i \le 2Q \qquad (7.1\text{-}10)$$

and $c_{-n} = c_n$

Combining (7.1-9) and (7.1-10), we obtain

$$H(z) = \sum_{i=0}^{2Q} a_i z^{-i} \qquad (7.1\text{-}11)$$

to be the transfer function of a DT filter that is physically realizable.

From (7.1-11) it is apparent that the FIR filter we have obtained has the following properties:

1. It has $(2Q + 1)$ impulse response coefficients, a_i, $0 \le i \le 2Q$.

2. The impulse response is symmetric about a_Q, as illustrated in Fig. 7.1-3 for the case $Q = 4$. This symmetry is due to the relation $a_i = c_{Q-i}$ in (7.1-10) and the fact that $c_{-n} = c_n$.

3. The duration of the impulse response is

    $$\tau = 2QT \qquad (7.1\text{-}12)$$

    where T is the sampling interval.

4. Its magnitude response and time-delay function can be found in the following way.

    First, from (7.1-7) it follows that

    $$\bar{H}(\nu) = e^{-jQ\pi\nu}\bar{H}_1(\nu) \qquad (7.1\text{-}13)$$

    which yields

    $$|\bar{H}(\nu)| = |\bar{H}_1(\nu)| \qquad (7.1\text{-}14)$$

since $|e^{-jQ\pi\nu}| = 1$.

Combining (7.1-4) and (7.1-14), there results

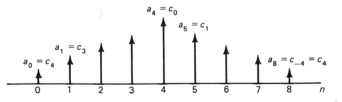

Fig. 7.1-3 Finite impulse response filter impulse response obtained via the Fourier series method, for $Q = 4$.

$$|\bar{H}(\nu)| \simeq |\bar{H}_d(\nu)|, \qquad |\nu| < 1 \qquad (7.1\text{-}15)$$

Equation (7.1-15) implies that the magnitude response of the filter we have designed approximates the desired magnitude response $|\bar{H}_d(\nu)|$.

Next, if $T_d(\nu)$ denotes the time-delay function of $\bar{H}(\nu)$ in (7.1-13), then

$$T_d(\nu) = \text{time delay of } e^{-jQ\pi\nu} + \text{time delay of } \bar{H}_1(\nu)$$

Clearly, the time delay of $e^{-jQ\pi\nu}$ is given by (5.5-27) to be $Q\pi$, while that of $\bar{H}_1(\nu)$ is zero, since $\bar{H}_1(\nu)$ is real. Hence

$$T_d(\nu) = Q\pi, \qquad |\nu| < 1 \qquad (7.1\text{-}16)$$

is a constant, in that it is independent of ν. Thus sinusoids of different frequencies are delayed by the same amount as they are processed by the filter we have designed. Consequently, this is a linear phase filter, which means that it does not introduce phase distortion.

This method is now illustrated by two examples.

7.2 EXAMPLES

Example 7.2-1: Design a low-pass FIR filter that approximates

$$\bar{H}_d(f) = \begin{cases} 1, & 0 \le f \le 1000 \text{ Hz} \\ 0, & \text{elsewhere in the range } 0 \le f \le f_N \end{cases} \qquad (7.2\text{-}1)$$

when the sampling frequency is 8000 sps. The impulse response duration is to be limited to 2.5 msec. Plot the resulting magnitude response.

Solution: Since the sampling frequency is 8000 sps, we have $f_N = 4000$Hz and $T = \dfrac{1}{8000}$ second. Again, $\tau = 2.5$ msec, and hence (7.1-12) yields

$$2.5 \times 10^{-3} = 2Q \left(\frac{1}{8000} \right)$$

which results in $Q = 10$.

The desired FIR filter thus has $21 (= 2Q + 1)$ coefficients a_i, $0 \le i \le 20$. These coefficients are obtained via the FS coefficients c_n in (7.1-5b). To obtain the FS coefficients, we use $\bar{H}_d(\nu)$ of the following sketch, which follows from (7.2-1) and the fact that $f_N = 4000$ Hz.

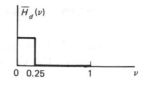

Thus (7.1-5b) yields

$$c_n = \int_0^{0.25} \cos{(n\pi\nu)}\, d\nu$$

$$= \left[\frac{\sin{(n\pi\nu)}}{\pi n} \right]_0^{0.25}$$

That is,

$$c_n = \frac{\sin{(0.25n\pi)}}{n\pi}, \qquad 0 \le n \le 10 \tag{7.2-2}$$

since $Q = 10$. Evaluation of (7.2-2) leads to the following FS coefficients:

$$c_0 = 0.25000, \quad c_1 = 0.22508, \quad c_2 = 0.15915$$
$$c_3 = 0.07503, \quad c_4 = 0, \quad c_5 = -0.04502 \tag{7.2-3}$$
$$c_6 = -0.05305, \quad c_7 = -0.03215, \quad c_8 = 0$$
$$c_9 = 0.02501, \quad c_{10} = 0.03183$$

Thus the transfer function $H(z)$ is given by (7.1-11) to be

$$H(z) = \sum_{i=0}^{20} a_i z^{-i} \tag{7.2-4}$$

where $a_i = c_{10-i}$, $0 \le i \le 20$, and $c_{-n} = c_n$.

The decibel magnitude response of $H(z)$ in (7.2-4) is plotted in Fig. 7.2-1, where

$$\text{dB} = 20 \log \left\{ \frac{|\bar{H}(f)|}{|\bar{H}(0)|} \right\} \tag{7.2-5}$$

From Fig. 7.2-1 it is apparent that the $H(z)$ we have obtained has a low-

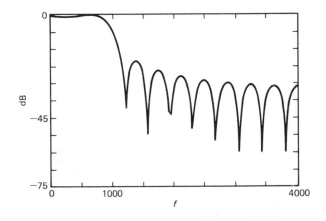

Fig. 7.2-1 Magnitude response of $H(z)$ in (7.2-4).

pass magnitude response with a cutoff of 1000 Hz, which approximates the desired magnitude response $\tilde{H}_d(f)$ in (7.2-1).

Example 7.2-2: Design a band-pass filter that approximates

$$\tilde{H}_d(f) = \begin{cases} 1, & 160 < f < 200 \text{ Hz} \\ 0, & \text{elsewhere in the range } 0 \le f \le f_N \end{cases} \quad (7.2-6)$$

Let the sampling frequency be 800 sps. Limit the duration of the impulse response to 50 msec, and plot the corresponding magnitude response.

Solution: With $\tau = 50$ msec and $T = \dfrac{1}{800}$, (7.1-12) yields

$$50 \times 10^{-3} = 2Q\left(\frac{1}{800}\right)$$

which results in $Q = 20$. The number of coefficients in the FIR filter we seek is thus 41 ($= 2Q + 1$).

Next the information in (7.2-6) is used to obtain the following sketch of $\tilde{H}_d(\nu)$, where $\nu = f/f_N$ and $f_N = 400$ Hz.

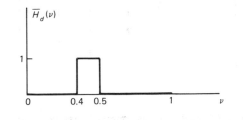

The FS coefficients are obtained via (7.1-5b) to be

$$c_n = \int_{0.4}^{0.5} \cos{(n\pi v)} \, dv$$

$$= \left[\frac{\sin{(n\pi v)}}{n\pi} \right]_{0.4}^{0.5}$$

or $c_n = \dfrac{\sin{(0.5n\pi)}}{n\pi} - \dfrac{\sin{(0.4n\pi)}}{n\pi},$ $0 \le n \le 20$ (7.2-7)

since $Q = 20$.

Hence (7.1-11) yields the transfer function

$$H(z) = \sum_{i=0}^{40} a_i z^{-i} \tag{7.2-8}$$

where $a_i = c_{20-i},$ $0 \le i \le 40$

with the c_n given by (7.2-7), and the relation $c_{-n} = c_n$.

The impulse response values so obtained are displayed in Fig. 7.2-2, while Fig. 7.2-3 shows the corresponding magnitude response, which implies that the preceding filter approximates the specified $\bar{H}_d(v)$ in (7.2-6).

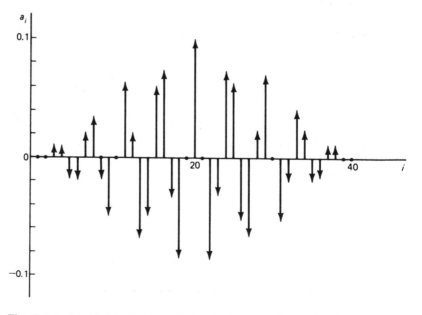

Fig. 7.2-2 Impulse response of filter designed in Example 7.2-2.

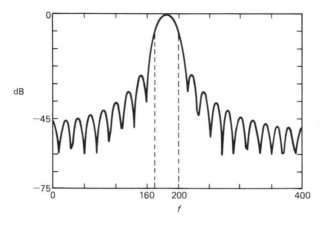

dB

−45

−75
0   160 200   400
f

Fig 7.2-3 Magnitude response of $H(z)$ in (7.2-8).

7.3 WINDOWING

In Chapters 1 and 4 we discussed the notion of windowing in the context of estimating the FT and PDS, respectively. We now show that windowing is also very useful for designing linear phase FIR filters.

The FIR filters that are obtained via the preceding FS method have a basic shortcoming, which results from the *abrupt truncation* of the FS expansion in (7.1-2); that is,

$$\bar{H}_d(\nu) = \sum_{n=-\infty}^{\infty} c_n e^{jn\pi\nu}, \qquad |\nu| < 1$$

to obtain

$$\bar{H}_1(\nu) = \sum_{n=-Q}^{Q} c_n e^{jn\pi\nu}, \qquad |\nu| < 1$$

in (7.1-5a), where Q is a finite positive integer. This type of truncation may result in poor convergence of the series in (7.1-5a), particularly in the vicinity of discontinuities. This aspect is illustrated in Fig. 7.3-1 for an assumed form of $\bar{H}_d(\nu)$.

Figure 7.3-1a illustrates the case when no truncation is used, while the effect of abrupt truncation is depicted in Fig. 7.3-1b, in which the presence of oscillations in $\bar{H}_1(\nu)$ due to poor convergence is apparent. The natural question that arises is: what causes these oscillations in $\bar{H}_1(\nu)$? The answer lies in realizing that an abrupt truncation of the infinite series in (7.1-2) is equivalent to *multiplying* it with the rectangular sequence

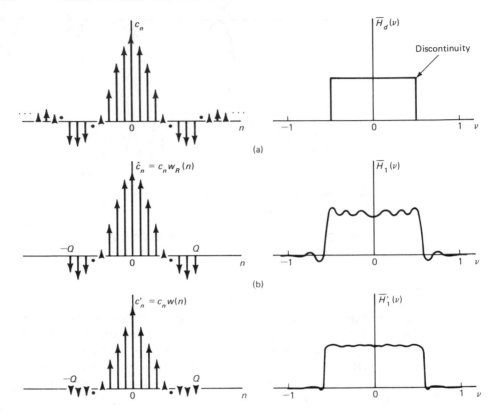

Fig. 7.3-1 Truncation using windowing.

$$w_R(n) = \begin{cases} 1, & |n| \leq Q \\ 0, & \text{elsewhere} \end{cases}$$

That is,

$$\hat{c}_n = c_n w_R(n)$$

as indicated in Fig. 7.3-1b. Now, since $\bar{H}_1(\nu)$, $\bar{H}_d(\nu)$, and $\bar{W}_R(\nu)$ are the FTs of the sequences $\hat{c}_n$, c_n, and $w_R(n)$, respectively, it follows that $\bar{H}_1(\nu)$ is the *convolution* of $\bar{H}_d(\nu)$ and $\bar{W}_R(\nu)$. Again, it can be shown that (Problem 7-15)

$$\bar{W}_R(\nu) = \sum_{n=-Q}^{Q} e^{jn\pi\nu} \tag{7.3-1a}$$

which can be expressed in the convenient form

$$\bar{W}_R(\nu) = \frac{\sin (Q + \tfrac{1}{2})\pi\nu}{\sin (\nu\pi/2)}, \qquad |\nu| < 1 \qquad (7.3\text{-}1b)$$

To illustrate, $\bar{W}_R(\nu)$ is plotted in Fig. 7.3-2 for $Q = 5$, where we observe that there is a main lobe and appreciably large side lobes. It is these large side lobes that cause the oscillations present in $\bar{H}_1(\nu)$ in Fig. 7.3-1b, when $\bar{W}_R(\nu)$ is convolved with $\bar{H}_d(\nu)$, as depicted in Fig. 7.3-3. The presence of such oscillations implies poor convergence properties and is known as the *Gibbs phenomenon*.

From the preceding discussion it follows that for a specified Q the size of the undesired oscillations in $\bar{H}_1(\nu)$ may be reduced by using window sequences whose FTs have large main lobes, but very small side lobes. To this end, we use the other window sequences we have defined in Chapters 1 and 4 (i.e., triangular, hanning, Hamming, and Kaiser). These sequences enable a more gradual truncation of the infinite expansion in (7.1-2), in contrast to the abrupt truncation realized via the rectangular window sequence.

Table 7.3-1 lists the triangular, hanning, Hamming, and Kaiser window sequences. We observe that each sequence consists of $2Q + 1$ nonzero samples that are symmetric about $n = 0$, and hence their FTs $\bar{W}(\nu)$ are real. These sequences are the same as the window sequences defined in Table 4.9-1, except that the latter are defined over the interval $0 \le m \le M - 1$. As an example, let us consider the Hamming window sequence, which is defined as

$$w_H(n) = \begin{cases} 0.54 + 0.46 \cos(n\pi/Q), & |n| \le Q \\ 0, & \text{elsewhere} \end{cases} \qquad (7.3\text{-}2)$$

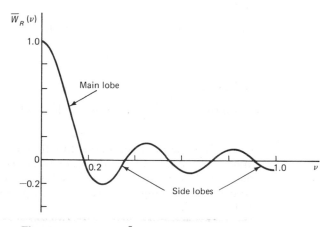

Fig. 7.3-2. Plot of $\bar{W}_R(\nu)$ for $Q = 5$ in (7.3 − 1b).

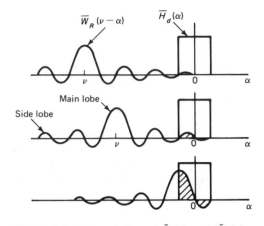

Fig. 7.3-3. Convolution of $\bar{H}_d(\nu)$ and $\bar{W}_R(\nu)$.

Table 7.3-1   Some Window Sequences for the Fourier Series Method[†]

1. Triangular

$$w_{tr}(n) = \begin{cases} 1 - \dfrac{|n|}{Q}, & |n| \le Q \\[2ex] 0, & \text{elsewhere} \end{cases}$$

2. Hanning

$$w_{ha}(n) = \begin{cases} 0.5\left(1 + \cos\dfrac{n\pi}{Q}\right), & |n| \le Q \\[2ex] 0, & \text{elsewhere} \end{cases}$$

3. Hamming

$$w_H(n) = \begin{cases} 0.54 + 0.46 \cos\left(\dfrac{n\pi}{Q}\right), & |n| \le Q \\[2ex] 0, & \text{elsewhere} \end{cases}$$

4. Kaiser

$$w_K(n) = \frac{I_0[\theta\sqrt{1 - (n/Q)^2}]}{I_0(\theta)}, \qquad |n| \le Q$$

where I_0 is the modified Bessel function of the first kind and zero order; the tradeoff between the main lobe width and side lobe level can be adjusted by varying the parameter θ; e.g., see p. 101 in [2].

[†]The material in this table is derived from [4].

We note that $w_H(n)$ is a "bell-shaped" sequence that is symmetric about $n = 0$, as depicted in Fig. 7.3-4 for $Q = 3$ and $Q = 6$. Thus if we multiply the FS coefficients c_n in (7.1-2) by $w_H(n)$, a gradual truncation of the related expansion will result. Also, from the decibel magnitude response of $w_H(n)$ displayed in Fig. 7.3-5a for $Q = 10$, it is apparent that its side lobes are less than -40 dB. In contrast, we see that the corresponding side lobes related to the rectangular window (Fig. 7.3-5b) are significantly larger.

From the preceding discussion, it follows that the magnitude response of the DT filter we design using the FS method can be improved by resorting to windowing. To do so, we simply multiply the FS coefficients c_n in (7.1-3) by one of the window sequences $w(n)$ in Table 7.3-1 to obtain

$$c'_n = c_n w(n), \qquad |n| \le Q \tag{7.3-3}$$

where the c'_n are called *modified FS coefficients*. As illustrated in Fig. 7.3-1c, the multiplication in (7.3-3) causes a gradual truncation of the c_n in Fig. 7.3-1a. The corresponding steady-state transfer function $\bar{H}'_1(\nu)$ is illustrated in Fig. 7.3-1c, and its time-delay function is zero since $\bar{H}'_1(\nu)$ is real. Hence, corresponding to (7.1-6) and (7.1-11), we have

$$H'_1(z) = \sum_{n=-Q}^{Q} c'_n z^n \tag{7.3-4}$$

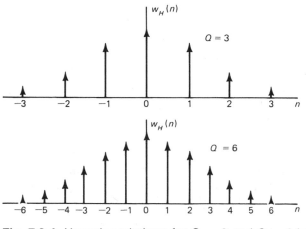

Fig. 7.3-4 Hamming windows for $Q = 3$ and $Q = 6$ in (7.3-2).

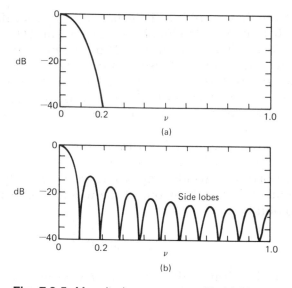

Fig. 7.3-5 Magnitude responses with (a) Hamming and (b) rectangular windows.

and

$$H'(z) = \sum_{i=0}^{2Q} a_i' z^{-i} \qquad (7.3\text{-}5a)$$

respectively. Here

$$a_i' = c_{Q-i}', \qquad 0 \le i \le 2Q \qquad (7.3\text{-}5b)$$

and $c'_{-n} = c'_n$, since $c_n = c_{-n}$ and $w(n)$ is an even function about $n = 0$.

To recapitulate, $H'(z)$ is the transfer function of the filter we obtain via the FS method with windowing. The magnitude response of this filter is an approximation to $|H_d(\nu)|$, which is desired (or specified), and it has the linear phase or constant time-delay property.

The preceding windowing process is now illustrated by two examples.

Example 7.3-1: Apply the windowing procedure to improve the lowpass filter magnitude response obtained in Example 7.2-1, using the Hamming window sequence. Plot the resulting magnitude response and compare it with that in Fig. 7.2-1.

Solution: First we obtain the Hamming window coefficients $w_H(n)$ in (7.3-2) for $1 \le n \le 10$, since $Q = 10$. This results in

$$w_H(0) = 1, \qquad w_H(1) = 0.97749, \qquad w_H(2) = 0.91215$$

$$w_H(3) = 0.81038, \qquad w_H(4) = 0.68215, \qquad w_H(5) = 0.54000 \qquad (7.3\text{-}6)$$

$$w_H(6) = 0.39785, \qquad w_H(7) = 0.26962, \qquad w_H(8) = 0.16785$$

$$w_H(9) = 0.10251, \qquad w_H(10) = 0.08000$$

Next the window sequence values in (7.3-6) are multiplied with the coefficients c_n in (7.2-3) to obtain the modified FS coefficients c'_n in (7.3-3). This yields

$$c'_0 = 0.25000, \qquad c'_1 = 0.22000, \qquad c'_2 = 0.14517, \qquad c'_3 = 0.06080$$

$$c'_4 = 0, \qquad c'_5 = 0.02431, \qquad c'_6 = -0.02111, \qquad c'_7 = -0.00867$$

$$c'_8 = 0, \qquad c'_9 = 0.00256, \qquad c'_{10} = 0.00255 \qquad (7.3\text{-}7)$$

Hence $H'(z)$ in (7.3-5) is given by

$$H'(z) = \sum_{i=0}^{20} a'_i z^{-i} \qquad (7.3\text{-}8)$$

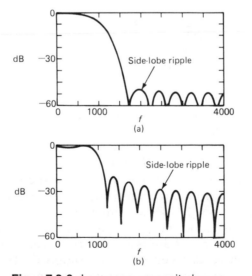

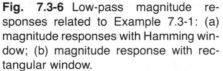

Fig. 7.3-6 Low-pass magnitude responses related to Example 7.3-1: (a) magnitude responses with Hamming window; (b) magnitude response with rectangular window.

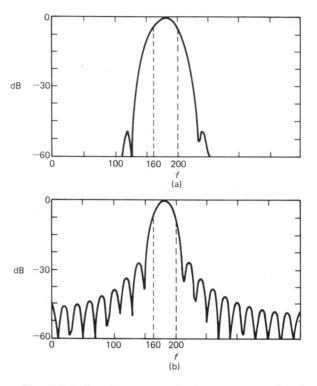

Fig. 7.3-7 Band-pass magnitude responses related to Example 7.3-2: (a) magnitude response with Hamming window; (b) magnitude response with rectangular window.

where $a'_i = c'_{Q-i},$ $0 \le i \le 20$

are obtained using the c'_n in (7.3-7) and the fact that $c'_{-n} = c'_n$.

The decibel magnitude response $|\bar{H}'(f)|_{dB}$ corresponding to $H'(z)$ in (7.3-8) is plotted in Fig. 7.3-6a. For the purposes of comparison, the decibel magnitude response obtained in Example 7.2-1 is plotted again in Fig. 7.3-6b. It is apparent that use of the Hamming window has resulted in a much better magnitude response, in that the sidelobe ripple indicated in Fig. 7.3-6 is substantially reduced.

The reader is encouraged to apply the Hamming window to design the band-pass filter considered in Example 7.2-2. The final results so obtained are summarized in Fig. 7.3-7, from which it is again apparent that the Hamming window provides a much better band-pass characteristic than that obtained in Example 7.2-2, which corresponds to a rectangular window.

7.4 IMPLEMENTATION CONSIDERATIONS

From the discussion in the previous section we see that a typical transfer function of the FIR filter we have designed is given by

$$H'(z) = \sum_{i=0}^{2Q} a_i' z^{-i} \tag{7.4-1}$$

where a_i' is the ith of $(2Q + 1)$ impulse response coefficients. This section will be devoted to discussing some aspects of implementing such filters. We start with the fundamental relation

$$Y(z) = H(z)X(z) \tag{7.4-2}$$

where $X(z)$ and $Y(z)$ are the ZTs of the filter input and output sequences $x(n)$ and $y(n)$, respectively. Substitution of (7.4-1) into (7.4-2) leads to

$$Y(z) = \sum_{i=0}^{2Q} a_i' X(z) z^{-i}$$

which yields the input–output relation

$$y(m) = \sum_{i=0}^{2Q} a_i' x(m - i), \qquad m \geq 0 \tag{7.4-3}$$

Equation (7.4-3) states that the output of an FIR filter at time m is simply the weighted sum of the current input sample $x(m)$ and the $2Q$ past input samples $x(m - i)$, $1 \leq m \leq 2Q$; see Fig. 7.4-1.

Letting $P = 2Q + 1$ and $h(i) = a_i'$, we note that (7.4-3) is merely the *linear convolution* relation we studied in Chapter 5; see (5.8-6). Thus

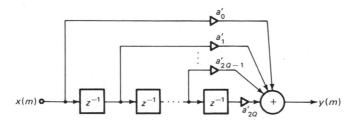

Fig. 7.4-1 Direct form realization of finite impulse response filters.

the entire discussion in Section 5.8 related to computing linear convolution is relevant to implementing FIR filters.

It is apparent from (7.4-3) that the direct approach for computing the output $y(m)$ is to carry out the products and sums involved in (7.4-3). However, this approach would require $(2Q + 1)$ multiplications and $2Q$ additions for *each* output sample $y(m)$ that is processed. Clearly, this becomes an inefficient approach as the filter length ($= 2Q + 1$) becomes large. In such cases, *fast convolution* techniques (which employ the FFT) are used to implement FIR filters (e.g., the fast convolution algorithm in Fig. 5.8-5, and the overlap-add method considered in the latter part of Section 5.8).

7.5 BRIEF DISCUSSION OF OTHER DESIGN METHODS

We conclude this chapter by briefly discussing some of the other methods that are available for designing FIR filters.

Frequency Sampling Method

Let $h(m)$, $0 \leq m \leq N - 1$, be the impulse response of the FIR filter we seek, and let $H(n)$, $0 \leq n \leq N - 1$, be the corresponding DFT coefficients. We note that the $H(n)$ can be expressed in the form

$$H(n) = H_n e^{j\psi_n}, \qquad 0 \leq n \leq N - 1 \tag{7.5-1}$$

where the H_n are *real numbers*, and the ψ_n are phase angles; for example, $1 + j1 = \sqrt{2}\, e^{j45°} = -\sqrt{2}\, e^{j225°}$. Our objective is to obtain the $h(m)$ by appropriately specifying H_n and ψ_n.

Since $H(n)$, $0 \leq n \leq N - 1$, are the DFT coefficients of $h(m)$, $0 \leq m \leq N - 1$, the definition of the IDFT in (4.5-4) yields

$$h(m) = \frac{1}{N} \sum_{n=0}^{N-1} H(n) W_N^{-nm}, \qquad 0 \leq m \leq N - 1 \tag{7.5-2}$$

where $W_N = e^{-j2\pi/N}$

Again, the transfer function $H(z)$ of an FIR filter with impulse response $h(m)$, $0 \leq m \leq N - 1$, is given by

$$H(z) = \sum_{m=0}^{N-1} h(m) z^{-m} \tag{7.5-3}$$

Substitution of (7.5-2) in (7.5-3) leads to

$$H(z) = \frac{1}{N} \sum_{n=0}^{N-1} H(n) \left[\sum_{m=0}^{N-1} (W_N^{-n} z^{-1})^m \right] \tag{7.5-4}$$

Since the inner summation in (7.5-4) is a geometric series with common ratio $W_N^{-n} z^{-1}$, it follows that (7.5-4) can be written in the form

$$H(z) = \frac{1}{N} \sum_{n=0}^{N-1} H(n) \frac{1 - z^{-N}}{1 - W_N^{-n} z^{-1}} \tag{7.5-5a}$$

which is equivalent to

$$H(z) = \frac{1}{N} \sum_{n=0}^{N-1} H(n) \frac{1 - z^{-N}}{1 - e^{j2\pi n/N} z^{-1}} \tag{7.5-5b}$$

since $W_N = e^{-j2\pi/N}$

Again, the substitution $z = e^{j\omega T}$ in (7.5-5) yields the steady-state transfer function of the FIR filter we seek; that is,

$$\bar{H}(\omega) = \frac{1}{N} \sum_{n=0}^{N-1} H(n) \frac{1 - e^{-jN\omega T}}{1 - e^{j2\pi n/N} e^{-j\omega T}}$$

$$= \frac{1}{N} \sum_{n=0}^{N-1} H(n) \frac{1 - e^{-jN(\omega T - 2\pi n/N)}}{1 - e^{-j(\omega T - 2\pi n/N)}}$$

That is,

$$\bar{H}(\omega) = \frac{1}{N} \sum_{n=0}^{N-1} H(n) \frac{e^{-jN(\omega T - 2\pi n/N)/2}}{e^{-j(\omega T - 2\pi n/N)/2}} \cdot \frac{\sin [N(\omega T - 2\pi n/N)/2]}{\sin [(\omega T - 2\pi n/N)/2]}$$

which results in

$$\bar{H}(\omega) = e^{-j(N-1)\omega T/2} \sum_{n=0}^{N-1} H(n) e^{j(N-1)n\pi/N} \frac{\sin [N(\omega T - 2\pi n/N)/2]}{N\sin [(\omega T - 2\pi n/N)/2]} \tag{7.5-6}$$

We leave it to the reader to show that $\bar{H}(\omega)$ in (7.5-6) is such that

$$\bar{H}(\omega_n) = H(n), \qquad 0 \leq n \leq N - 1 \tag{7.5-7}$$

where $\omega_n = 2\pi n/(NT)$ is the nth DFT frequency component, which means that (7.5-6) yields the desired frequency response at the sampling frequencies ω_n. As such, this approach for designing FIR filters is called the *frequency sampling method*. At frequencies other than ω_n, the DT filter so obtained provides an interpolation between adjacent frequencies. This interpolation is due to the sin (Nx)/sin (x) functions in (7.5-6), with $x = (\omega T - 2\pi n/N)/2$.

OTHER ISSUES. We must now choose $H(n)$ so that (1) the filter has the linear phase (constant time delay) property in (5.5-27), and (2) the impulse response sequence $h(m)$ is real valued.

In (7.5-6) we note that $H(n) = H_n e^{j\psi_n}$, which means that if the phase angles are chosen as

$$\psi_n = -\frac{(N-1)n\pi}{N}, \qquad 0 \leq n \leq N - 1 \qquad (7.5\text{-}8)$$

the quantity in the summation is real. Then the resulting filter will have the linear phase property, and its time-delay function will be equal to $(N - 1)/2$ for $0 \leq \omega T \leq \pi$.

Next, to ensure that the $h(m)$ in (7.5-2) are real, we note that W_N has the property

$$\widetilde{W}_N^{-n} = W_N^{-(N-n)} \qquad (7.5\text{-}9)$$

for $1 \leq n \leq N/2 - 1$ when N is even, and for $1 \leq n \leq (N - 1)/2$ when N is odd, where the symbol $\sim$ denotes complex-conjugate.

In view of (7.5-9), let us choose the $H(n)$ so that

$$\widetilde{H}(n) = H(N - n) \qquad (7.5\text{-}10)$$

for $1 \leq n \leq (N/2) - 1$ when N is even, and for $1 \leq n \leq (N - 1)/2$ when N is odd. Then, substitution of (7.5-9) and (7.5-10) in (7.5-2) results in

$$h(m) = \frac{1}{N}\left\{ H(0) + \sum_{k=1}^{(N/2)-1} 2 \operatorname{Re}\left[H(k)W^{-km}\right] - H\left(\frac{N}{2}\right) \right\} \qquad (7.5\text{-}11\text{a})$$

when N is even, and

$$h(m) = \frac{1}{N}\left\{ H(0) + \sum_{k=1}^{(N-1)/2} 2 \operatorname{Re}\left[H(k)W^{-km}\right] \right\} \qquad (7.5\text{-}11\text{b})$$

when N is odd.

From (7.5-11a) it is clear that if we choose

$$H\left(\frac{N}{2}\right) = 0, \quad \text{for } N \text{ even} \tag{7.5-12}$$

then

$$h(m) = \frac{1}{N}\left\{H(0) + \sum_{k=1}^{(N/2)-1} 2 \operatorname{Re}\left[H(k)W^{-km}\right]\right\} \tag{7.5-13}$$

when N is even.

Furthermore, since H_0 is a real number and $\psi_0 = 0$ [see (7.5-8)], $H(0)$ is always real. Thus from (7.5-11b) and (7.5-13) we conclude that the impulse response values of the filter we seek are real, if the DFT coefficients $H(n)$ satisfy the condition in (7.5-10) and the additional condition in (7.5-12) when N is even.

Finally, we combine the requirements in (7.5-8) and (7.5-10). To this end, we substitute (7.5-8) in (7.5-10) to obtain

$$H_n e^{j(N-1)n\pi/N} = H_{N-n}e^{-j(N-1)(N-n)\pi/N}, \quad 0 \le n \le N - 1$$

This equation simplifies to yield

$$H_n = H_{N-n}e^{-j(N-1)\pi}, \quad 0 \le n \le N - 1$$

That is,

$$H_n = (-1)^{N-1}H_{N-n}$$

or

$$H_n = -H_{N-n}, \quad \text{for even } N$$

and

$$H_n = H_{N-n}, \quad \text{for odd } N$$

Thus, in summary, $H(n) = H_n e^{j\psi_n}, 0 \le n \le N - 1$ must be chosen to satisfy the following constraints:

$$H_n = -H_{N-n} \quad \text{for } N \text{ even} \tag{7.5-14a}$$

$$H\left(\frac{N}{2}\right) = 0, \quad \text{for } N \text{ even} \tag{7.5-14b}$$

$$H_n = H_{N-n} \text{ for } N \text{ odd}, \tag{7.5-14c}$$

and

$$\psi_n = -(N - 1)n\pi/N, \tag{7.5-14d}$$

where H_0 is any finite real number.

Example 7.5-1: Use the frequency sampling method to design an 11-coefficient FIR low-pass filter whose cutoff frequency f_c equals $2f_s/11$, where f_s is the sampling frequency. Plot the magnitude response of the resulting filter.

Solution: Since $N = 11$, the fundamental DFT frequency is given by $f_0 = 1/11T = f_s/11$, where T is the sampling interval. Hence $n = 2$ corresponds to the cutoff frequency. According to (7.5-14c) and (7.5-14d), we may choose

$$H_n = \begin{cases} 1, & \text{for } n = 1, 2 \text{ and } n = 9, 10 \\ 0, & \text{for } 3 \le n \le 8 \end{cases} \qquad (7.5\text{-}15a)$$

and $\psi_n = -10n\pi/11, \qquad 0 \le n \le 10$

Also, let $H_0 = 1$. Then substitution of this information in (7.5-1) results in

$$H(n) = \begin{cases} e^{-j10n\pi/11}, & \text{for } n = 0, 1, 2 \text{ and } n = 9, 10 \\ 0, & 3 \le n \le 8 \end{cases} \qquad (7.5\text{-}15b)$$

Next, substituting (7.5-15b) in (7.5-11b), we obtain the following impulse response:

$$h(0) = 0.06942, \qquad h(1) = -0.05403, \qquad h(2) = -0.10945,$$
$$h(3) - 0.04733, \qquad h(4) = 0.31938, \qquad h(5) - 0.45455,$$
$$h(6) = 0.31938, \qquad h(7) = 0.04733, \qquad h(8) = -0.10945,$$
$$h(9) = -0.05403, \qquad h(10) = 0.06942$$

The corresponding magnitude response is displayed in Fig. 7.5-1 (thick line), using the computer program in Appendix 5.1. As was the case with the FS method, we observe that there is a large ripple in the vicinity of the discontinuity (jump) that occurs at the normalized cutoff frequency ν_c. The value of ν_c is $\dfrac{4}{11}$, since $\nu_0 = \dfrac{2}{11}$, and the cutoff frequency corresponds to the second harmonic, that is, n_0 with $n = 2$. We also note that the magnitude response is 1 at $n = 1, 2$ and zero at $n = 3, 4$, and 5, as specified by the H_n in (7.5-15a). The ripple effect indicated in Fig. 7.5-1 can be reduced by allowing a *smoother* transition from the passband to the stopband. To this end, let us choose

$$H_n = \begin{cases} 1, & \text{for } n = 1, 2 \text{ and } n = 9, 10 \\ 0.5, & \text{for } n = 3 \text{ and } n = 8 \\ 0, & \text{for } 4 \le n \le 7 \end{cases}$$

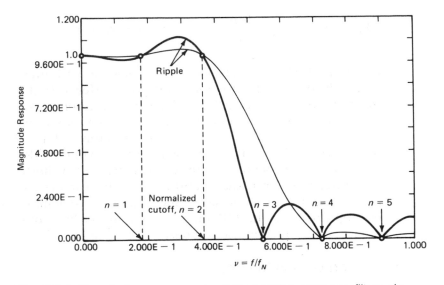

Fig. 7.5-1 Magnitude responses for finite impulse response filters designed in Example 7.5-1.

Also let $H_0 = 1$. Then, corresponding to (7.5-15b), we obtain

$$H(n) = \begin{cases} e^{-j10n\pi/11}, & \text{for } n = 0, 1, 2 \text{ and } n = 9, 10 \\ 0.5e^{-jn\pi/11}, & \text{for } n = 3 \text{ and } n = 8 \\ 0, & \text{for } 4 \le n \le 7 \end{cases}$$

The resulting filter then has the following impulse response:

$$h(0) = 0.00987, \quad h(1) = 0.02244, \quad h(2) = -0.07168$$
$$h(3) = -0.03989, \quad h(4) = 0.30643, \quad h(5) = 0.54545$$
$$h(6) = 0.30643, \quad h(7) = -0.03989, \quad h(8) = -0.07168$$
$$h(9) = 0.02244, \quad h(10) = 0.00987$$

The corresponding magnitude response is also displayed in Fig. 7.5-1 and indicated by the thin line. This response shows a significant reduction in the ripple magnitude, but only at the expense of a wider transition band (i.e., a larger bandwidth).

In conclusion, we note that $H(z)$ in (7.5-5a) can be written as

$$H(z) = \frac{1 - z^{-N}}{N} \sum_{n=0}^{N-1} G_n(z) \qquad (7.5\text{-}16)$$

where $\quad G_n(z) = \dfrac{H(n)}{1 - W_N^{-n}z^{-1}}, \qquad 0 \leq n \leq N - 1$

is the transfer function of a first-order resonator (IIR filter), whose poles lie on the unit circle at equidistant points. Hence $H(z)$ can be implemented as depicted in Fig. 7.5-2. Furthermore, resonators with complex-conjugate poles can be combined into second-order sections whose coefficients are real (see Problem 7-18). The structure in Fig. 7.5-2 is intriguing in that the frequency response information of the FIR filter appears directly in the resonators via the coefficients $H(n)$. A shortcoming of this structure, however, is that stability problems can arise because of the poles present in each resonator. Such stability problems can be avoided by implementing the FIR filter we obtain via fast convolution techniques, using the impulse response $h(m)$ given by (7.5-11b) or (7.5-13).

Computer-Aided Design Methods

Filters obtained via the FS and frequency sampling methods usually have very "good" frequency responses, but are not optimum with respect to commonly used criteria. One such criterion is known as minimization of the Chebyshev norm. This criterion is concerned with minimizing the peak (maximum) error over the Nyquist band. To this end, iterative numerical techniques are used to seek optimum solutions. Such techniques are collectively referred to as *computer-aided design methods*.
    A number of important contributions have been made in the area

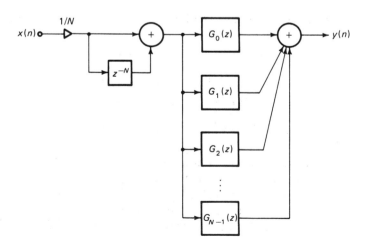

Fig. 7.5-2 Recursive form implementation for finite impulse response filter with transfer function $H(z)$ in (7.5-5a).

of computer-aided design methods for FIR filters (e.g., see [8 to 12]). Of these, the method due to McClellan and Parks [11, 12] is the most effective for designing FIR linear phase filters. A FORTRAN program that implements this computer-aided design technique is available; see [13]. Excellent discussions of computer-aided design methods are also available in advanced texts such as references [1] and [2].

7.6 SUMMARY

This chapter has been devoted to the development of the FS method, which is a simple but effective method for designing FIR filters. The filters so obtained have magnitude responses that approximate desired (or specified) magnitude responses and are guaranteed to have the linear phase (or constant time delay) property. It was also shown that a substantial improvement in the magnitude response can be realized by resorting to windowing techniques. A computer program for designing FIR low-pass, high-pass, band-pass, and band-stop filters via the FS method is listed in Appendix 7.1.

While developing the FS method, we assumed that $\bar{H}_d(\nu)$ is a real function in the interval $|\nu| < 1$. The same procedure can be used for a given $\bar{H}_d(\nu)$, which is a complex function in the interval $|\nu| < 1$. For example, see Problem 7-11. An introductory discussion of the frequency sampling method for designing FIR filters was also included.

Linear phase FIR filters are used in applications where phase distortion is to be avoided (e.g., filtering speech data). Phase distortion is avoided in such filters since they possess the linear phase or constant time-delay property.

PROBLEMS

7–1   Starting with the formula for c_n given in (7.1-2), derive (7.1-3).

7–2   An FIR filter has the transfer function

$$G(z) = \tfrac{1}{4} [0.5 + 1.3z^{-1} + 0.4z^{-2} + 1.3z^{-3} + 0.5z^{-4}]$$

Plot the decibel magnitude response and also the phase response.

7–3   Consider an FIR filter whose transfer function is given by

$$H(z) = \sum_{m=0}^{2M} h(m) \, z^{-m}$$

where the impulse response coefficients $h(m)$, $0 \leq m \leq 2M$, are real, and such that $h(m) = h(2M - m)$, $0 \leq m \leq M - 1$. Show that

$$|\bar{H}(\nu)| = |2 \sum_{m=0}^{M-1} h(m) \cos [(M - m)\pi\nu] + h(M)|,$$

and $T_d(\nu) = M\pi, \quad$ for $0 \leq \nu \leq 1$

7–4   In Problem 7-3, suppose the impulse response coefficients are such that $h(m) = -h(2M - m)$, $0 \leq m \leq M - 1$, and $h(M) = 0$. Show that

$$|\bar{H}(\nu)| = |2 \sum_{m=0}^{M-1} h(m) \sin [(M - m)\pi\nu]|$$

and $T_d(\nu) = M\pi, \quad$ for $0 \leq \nu \leq 1$

7–5   In Problem 7-3, consider $M = 2$. Then we have the transfer function

$$\bar{H}(\nu) = 2h(0) \cos (2\pi\nu) + 2h(1) \cos (\pi\nu) + h(2), \quad 0 \leq \nu \leq 1$$

If $\bar{H}(\nu)$ is to satisfy the conditions $\bar{H}(0) = 1$ and $\bar{H}(1) = 0$, show that the impulse response coefficients are as follows:

$$h(1) = \frac{1}{4} \quad \text{and} \quad h(2) = \left[\frac{1}{2} - 2h(0) \right]$$

where $h(0)$ is arbitrary.

7–6   In Problem 7-5, obtain three plots of $|\bar{H}(\nu)|$, $0 \leq \nu \leq 1$, for the following cases: $h(0) = 0$, $h(0) = \frac{1}{4}$, and $h(0) = \frac{1}{8}$.

7–7   In Problem 7-5, find $h(0)$, $h(1)$, and $h(2)$ such that $\bar{H}(\nu)$ satisfies the following conditions: $\bar{H}(0) = 1$, $\bar{H}(0.5) = 1$, and $\bar{H}(1) = 0$.

7–8   Design an FIR low-pass DT filter for which $\bar{H}_d(f)$ is as follows:

$$\bar{H}_d(f) = \begin{cases} 1, & 0 \leq f \leq 5 \text{ Hz} \\ 0, & \text{elsewhere in the range } 0 \leq f \leq f_N \end{cases}$$

The sampling frequency is 20 sps, and the impulse response is to have a duration of 1 sec. Use (a) a rectangular window, and (b) a Hamming window. Plot the resulting decibel magnitude response.

7–9   Repeat Problem 7-8 for a band-pass filter with

$$\tilde{H}_d(f) = \begin{cases} 1, & 2 \le f \le 3 \text{ Hz} \\ 0, & \text{elsewhere in the range } 0 \le f \le f_N \end{cases}$$

7–10   Given $\tilde{H}_d(\nu)$ as shown in the following sketch.

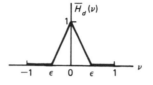

Show that the FS coefficients c_n given by (7.1-5b) are as follows:

$$c_0 = 0.5\epsilon$$

$$c_n = \frac{1}{n^2 \pi^2 \epsilon} [1 - \cos(n\pi\epsilon)], \qquad 1 \le n \le Q$$

Design a FIR filter for the case $\epsilon = 0.4$, using rectangular and Hamming windows for $Q = 10$. Plot the resulting decibel magnitude responses.

7–11 $\tilde{H}_d(\nu)/j$ in the following sketch is that for an ideal differentiator is shown below.

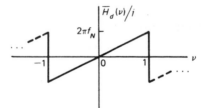

If $\tilde{H}_d(\nu)$ is expanded in a FS, then, using (7.1-2), show that the nth FS coefficient c_n is given by

$$c_n = 2\pi f_N \int_0^1 \nu \sin(n\pi\nu) \, d\nu, \qquad 1 \le n \le Q$$

with $c_0 = 0$ and $c_{-n} = -c_n$. Show that

$$c_n = \frac{(-1)^{n+1}}{nT}, \qquad 1 \le n \le Q$$

Use this information to design an FIR differentiator for $Q = 10$ with rectangular and Hamming window sequences. Plot the resulting decibel magnitude responses.

7–12   The purpose of this problem is to compare the performances of the triangular and hanning windows (see Table 7.3-1) in connection with Example 7.2-1. This is to be done by plotting the decibel magnitude responses of the filters obtained using the FS method with these window sequences, and comparing them with the plots shown in Fig. 7.3-6.

7–13   It is known that the Bessel function I_0 associated with the Kaiser window can be evaluated via the power series

$$I_0(x) = 1 + \sum_{k=1}^{\infty} \frac{(x/2)^k}{k!}$$

A FORTRAN subroutine to evaluate $I_0(x)$ is available; see page 103 of [2]. Use this program to generate a Kaiser window sequence for $Q = 10$, and hence rework Example 7.2-1. Compare the resulting magnitude response with those in Fig. 7.3-6 and the responses obtained in Problem 7-12.

7–14   Use the FS method with the Hamming window to design a 21-coefficient FIR *band-stop* filter with the following specifications:

$$\bar{H}_d(f) = \begin{cases} 0, & 20 \le f \le 30 \text{ Hz} \\ 1, & \text{elsewhere in the range } 0 \le f \le f_N \end{cases}$$

The sampling frequency is 200 sps.

7–15   Starting with (7.3-1a), derive (7.3-1b). *Hint:* First show that

$$\bar{W}_R(\nu) = e^{-jQ\pi\nu} \left[\frac{1 - e^{j(2Q+1)\pi\nu}}{1 - e^{j\pi\nu}} \right]$$

7–16   Use the frequency sampling method to design a linear phase low-pass FIR filter with 10 coefficients, whose cutoff frequency is 100 Hz. Assume a sampling frequency of 1000 sps. Plot the magnitude response of the resulting filter.

7–17   Design a 21-coefficient band-pass filter with linear phase, using the frequency sampling method. Let the lower and upper cutoff frequencies be 100 and 200 Hz, respectively. Assume a sampling frequency of 2100 sps. Plot its magnitude response.

7–18   For example, consider the case when N is odd. Then (7.5-16) can be written as

$$H(z) = \frac{(1 - z^{-N})}{N} \left[\frac{H_0}{1 - z^{-1}} + \sum_{k=1}^{(N-1)/2} [G_k(z) + G_{N-k}(z)] \right]$$

where $G_k(z) = \dfrac{H(k)}{1 - W_N^{-k}z^{-1}} = \dfrac{H_k e^{j\psi_k}}{1 - W_N^{-k}z^{-1}}$

and the ψ_n are given by (7.5-8). Using the properties that

$H_{N-k} = H_k$ from (7.5-14c)

$W_N^{-(N-k)} = \widetilde{W}_N^{-k}$ from (7.5-9) where "$\sim$" denotes complex conjugate

$e^{j\psi_{N-k}} = e^{-j\psi_k}$ from (7.5-8)

show that $H(z)$ can be expressed as follows:

$$H(z) = \frac{1 - z^{-N}}{N} \left[\frac{H_0}{1 - z^{-1}} + \sum_{k=1}^{(N-1)/2} H_k(z) \right]$$

where $H_k(z) = \dfrac{2(-1)^k H_k \cos(k\pi/N)(1 - z^{-1})}{1 - 2\cos(2\pi k/N)z^{-1} + z^{-2}}$

APPENDIX 7.1[†] COMPUTER PROGRAM FOR DESIGNING FINITE IMPULSE RESPONSE FILTERS

The program listed in this appendix enables one to design FIR filters via the FS method, using the rectangular or Hamming windows. Low-pass, high-pass, band-pass, and band-stop (or band-reject) filters can be designed. To illustrate, suppose we wish to approximate the band-pass response given by (7.2-6) using a 41-coefficient FIR filter. Then the pertinent queries and related responses are as follows:

```
DO YOU WANT A BAND-REJECT FILTER (Y/N=BAND-PASS)? N
VALUE OF Q? 20
LOWER CUTOFF, V1? 0.4
UPPER CUTOFF, V2? 0.5
DO YOU WANT HAMMING? Y
STOP ***NORMAL TERMINATION***
```

[†]Program developed by David Martz and Wayne Blasi.

We note that use of a Hamming window is requested. The resulting impulse response values a_i are listed next, and the corresponding decibel magnitude response is displayed in Fig. 7.3-7a.

```
A(  0)=     .0000
A(  1)=    -.0001
A(  2)=     .0011
A(  3)=     .0010
A(  4)=    -.0032
A(  5)=    -.0046
A(  6)=     .0058
A(  7)=     .0129
A(  8)=    -.0062
A(  9)=    -.0264
A( 10)=     .0000
A( 11)=     .0422
A( 12)=     .0160
A( 13)=    -.0541
A( 14)=    -.0409
A( 15)=     .0551
A( 16)=     .0690
A( 17)=    -.0415
A( 18)=    -.0914
A( 19)=     .0155
A( 20)=     .1000
A( 21)=     .0155
A( 22)=    -.0914
A( 23)=    -.0415
A( 24)=     .0690
A( 25)=     .0551
A( 26)=    -.0409
A( 27)=    -.0541
A( 28)=     .0160
A( 29)=     .0422
A( 30)=     .0000
A( 31)=    -.0264
A( 32)=    -.0062
A( 33)=     .0129
A( 34)=     .0058
A( 35)=    -.0046
A( 36)=    -.0032
A( 37)=     .0010
A( 38)=     .0011
A( 39)=    -.0001
A( 40)=     .0000
```

```
C*******************************************************************************
C
        COMPILER FREE
        DIMENSION A(0:512),C(0:256)
        INTEGER P
1       FORMAT ('<12>DO YOU WANT HAMMING? ',Z)
2       FORMAT (S1)
3       FORMAT ('<12>DO YOU WANT A BAND REJECT FILTER (Y/N=BANDPASS)? ',Z)
```

```
    4      FORMAT ('<12>A(',I3,')=',F10.4)
    5      FORMAT ('<23><14>')
C
C          CALCULATE FIR COEFFICIENTS
C
           TYPE'<7>        F I R   F I L T E R   D E S I G N'
           IBR=0
           WRITE(10,3)
           READ(11,2) IB
           IF(IB.EQ.'Y<0>')IBR=1
           ACCEPT'<12>VALUE OF P ?',P
           ACCEPT'<12>LOWER CUTOFF, V1? ',V1
           ACCEPT'<12>UPPER CUTOFF, V2? ',V2
           PI=3.14159
           C(0)=V2-V1
           IF(IBR.EQ.1)C(0)=1-C(0)
           DO 50 N=1,P
           C(N)=(SIN(V2*N*PI))/(N*PI)-(SIN(V1*N*PI))/(N*PI)
           IF (IBR.EQ.1)C(N)=-C(N)
    50     CONTINUE
C
C          CHECK TO SEE IF HAMMING DESIRED
C
           WRITE(10,1)
           READ(11,2)IHAM
           IF (IHAM.NE.'Y<0>')GO TO 70
           DO 60 N=0,P
           C(N)=C(N)*(0.54+0.46*(COS(N*PI/P)))
    60     CONTINUE
C
C          GENERATE COEFFICIENT ARRAY
C
    70     DO 80 N=0,P
           A(N)=C(P-N)
    80     CONTINUE
C
C          GENERATE CONJUGATE POINTS
C
           DO 85 IC=1,P
           A(P+IC)=A(P-IC)
    85     CONTINUE
C
C          PRINT FILTER WEIGHTS
C
           DO 90 IPO=0,2*P
           WRITE (12,4) IPO,A(IPO)
    90     CONTINUE
           STOP***NORMAL TERMINATION***
           END
```

REFERENCES

1. A. OPPENHEIM and R. SCHAFER, *Digital Signal Processing*, Prentice-Hall, Englewood Cliffs, N.J., 1975.

2. L. R. RABINER and B. GOLD, *Theory and Application of Digital Signal Processing*, Prentice-Hall, Englewood Cliffs, N.J., 1975.

3. S. D. STEARNS, *Digital Signal Analysis*, Hayden, Rochelle Park, N.J., 1975.

4. W. D. STANLEY, *Digital Signal Processing*, Reston, Reston, Va., 1975.

5. R. W. Hamming, *Digital Filters*, Prentice-Hall, Englewood Cliffs, N.J., 1975.

6. S. A. Tretter, *Discrete-Time Signal Processing*, Wiley, New York, 1976.

7. M. T. Jong, *Methods of Discrete Signal and System Analysis*, McGraw-Hill, New York, 1982.

8. L. R. Rabiner, "Linear Program Design of Finite Impulse Response (FIR) Filters," *IEEE Trans. Audio and Electroacoust.*, Vol. AU-20, 1972, pp. 280–288.

9. O. Herrmann, "Design of Nonrecursive Filters with Linear Phase," *Electronics Letters*, Vol. 6, No. 11, May 28, 1970, pp. 328–329.

10. E. Hofstetter, A. V. Oppenheim, and J. Siegel, "A New Technique for the Design of Nonrecursive Digital Filters," Proc. of the Fifth Annual Princeton Conference on Information Sciences and Systems, March 1971, pp. 64–72.

11. T. W. Parks and J. H. McClellan, "Chebyshev Approximation for Nonrecursive Digital Filters with Linear Phase," *IEEE Trans. Circuit Theory*, Vol. CT-29, 1972, pp. 189–194.

12. J. H. McClellan and T. W. Parks, "A Unified Approach to the Design of Optimum FIR Linear Phase Digital Filters," *IEEE Trans. Circuit Theory*, Vol. CT-20, 1973, pp. 697–701.

13. J. H. McClellan, T. W. Parks, and L. R. Rabiner, "A Computer Program for Designing Optimum FIR Digital Filters," *IEEE Trans. Audio and Electroacoust.*, Vol. AU-21, 1973, pp. 506–526.

8

Some Implementation Considerations

We recall that the general difference equation that describes a DT system is the convolution relation in (5.1-1). For convenience, we repeat (5.1-1) below as (8.0-1):

$$y(n) = \sum_{i=0}^{k} a_i x(n - i) - \sum_{j=1}^{m} b_j y(n - j), \quad n \geq 0 \qquad (8.0\text{-}1)$$

where the a_i and b_j are the system coefficients, and $x(n)$, $y(n)$ are the input and output sequences, respectively.

In (8.0-1) we must appreciate that the a_i, b_j, $x(n)$, and $y(n)$ are all considered to take a *continuum* of values. Hence an *infinite* precision is assumed. However, when such DT systems are implemented in *digital* (binary) form, only *finite* precision is available. This is because the related implementations are in terms of computer algorithms or special-purpose digital hardware, where the pertinent word lengths are finite. Thus such implementations of DT systems (filters) may be referred to as *digital systems* (filters).

Our main goal in this chapter is to address some issues related to implementing a DT system in special-purpose hardware. The issues we shall consider are the following:

1. Input quantization error introduced by using a finite number of bits to represent (encode) the input sequence $x(n)$.

2. Coefficient inaccuracy error that results from using a finite number of bits to represent each of the system coefficients a_i, b_j.

3. Roundoff error that occurs each time a product $a_i x(n - i)$ or $b_j y(n - j)$ in (8.0-1) is rounded so that it can be accommodated in a prescribed word length.

4. A scaling procedure that helps to avoid overflow of the adders used in an implementation.

5. Certain types of undesired oscillations that may occur due to limited accuracy and adder overflows.

6. A table lookup implementation where no multiplications are necessary.

In practice one is usually concerned with these issues in connection with IIR systems. Hence this chapter is for the most part pertinent to the implementation of IIR systems. In Chapter 7 it was pointed out that FIR systems can be implemented in an efficient manner via the FFT (i.e., using fast convolution techniques).

An in-depth treatment of the issues listed are beyond the scope of this book. Our discussion will serve to provide the reader with an appreciation and some working knowledge of these issues. With this introduction and a background in random-signal analysis, the reader can pursue a rigorous treatment of the subject matter in more advanced textbooks (e.g., see [1 to 3]).

It is well known that digital systems are associated with two types of arithmetic, *fixed point* and *floating point*.[†] For real-time applications, fixed-point arithmetic is widely used since it results in more economical and faster hardware than floating-point arithmetic. On the other hand, floating-point arithmetic leads to increased dynamic range and improved accuracy of processing. Hence one is usually not concerned with the listed issues when floating-point arithmetic is used (e.g., in non-real-time applications where general-purpose computers are used). Thus our discussion in this chapter will be restricted to the listed issues as they relate to implementations that employ fixed-point arithmetic. To this end, we commence with a review of fixed-point arithmetic.

8.1 FIXED-POINT ARITHMETIC

We assume that all data values x are scaled and that $-1 < x < 1$. Then a given fractional number x has a fixed-point representation as follows:

[†]The interested reader may refer to Appendix 8.1 for a brief review of fixed-point and floating-point arithmetic.

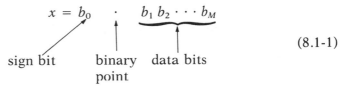

$$\text{(8.1-1)}$$

where $b_0 = \begin{cases} 0, & \text{for } x \geq 0 \\ 1, & x < 0 \end{cases}$

and M is the number of *data bits*. The corresponding *word length* (say B) is $(M + 1)$ bits; that is, $B = M + 1$.

Given a fractional decimal number $(x)_{10}$, 2's complement notation is used to obtain the representation in (8.1-1). The procedure to do so depends upon whether $(x)_{10}$ is positive or negative. Thus we consider two cases.

CASE 1. $(x)_{10}$ is Positive. Here $b_0 = 0$ and b_n, $1 \leq n \leq M$, are obtained via the straightforward decimal-to-binary conversion process. For example, suppose $(x)_{10} = 0.625$ and $M = 4$. Then $b_1 = b_3 = 1$ and $b_2 = b_4 = 0$, as illustrated in Fig. 8.1-1. Thus the fixed-point representation of $(x)_{10} = 0.625$ is

$$x = 0.1010 \qquad \text{(8.1-2)}$$

CASE 2. $(x)_{10}$ is Negative. The desired fixed-point representation is obtained via two steps, as follows:

1. Obtain the fixed-point representation of $|(x)_{10}|$ as discussed in case 1, and denote the resulting number by $\hat{x}$.

2. Copy the bits of $\hat{x}$ starting with the rightmost (least significant) bit, through the first bit that is a 1. Then complement the rest of the bits.

$$
\begin{array}{rl}
& \boxed{0}.625 \\
& \quad \times 2 \\
\hline
b_1 \rightarrow & \boxed{1}.250 \\
& \quad \times 2 \\
\hline
b_2 \rightarrow & \boxed{0}.500 \\
& \quad \times 2 \\
\hline
b_3 \rightarrow & \boxed{1}.000 \\
& \quad \times 2 \\
\hline
b_4 \rightarrow & \boxed{0}.000
\end{array}
$$

Fig. 8.1-1 Decimal-to-binary conversion procedure, for $(x)_{10} = 0.625$ and $M = 4$.

To illustrate, let $(x)_{10} = -0.625$, which implies that $|(x)_{10}| = 0.625$. From the example considered in case 1, it is apparent that the fixed-point representation of $|(x)_{10}| = 0.625$ as given in (8.1-2). Thus

$$\hat{x} = 0.1010 \tag{8.1-3}$$

Then step 2 yields

$$x = 1.0110 \tag{8.1-4}$$
$$\uparrow$$

where the symbol $\uparrow$ indicates the occurrence of the first bit in $\hat{x}$ that is a 1, starting with the rightmost bit. The rest of the bits that appear in $\hat{x}$ are then complemented. Thus x in (8.1-4) is the desired fixed-point representation of $(x)_{10} = -0.625$.

Next, if the sign and data bits in (8.1-1) are given, it is easy to evaluate the corresponding decimal number $(x)_{10}$ using the formula

$$(x)_{10} = -b_0 + \sum_{n=1}^{M} b_n 2^{-n} \tag{8.1-5}$$

For example, let $x = 1.00101$, which implies that $b_0 = 1$, $b_1 = b_2 = b_4 = 0$, and $b_3 = b_5 = 1$. Substitution of these values for b_0 and b_n, $1 \leq n \leq 5$, in (8.1-5) leads to

$$(x)_{10} = -1 + (1)2^{-3} + (1)2^{-5}$$
$$= -0.84375$$

Truncation and Rounding

Truncation of a fixed-point number to K data bits, where $K < M$, is the process of dropping all bits beyond b_K in (8.1-1). Let x_T denote the resulting fixed-point number and $(x_T)_{10}$ denote its decimal equivalent. Then, corresponding to (8.1-1) and (8.1-5), we have

$$x_T = b_0 \cdot b_1 b_2 \cdots b_K \tag{8.1-6}$$

and

$$(x_T)_{10} = -b_0 + \sum_{n=1}^{K} b_n 2^{-n} \tag{8.1-7}$$

On the other hand, *rounding* of a fixed-point number to K data bits is the process of adding a 1 to the $(K + 1)$th data bit of x in (8.1-1) and truncating to K data bits. For example, if

$$x = 0.11010111 \qquad (8.1\text{-}8)$$

and we wish to round x to 7 bits, then we form

$$x' = 0.11010111 + 0.00000001 \qquad (8.1\text{-}9)$$
$$= 0.11011000$$

Next, we truncate x' to 7 bits to obtain

$$x_R = 0.1101100 \qquad (8.1\text{-}10)$$

which is the fixed-point number corresponding to x in (8.1-8) when the latter is rounded to 7 data bits.

The fact that x_R in (8.1-10) is a reasonable answer can be verified by converting it and x in (8.1-8) to their decimal equivalents via (8.1-7) with $K = 8$ and 7, respectively. This yields

$$(x)_{10} = 0.83984375$$

and $\qquad\qquad\quad (x_R)_{10} = 0.84375$

from which we see that $(x_R)_{10}$ compares quite closely with $(x)_{10}$.

Decimal-to-Binary Accuracy Considerations [4]

Suppose we are given a positive fractional *decimal* number $(x)_{10}$ that consists of D digits; that is,

$$(x)_{10} = \sum_{n=1}^{D} d_n 10^{-n}, \qquad d_n \epsilon \{0, 1, \cdots, 9\} \qquad (8.1\text{-}11)$$

The representation in (8.1-11) implies that the *accuracy* with which $(x)_{10}$ can be specified is given by

$$(x)_{10} - \Delta x \le (x)_{10} \le (x)_{10} + \Delta x \qquad (8.1\text{-}12)$$

where $\quad \Delta x = \dfrac{10^{-D}}{2}$

Now suppose we wish to approximate $(x)_{10}$ by the *binary* number y, which consists of M data bits; that is,

$$(x)_{10} \simeq (y)_{10} = \sum_{n=1}^{M} b_n 2^{-n} \qquad (8.1\text{-}13)$$

where $(y)_{10}$ is the decimal number corresponding to y.

The *accuracy* associated with the representation in (8.1-13) is

$$(y)_{10} - \Delta y \le (y)_{10} \le (y)_{10} + \Delta y \qquad (8.1\text{-}14)$$

where $\quad \Delta y = \dfrac{2^{-M}}{2}$

Our objective is to choose M so that the accuracy related to $(y)_{10}$ is *at least* equal to that related to $(x)_{10}$. Clearly, the condition that has to be satisfied is

$$\Delta y \le \Delta x \qquad (8.1\text{-}15)$$

where Δx and Δy are defined in (8.1-12) and (8.1-14), respectively. Hence (8.1-15) is equivalent to

$$\frac{2^{-M}}{2} \le \frac{10^{-D}}{2}$$

which yields

$$M \ge D \log_2 10 \qquad (8.1\text{-}16)$$

or $\qquad\qquad M \ge 3.3D$

Equation (8.1-16) is the desired result, which can be used to obtain M for a specified D. To illustrate, if $D = 2$, then (8.1-16) implies we must use 7 data bits or more to encode decimal numbers given by (8.1-11) so that the corresponding binary representation is at least as accurate as the 2-digit decimal representation.

8.2 INPUT QUANTIZATION ERROR

A given CT input signal is converted into digital form by means of an analog-to-digital (A/D) converter, as illustrated in Fig. 8.2-1. The input

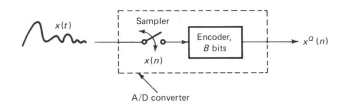

Fig. 8.2-1 Analog-to-digital conversion.

signal is sampled to obtain the sequence $x(n)$. Each value $x(n)$ is encoded using *word lengths* of B bits to obtain the quantized output $x^Q(n)$. Each word consists of M data bits and one sign bit.

We assume that the sequence $x(n)$ is scaled such that $|x(n)| < 1$. This implies that the pertinent dynamic range is 2. Since the encoder employs B bits, the number of levels available for quantizing $x(n)$ is 2^B. Thus the interval between successive levels, q, is given by

$$q = \frac{2}{2^B} = 2^{-B+1} \tag{8.2-1}$$

and is called the quantization step size. The input value $x(n)$ is then rounded to the nearest level, as illustrated in Fig. 8.2-2, for $B = 4$ and

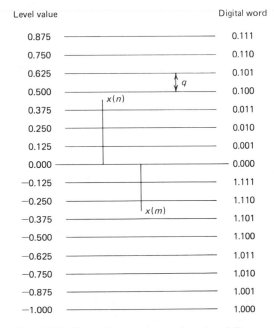

Fig. 8.2-2 Encoding process related to A/D conversion.

$q = 0.125$; see (8.2-1). The output of the A/D converter is thus a 4-bit word where a fixed-point representation is assumed. To illustrate, suppose $x(n)$ shown in Fig. 8.2-2 equals 0.45. Then, *rounding*[†] to the nearest level, we obtain $x^Q(n) = 0.5$ since

$$|x(n) - x^Q(n)| = 0.05 < \frac{q}{2} \qquad (8.2\text{-}2)$$

where $q = 0.125$. The corresponding output of the A/D converter is the 4-bit word 0.100. Similarly, if $x(m) = -0.28$, then $x^Q(m) = -0.25$, since

$$|x(m) - x^Q(m)| = 0.03 < \frac{q}{2} \qquad (8.2\text{-}3)$$

where $q = 0.125$. This results in the output word 1.110.

From (8.2-2) and (8.2-3), it follows that we can view the A/D converter output as being the sum of the actual input $x(n)$ and an *error (noise)* component $e(n)$; that is,

$$x^Q(n) = x(n) + e(n) \qquad (8.2\text{-}4)$$

In (8.2-4) we note that $e(n)$ is a *random* quantity in that its value cannot be predicted exactly. Thus the values it takes can only be described via statistical methods. In particular, it is said to be a uniformly distributed *random variable* in the interval $(-q/2, q/2)$. In other words, it can take any value between $-q/2$ and $+q/2$ with equal probability.

A rigorous analysis of the additive noise model in (8.2-4) requires a background in random-signal analysis. For our purposes it suffices to merely use the final results one obtains via such an analysis. The overall effect of $e(n)$ is assessed in terms of its *variance* (or *power*), which is denoted by σ^2_e. It can be shown that[††]

$$\sigma^2_e = \frac{q^2}{12}$$

$$\qquad (8.2\text{-}5)$$

or
$$\sigma^2_e = \frac{2^{-2B}}{3}$$

since (8.2-1) implies that $q = 2^{-B+1}$.

[†]Throughout this discussion we shall restrict our attention to rounding.

[††]The reader with a background in random signals may refer to Appendix 8.4 for the related derivation.

We shall refer to σ^2_e in (8.2-5) as being the steady-state *noise power due to input quantization*. From (8.2-5) it is clear that σ^2_e tends to zero as B tends to infinity. This is what one would expect, since increasing values of B represent larger A/D converter word lengths. The condition B tending to infinity would thus imply that the A/D converter has infinite precision. In practice, typical A/D converter word lengths are 8, 12, and 16.

Output Noise Power

Now suppose the output of the A/D converter, $x^Q(n)$, becomes the input to a DT system whose transfer function is $H(z)$; see Fig. 8.2-3. Then from (8.2-4) it is apparent that the output $y'(n)$ will consist of two parts. The first is due to $x(n)$, and the other is due to the noise component $e(n)$. Thus we have

$$y'(n) = y(n) + \epsilon(n) \qquad (8.2\text{-}6)$$

where $y(n)$ and $\epsilon(n)$ are the outputs of $H(z)$ due to the inputs $x(n)$ and $e(n)$, respectively. The quantity $\epsilon(n)$ is again a random variable in the sense that its value cannot be obtained exactly. As in the case of $e(n)$, we assess the influence of $\epsilon(n)$ in terms of its variance (or power). It can be shown that[†]

$$\sigma^2_\epsilon = \sigma^2_e \cdot \frac{1}{2\pi j} \oint_{C_u} z^{-1} H(z) H(z^{-1})\, dz \qquad (8.2\text{-}7)$$

where $\oint_{C_u}$ denotes integration *around the unit circle* $|z| = 1$ in the counterclockwise direction.

The integral

$$I = \frac{1}{2\pi j} \oint_{C_u} z^{-1} H(z) H(z^{-1})\, dz \qquad (8.2\text{-}8)$$

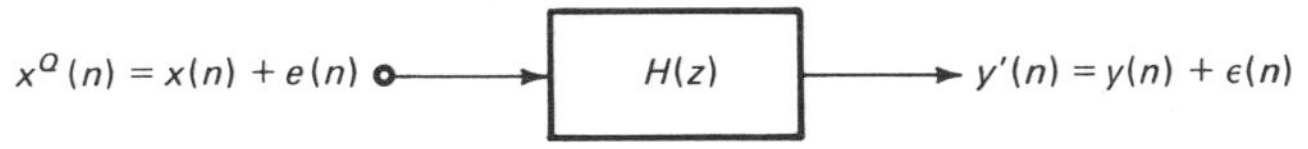

$$x^Q(n) = x(n) + e(n) \; \circ\!\!\longrightarrow \boxed{H(z)} \longrightarrow y'(n) = y(n) + \epsilon(n)$$

Fig. 8.2-3 Output of discrete-time system when subjected to quantized input $x^Q(n)$.

[†]The reader with a background in random signals may refer to Appendix 8.4 for a derivation of this result.

can be evaluated via the notion of residues, as was the case with the evaluation of the IZT in (3.5-12). In particular, we have

$$
\begin{aligned}
I = \text{ sum of residues of } L(z) = z^{-1}H(z)H(z^{-1}) \\
\text{ at the poles of } L(z) \text{ that lie } within \text{ the} \\
\text{ the unit circle } |z| = 1
\end{aligned}
\tag{8.2-9}
$$

To illustrate, suppose

$$
H(z) = \frac{z}{z - \alpha}
\tag{8.2-10}
$$

where $|\alpha| < 1$. Then

$$
L(z) = z^{-1}H(z)H(z^{-1})
\tag{8.2-11}
$$

yields

$$
L(z) = \frac{z^{-1}}{(z - \alpha)(z^{-1} - \alpha)}
$$

$$
= \frac{1}{(z - \alpha)(1 - \alpha z)}
\tag{8.2-12}
$$

It is apparent that $L(z)$ in (8.2-12) has two poles, at $z_1 = \alpha$ and $z_2 = 1/\alpha$. The pole at $z = z_2$ lies *outside* the unit circle. Hence we need only to evaluate the residue at $z = z_1$, which is

$$
R_{z=\alpha} = (z - \alpha)L(z)\big|_{z=\alpha}
\tag{8.2-13}
$$

$$
= \frac{1}{1 - \alpha^2}
$$

Thus from (8.2-8), (8.2-9), and (8.2-13), we obtain

$$
I = \frac{1}{2\pi j} \oint_{C_u} \frac{z^{-1}\, dz}{(z - \alpha)(z^{-1} - \alpha)} = \frac{1}{1 - \alpha^2}
\tag{8.2-14}
$$

Example 8.2-1: Suppose the A/D conversion indicated in Fig. 8.2-1 employs 8 bits (i.e., $B = 8$). The resulting $x^Q(n)$ are then processed through the first-order IIR system whose transfer function is

$$
H(z) = \frac{z}{z - 0.999}
\tag{8.2-15}
$$

Find the steady-state noise power due to quantization that occurs at the output of $H(z)$.

Solution: The noise power due to quantization at the input to this system is σ^2_e in (8.2-5) with $B = 8$. Hence

$$\sigma^2_e = \frac{2^{-16}}{3} \tag{8.2-16}$$

Next the noise power at the output of the system is σ^2_ϵ in (8.2-7), with σ^2_e in (8.2-16) and $H(z)$ in (8.2-15); that is,

$$\sigma^2_\epsilon = \frac{2^{-16}}{3}I \tag{8.2-17}$$

where $I = \frac{1}{2\pi j}\oint_{C_u} \frac{z^{-1}\,dz}{(z - 0.999)(z^{-1} - 0.999)}$

It follows that the value of I in (8.2-17) is obtained by substituting $\alpha = 0.999$ in (8.2-14). Thus the desired answer is

$$\sigma^2_\epsilon = \frac{2^{-16}}{3}(500.25)$$

$$= 0.00254$$

8.3 COEFFICIENT INACCURACY ERROR

The general transfer function of a DT system we wish to implement is given by (5.1-3) to be

$$H(z) = \frac{\displaystyle\sum_{i=0}^{k} a_i z^{-i}}{1 + \displaystyle\sum_{j=1}^{m} b_j z^{-j}}$$

In this $H(z)$ the system coefficients a_i and b_j can only be approximated since they have to be encoded in binary form using a finite number of bits, say C. Let a_i^Q and b_j^Q denote the quantized values corresponding to a_i and b_j. Then we have

$$a_i^Q = a_i + \alpha_i \tag{8.3-1}$$

and

$$b_j^Q = b_j + \beta_j$$

where α_i and β_j represent error terms due to quantization. Since C bits are used for quantization purposes, α_i and β_j are such that

$$|\alpha_i| \le 2^{-C+1} \tag{8.3-2}$$

and $\qquad\qquad\qquad |\beta_j| \le 2^{-C+1}$

assuming one sign bit and one integer bit before the binary point. An appropriate modification is needed if some coefficients need more than one integer bit. As a result, the transfer function that is actually implemented is not $H(z)$ in (5.1-3), but $H^Q(z)$, which is as follows:

$$H^Q(z) = \frac{\displaystyle\sum_{i=0}^{k} a_i^Q z^{-i}}{1 + \displaystyle\sum_{j=1}^{n} b_j^Q z^{-j}} \tag{8.3-3}$$

where a_i^Q and b_j^Q are defined in (8.3-1).

Now, if the word length C is not sufficiently large, certain undesirable effects occur. For example, the frequency characteristics (e.g., magnitude, phase, bandwidth) of $H^Q(z)$ may differ appreciably from those of $H(z)$. Also, if the poles of $H(z)$ are close to the unit circle, then those of $H^Q(z)$ may lie just outside the unit circle, resulting in an unstable implementation. Kaiser [5] has shown that such undesirable effects due to coefficient inaccuracy are far more pronounced when high-order systems are directly implemented, using the direct form 1 or direct form 2 realizations in Figs. 5.6-1 and 5.6-2, respectively. Hence the *cascade* and *parallel* form realizations in Figs. 5.6-3 and 5.6-4, respectively, are preferred, where each $H_i(z)$ is a *first-* or *second-order* section. This aspect is now illustrated by the following simple example.

Consider the second-order IIR system

$$H(z) = \frac{1}{(1 - 0.5z^{-1})(1 - 0.4z^{-1})} \tag{8.3-4}$$

whose poles are located at $z = 0.5$ and $z = 0.4$.

It is clear that $H(z)$ can be realized in at least two ways; that is, direct form 1 or cascade. The transfer function pertinent to the direct form 1 realization in Fig. 8.3-1a is

$$H(z) = \frac{1}{1 + b_1 z^{-1} + b_2 z^{-2}} \tag{8.3-5a}$$

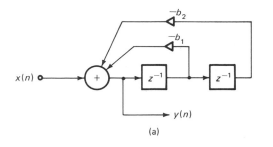

(a)

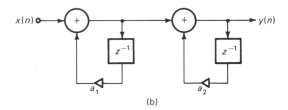

(b)

Fig. 8.3-1 (a) Direct form 1 and (b) cascade realizations of *H(z)* in (8.3-4).

where $b_1 = -0.9$ and $b_2 = 0.2$.
Again, the transfer function pertinent to the cascade realization in Fig.
8.3-1b is in the form

$$H(z) = H_1(z)H_2(z) \qquad (8.3\text{-}5b)$$

where $H_1(z) = \dfrac{1}{1 - a_1 z^{-1}}$

$H_2(z) = \dfrac{1}{1 - a_2 z^{-1}}$

with $a_1 = 0.5$ and $a_2 = 0.4$.

We now ask the following question [6]: are the pole locations $z = 0.5$ and $z = 0.4$ more *sensitive* to changes in b_1 and b_2 (i.e., direct form 1 realization), when compared to corresponding changes in a_1 and a_2 (i.e., cascade realization)? To answer this question, let us assume that a_1, a_2 and b_1, b_2 are encoded using a sign bit and 3 data bits each. Then using truncation, corresponding to $H(z)$ in (8.3-5a) and (8.3-5b), we obtain (see Problem 8-5)

$$H'(z) = \frac{1}{1 - 0.875z^{-1} + 0.125z^{-2}} \qquad (8.3\text{-}6a)$$

and
$$H''(z) = \frac{1}{1 - 0.5z^{-1}} \cdot \frac{1}{1 - 0.375z^{-1}} \qquad (8.3\text{-}6b)$$

For the purposes of comparison, we now summarize the pole locations of the given transfer function $H(z)$ in (8.3-4) and those corresponding to its direct form 1 and cascade realization given by $H'(z)$ and $H''(z)$, respectively.

Transfer Function	Pole Locations
$H(z)$	$z_1 = 0.5,\quad z_2 = 0.4$
$H'(z)$	$z_1' = 0.695,\quad z_2' = 0.18$
$H''(z)$	$z_1'' = 0.5,\quad z_2'' = 0.375$

From this table it is clear that the poles of $H''(z)$ are closer to those of $H(z)$, which means that the cascade realization is less sensitive to the process of quantizing each filter coefficient using 3 data bits. Clearly, larger word lengths will be required to make the direct form 1 realization less sensitive. This implies that the cascade realization is more cost effective.

8.4 PRODUCT ROUNDOFF ERROR

In implementing a DT filter we are required to form products of the form

$$p(n) = \alpha d(n) \qquad (8.4\text{-}1)$$

where $p(n)$ is the desired product, α is a filter coefficient, and $d(n)$ is a data value. Let the word lengths associated with α and $d(n)$ be C and B bits, respectively. Then a direct product evaluation yields $B + C$ bits for $p(n)$. This product will have to be stored as another data value in a B-bit word. As such, we have two choices. The $(B + C)$-bit product can either be truncated (chopped) or rounded to B bits. We shall restrict our attention to the rounding case.

The process of rounding a $(B + C)$-bit product to B bits is modeled as depicted in Fig. 8.4-1. Here $p(n)$ is the exact product of α and $d(n)$ if a multiplier of infinite precision were available, and $p^Q(n)$ is the corresponding quantized product that we actually obtain. We relate $p(n)$ and $p^Q(n)$ via the error (noise) source whose steady-state power (variance) is σ_r^2. The question is: what is the value of σ_r^2? The answer lies in

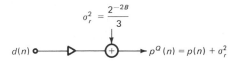

Fig. 8.4-1 Roundoff error noise model.

realizing that the process of rounding a $(B + C)$-bit number to a B-bit number is very similar to that of quantizing input samples $x(n)$ using B bits each. This aspect was discussed in Section 8.2, where each input sample $x(n)$ was rounded to B bits during the A/D conversion process; see σ_e^2 in (8.2-5). Hence product roundoff error is modeled as

$$p^Q(n) = p(n) + \sigma_r^2 \qquad (8.4\text{-}2)$$

where σ_r^2 is the steady-state noise power due to rounding $(B + C)$-bit products to B bits, and is given by

$$\sigma_r^2 = \frac{2^{-2B}}{3} \qquad (8.4\text{-}3)$$

To illustrate, consider the first-order DT system

$$y(n) = a_1 y(n - 1) + x(n), \qquad n \geq 0 \qquad (8.4\text{-}4)$$

with $|a_1| < 1$, the realization of which is shown in Fig. 8.4-2. The noise source 1 indicated in Fig. 8.4-2 accounts for rounding the product $a_1 y(n - 1)$ of $B + C$ bits to B bits. Next, the effect of introducing source 1 is to add noise to the output $y(n)$. The output noise power is denoted in Fig. 8.4-2 by σ_ϵ^2 and is given by (8.2-7) to be

$$\sigma_\epsilon^2 = \sigma_e^2 \frac{1}{2\pi j} \oint_{C_u} z^{-1} P(z) P(z^{-1}) \, dz \qquad (8.4\text{-}5)$$

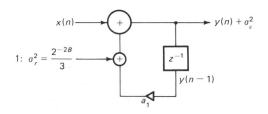

Fig. 8.4-2 Implementation of system in (8.4-4) and accounting for rounding the $(B + C)$-bit product $a_1 y(n - 1)$ to B bits.

where $P(z)$ is the transfer function between source 1 and the output.

From Fig. 8.4-2 it is apparent that $P(z)$ is merely the transfer function $H(z) = Y(z)/X(z)$; that is,

$$P(z) = H(z) = \frac{z}{z - a_1} \tag{8.4-6}$$

The value of the contour integral in (8.4-5) is given by (8.2-14) with $\alpha = a_1$. That is,

$$\sigma_\epsilon^2 = \frac{2^{-2B}}{3(1 - a_1^2)} \tag{8.4-7}$$

Thus, for a given B and a_1, we can use (8.4-7) to evaluate σ_ϵ^2, which is the steady-state output noise power due to rounding the product $a_1 y(n - 1)$ in (8.4-4).

Next we work an assortment of examples where the word lengths associated with the filter coefficients and data values are assumed to be C and B bits, respectively. Each product is then rounded to B bits.

Example 8.4-1: Consider the IIR system

$$H(z) = H_1(z)H_2(z) \tag{8.4-8}$$

where $H_1(z) = \dfrac{1}{1 - a_1 z^{-1}}$

$H_2(z) = \dfrac{1}{1 - a_2 z^{-1}}$

with $|a_1| < 1$ and $|a_2| < 1$. The corresponding cascade realization is shown in Fig. 8.4-3a. Find the steady-state output noise power due to product roundoff.

Solution: From Fig. 8.4-3a it is apparent that there are two products, $a_1 y_1(n - 1)$ and $a_2 y(n - 1)$. Rounding these products results in noise sources each with variances $2^{-2B}/3$, as indicated in Fig. 8.4-3b.

The steady-state output noise power σ_ϵ^2 consists of two contributions due to the noise sources 1 and 2, respectively. The transfer function "seen" by source 2 is $H_2(z)$, while that seen by source 1 is $H(z)$

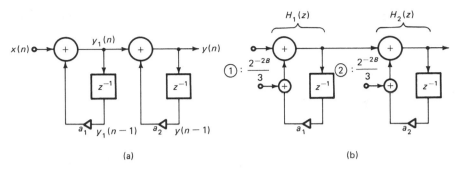

Fig. 8.4-3 (a) Cascade realization and (b) noise sources for Example 8.4-1.

$= H_1(z)H_2(z)$. We assume that these two contributions can be added.[†]
Thus (8.2-7) yields σ_ϵ^2 to be

$$\sigma_\epsilon^2 = \frac{2^{-2B}}{3} [I_1 + I_2] \tag{8.4-9}$$

where $\quad I_2 = \dfrac{1}{2\pi j} \displaystyle\oint_{C_u} \underbrace{\left[\dfrac{1}{1 - a_2 z^{-1}} \right]}_{H_2(z)} \underbrace{\left[\dfrac{1}{1 - a_2 z} \right]}_{H_2(z^{-1})} z^{-1}\, dz$

and

$$I_1 = \frac{1}{2\pi j} \oint_{C_u} \underbrace{\left[\frac{1}{1 - a_1 z^{-1}} \right]}_{H_1(z)} \underbrace{\left[\frac{1}{1 - a_1 z} \right]}_{H_1(z^{-1})} \underbrace{\left[\frac{1}{1 - a_2 z^{-1}} \right]}_{H_2(z)} \underbrace{\left[\frac{1}{1 - a_2 z} \right]}_{H_2(z^{-1})} z^{-1}\, dz$$

The integrals I_1 and I_2 in (8.4-9) can be evaluated via (8.2-8) and (8.2-9). It can be shown that (Problem 8-6)

$$I_2 = \frac{1}{1 - a_2^2} \tag{8.4-10a}$$

and $\quad I_1 = \dfrac{a_1}{(1 - a_1^2)(a_1 - a_2)(1 - a_2 a_1)}$

$$+ \frac{a_2}{(1 - a_2^2)(a_2 - a_1)(1 - a_1 a_2)} \tag{8.4-10b}$$

[†]The reader who is familiar with random signal theory will recognize that sources 1 and 2 are considered to be *uncorrelated*.

Substitution of (8.4-10) in (8.4-9) leads to the desired result:

$$\sigma_\epsilon^2 = \frac{2^{-2B}}{3}\left[\frac{1}{1 - a_2^2} + \frac{1 + a_1 a_2}{1 - a_1 a_2}\frac{1}{(1 - a_1^2)(1 - a_2^2)}\right] \qquad (8.4\text{-}11)$$

Example 8.4-2: Given the IIR system

$$H(z) = \frac{1}{1 - 2\gamma \cos(\omega_0 T)z^{-1} + \gamma^2 z^{-2}} \qquad (8.4\text{-}12)$$

whose complex-conjugate poles are given by

$$z_1 = \gamma e^{j\omega_0 T} \quad \text{and} \quad z_2 = \gamma e^{-j\omega_0 T}$$

where $|\gamma| < 1$ for stability.

This system can be implemented via the direct form 2 realization in Fig. 5.6-2, as depicted in Fig. 8.4-4a. Examination of Fig. 8.4-4a shows that the process of rounding following the products $2\gamma \cos(\omega_0 T)y(n - 1)$ and $\gamma^2 y(n - 2)$ introduces the noise sources 1 and 2 indicated in Fig. 8.4-4b. It is seen that each of the noise sources sees the transfer function $H(z)$. As in Example 8.4-1, we assume that these two contributions can be added. Hence (8.2-7) yields σ_ϵ^2 to be

$$\sigma_\epsilon^2 = 2\left[\frac{2^{-2B}}{3}\right] I \qquad (8.4\text{-}13)$$

where

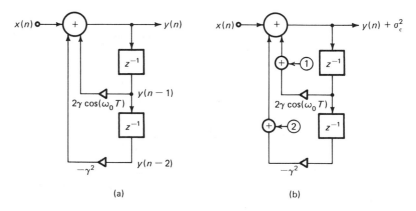

(a) (b)

Fig. 8.4-4 (a) Direct form 2 realization and (b) noise sources for Example 8.4-2.

$$I = \frac{1}{2\pi j} \oint_{C_u} \left[\frac{1}{(1 - z_1 z^{-1})(1 - z_2 z^{-1})} \right] \left[\frac{1}{(1 - z_1 z)(1 - z_2 z)} \right] z^{-1} \, dz$$

with $z_1 = \gamma e^{j\omega_0 T}$ and $z_2 = \gamma e^{-j\omega_0 T}$. Examination of I reveals that the poles *inside* the unit circle $|z| = 1$ are z_1 and z_2. Thus (8.2-8) and (8.2-9) imply that

$$I = R_{z=z_1} + R_{z=z_2} \tag{8.4-14}$$

The desired residues are given by

$$R_{z=z_1} = \frac{z_1}{(z_1 - z_2)(1 - z_1^2)(1 - z_1 z_2)} \tag{8.4-15}$$

and

$$R_{z=z_2} = \frac{z_2}{(z_2 - z_1)(1 - z_2^2)(1 - z_1 z_2)}$$

Substitution of (8.4-15) in (8.4-14) and subsequent manipulation of terms results in

$$I = \frac{1 + z_1 z_2}{1 - z_1 z_2} \frac{1}{(1 - z_1^2)(1 - z_2^2)} \tag{8.4-16}$$

With $z_1 = \gamma e^{j\omega_0 T}$ and $z_2 = \gamma e^{-j\omega_0 T}$, it is easy to verify that

$$1 + z_1 z_2 = 1 + \gamma^2$$
$$1 - z_1 z_2 = 1 - \gamma^2$$

and

$$(1 - z_1^2)(1 - z_2^2) = (1 - \gamma^2 e^{j2\omega_0 T})(1 - \gamma^2 e^{-j2\omega_0 T})$$
$$= 1 - 2\gamma^2 \cos(2\omega_0 T) + \gamma^4$$

Thus (8.4-16) becomes

$$I = \frac{1 + \gamma^2}{(1 - \gamma^2)(1 + \gamma^4 - 2\gamma^2 \cos(2\omega_0 T))} \tag{8.4-17}$$

Substituting (8.4-17) in (8.4-13), we obtain the desired steady-state output noise power as

$$\sigma_\epsilon^2 = 2 \left[\frac{2^{-2B}}{3} \right] \left\{ \frac{1 + \gamma^2}{1 - \gamma^2} \frac{1}{[1 + \gamma^4 - 2\gamma^2 \cos(2\omega_0 T)]} \right\} \tag{8.4-18}$$

Example 8.4-3: Consider the transfer function

$$H(z) = H_1(z)H_2(z) \tag{8.4-19}$$

where $H_1(z) = \dfrac{0.4z^{-1}}{1 - 0.8z^{-1}}$

$H_2(z) = \dfrac{z^{-1}}{1 - 0.9z^{-1}}$

Cascade realizations for $H(z) = H_1(z)H_2(z)$ and $H(z) = H_2(z)H_1(z)$ are shown in Fig. 8.4-5. Find the corresponding steady-state output noise powers, $\sigma_{\epsilon_1}^2$ and $\sigma_{\epsilon_2}^2$.

Solution: The noise sources that account for rounding following multiplication are indicated in Fig. 8.4-6. As before, we assume that the individual contributions of these sources can be added to obtain the output noise powers $\sigma_{\epsilon_1}^2$ and $\sigma_{\epsilon_2}^2$.

From Fig. 8.4-6a it is apparent that the noise source 1 sees the transfer function $H_1(z)H_2(z)$, while that seen by sources 2 and 3 is $H_2(z)$. Thus (8.2-7) yields

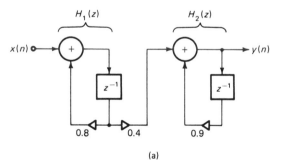

(a)

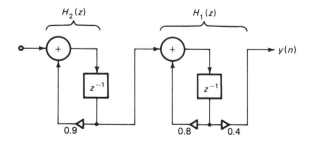

Fig. 8.4-5 Cascade realizations of $H(z)$ in (8.4-19).

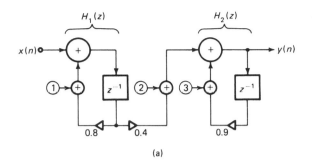

(a)

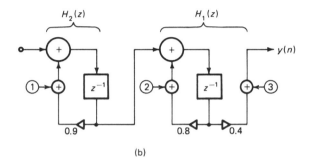

(b)

Fig. 8.4-6. Noise sources for Example 8.4-3.

$$\sigma_{\epsilon_1}^2 = \left[\frac{2^{-2B}}{3}\right] I_1 + 2\left[\frac{2^{-2B}}{3}\right] I_2 \qquad (8.4\text{-}20)$$

where $I_1 = \dfrac{1}{2\pi j} \oint_{C_u} z^{-1} H_1(z) H_2(z) H_1(z^{-1}) H_2(z^{-1})\, dz$

$I_2 = \dfrac{1}{2\pi j} \oint_{C_u} z^{-1} H_2(z) H_2(z^{-1})\, dz$

It is left as an exercise to show that (Problem 8-7)

$$I_1 = 14.37 \qquad (8.4\text{-}21)$$

and

$$I_2 = 5.26$$

Substituting (8.4-21) in (8.4-20), there results

$$\sigma_{\epsilon_1}^2 = 25.01\left[\frac{2^{-2B}}{3}\right] \qquad (8.4\text{-}22)$$

Similarly, from Fig. 8.4-6b we have

$$\sigma_{\epsilon_2}^2 = \left[\frac{2^{-2B}}{3}\right] I_1 + \left[\frac{2^{-2B}}{3}\right] I_3 + \frac{2^{-2B}}{3} \tag{8.4-23}$$

where the three contributions are due to the noise sources 1, 2, and 3, respectively. I_1 is given by (8.4-21), and

$$I_3 = \frac{1}{2\pi j} \oint_{C_u} H_1(z)H_1(z^{-1})z^{-1}\, dz \tag{8.4-24}$$

It can be shown that (Problem 8-8)

$$I_3 = 0.44$$

Substituting $I_1 = 14.37$ and $I_3 = 0.44$ in (8.4-23), we get

$$\sigma_{\epsilon_2}^2 = 15.81\left[\frac{2^{-2B}}{3}\right] \tag{8.4-25}$$

From (8.4-22) and (8.4-25) it is clear that $\sigma_{\epsilon_2}^2$ is less than $\sigma_{\epsilon_1}^2$. This observation implies that the *order* in which individual sections are cascaded also influences the output noise power due to roundoff.

8.5 SCALING CONSIDERATIONS

The objective of using scaling is to avoid overflows in a prescribed adder of a DT system's fixed-point implementation. It is important to avoid overflows since they result in undesired effects such as oscillations of large amplitude [1 to 6]. A detailed analysis related to scaling is beyond the scope of this book. An excellent discussion of the same is available in a more advanced text [1]. Our goal in this section is to acquaint the reader with a class of scaling formulas and to illustrate their use via examples.

The notion of scaling is best introduced by referring to the *single* second-order section shown in Fig. 8.5-1. We observe that there are two adders, 1 and 2. The reason for introducing a scale factor s_0 at the input is to prevent overflow at the output of the adder 1. Thus, if the output of adder 1 is the sequence $v(n)$, then from Fig. 8.5-1 it follows that

$$H'(z) = \frac{V(z)}{X(z)} = \frac{s_0}{1 + b_1 z^{-1} + b_2 z^{-2}} \tag{8.5-1}$$

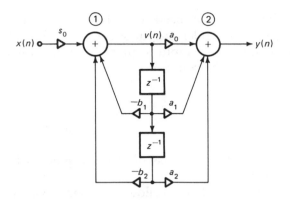

Fig. 8.5-1 Scaling the input by s_0 in order to prevent overflow in adder 1.

Now the overall input–output transfer function of the implementation in Fig. 8.5-1 is given by

$$H(z) = s_0 \left[\frac{a_0 + a_1 z^{-1} + a_2 z^{-2}}{1 + b_1 z^{-1} + b_2 z^{-2}} \right]$$

That is,

$$H(z) = s_0 \frac{N(z)}{D(z)} \qquad (8.5\text{-}2)$$

where $N(z) = a_0 + a_1 z^{-1} + a_2 z^{-2}$

From (8.5-2) it follows that (8.5-1) can be expressed as

$$H'(z) = \frac{s_0}{D(z)} \qquad (8.5\text{-}3)$$

It has been shown that a reasonable value for s_0 in order to avoid overflow in adder 1 in Fig. 8.5-1 is given by[†]

$$s_0^2 = \frac{1}{I_1} \qquad (8.5\text{-}4)$$

where $I_1 = \dfrac{1}{2\pi j} \displaystyle\oint_{C_u} \dfrac{z^{-1} \, dz}{D(z) \, D(z^{-1})}$

[†] The interested reader may refer to Appendix 8.4 for a derivation of this result.

The basic strategy used to derive (8.5-4) is to force the instantaneous energy in the adder output sequence $v(n)$ in Fig. 8.5-1, to be less than the energy in the input sequence $x(n)$. In other words, the value of s_0 ensures that the condition

$$v^2(n) < \sum_{n=0}^{\infty} x^2(n) \tag{8.5-5}$$

is satisfied, for the class of input sequences for which the right-hand side of (8.5-5) is finite.

We note that overflow in adder 2 was not addressed in the preceding discussion. This is because the output, $y(n)$, of adder 2 is not used as input to another stage, since the second-order section in Fig. 8.5-1 is assumed to be by itself. This will not be the case, however, when the section in Fig. 8.5-1 is cascaded with another, as will be discussed later. To illustrate the use of the scaling formula in (8.5-4), we now work two examples.

Example 8.5-1: Given:

$$H(z) = \frac{0.245 + 0.245z^{-1}}{1 - 0.509z^{-1}} = \frac{N(z)}{D(z)} \tag{8.5-6}$$

is the transfer function of a low-pass Butterworth DT filter whose cutoff frequency is 1 kHz at a sampling frequency of 10 kHz. Find the scaling factor s_0 to avoid overflow in adder 1 of the implementation shown in Fig. 8.5-2.

Solution: From (8.5-4) and (8.5-6) we have

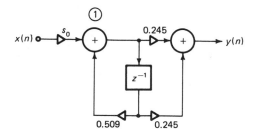

Fig. 8.5-2 Implementation of $H(z)$ in (8.5-6), where s_0 is the scaling factor to avoid over-flow in adder 1.

$$s_0^2 = \frac{1}{I_1} \qquad (8.5\text{-}7)$$

where $I_1 = \dfrac{1}{2\pi j} \displaystyle\oint_{C_u} \dfrac{z^{-1}\,dz}{(1 - 0.509z^{-1})(1 - 0.509z)}$

It is clear that I_1 is identical to I_2 in (8.4-9) with $a_2 = 0.509$. Thus (8.4-10a) yields the value of I_1 to be $1/[1 - (0.509)^2] = 1.3496$. With $I_1 = 1.3504$ in (8.5-7), the desired scaling factor is

$$s_0 = 0.86$$

Example 8.5-2: Consider a DT low-pass Butterworth filter whose transfer function is given by

$$H(z) = \frac{0.0676(1 + 2z^{-1} + z^{-2})}{1 - 1.1422z^{-1} + 0.4124z^{-2}} = \frac{N(z)}{D(z)} \qquad (8.5\text{-}8)$$

Find the scaling factor s_0 to avoid overflow in adder 1 of the implementation shown in Fig. 8.5-3.

Solution: The desired value of s_0 is obtained by evaluating (8.5-4) with $D(z)$ as given in (8.5-8); that is,

$$s_0^2 = \frac{1}{I_1} \qquad (8.5\text{-}9)$$

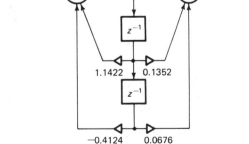

Fig. 8.5-3 Implementation of $H(z)$ in (8.5-8), where s_0 is the scaling factor to avoid overflow in adder 1.

where $I_1 = \dfrac{1}{2\pi j} \displaystyle\oint_{C_u} \dfrac{z^{-1}\,dz}{D(z)D(z^{-1})}$

with $D(z) = 1 - 1.1422z^{-1} + 0.4124z^{-2}$. The integral I_1 in (8.5-9) is readily evaluated via (8.4-17). This is because $D(z)$ has complex-conjugate roots, since $(-1.1422)^2 < 4(0.4124)$. Thus, comparing the coefficients of $D(z)$ with those of the denominator of $H(z)$ in (8.4-12), we obtain

$$\gamma^2 = 0.4124 \tag{8.5-10}$$

and
$$2\gamma \cos(\omega_0 T) = 1.1422$$

From (8.5-10) it follows that

$$\gamma = 0.642$$

and
$$\omega_0 T = 0.4743$$

which means that $\cos(2\omega_0 T) = 0.5828$.

Thus, substitution of $\gamma = 0.642$ and $\cos(2\omega_0 T) = 0.5828$ in (8.4-17) leads to $I_1 = 3.484$. With this value for I_1, (8.5-9) yields the scaling factor to be $s_0 = 0.536$.

Cascade of Two Stages

We now consider a cascade of two second-order sections as illustrated in Fig. 8.5-4. The transfer functions of the two stages are given by

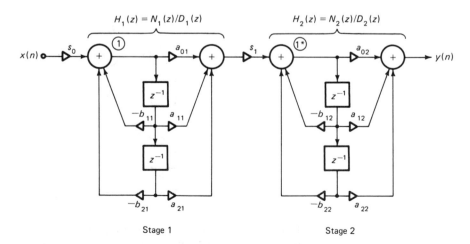

Fig. 8.5-4 Cascade of two second-order stages, where s_0 and s_1 avoid overflows in adders 1 and 1*, respectively.

$$H_1(z) = \frac{a_{01} + a_{11}z^{-1} + a_{21}z^{-2}}{1 + b_{11}z^{-1} + b_{21}z^{-2}} = \frac{N_1(z)}{D_1(z)}$$

(8.5-11)

and
$$H_2(z) = \frac{a_{02} + a_{12}z^{-1} + a_{22}z^{-2}}{1 + b_{12}z^{-1} + b_{22}z^{-2}} = \frac{N_2(z)}{D_2(z)}$$

where the subscript j in a_{ij} and b_{ij} represents the jth stage, $j = 1, 2$.

The basic idea is to use the scaling factor s_0 to avoid overflow in the adder 1 by scaling the input to stage 1. Next, the input to stage 2 is scaled by s_1 so that overflow in the adder 1* in stage 2 is avoided. We note that adders 1 and 1* are the first (leftmost) adders in each of the two stages.

From the discussion related to the single stage considered earlier, it is apparent that s_0 can be evaluated via (8.5-4) using the formula

$$s_0^2 = \frac{1}{I_1}$$

(8.5-12)

where $I_1 = \dfrac{1}{2\pi j} \oint_{C_u} \dfrac{z^{-1}\,dz}{D_1(z)D_1(z^{-1})}$

where $D_1(z)$ is the denominator polynomial of $H_1(z)$.

It is clear from Fig. 8.5-4 that the input is first scaled, processed through stage 1, and then enters stage 2. Thus it is reasonable to expect that the scaling formula for s_1 depends on s_0, $H_1(z)$, and $1/D_2(z)$, which is the input–output transfer function of adder 1*. It can be shown that[†]

$$s_1^2 = \frac{1}{s_0^2 I_2}$$

(8.5-13)

where $I_2 = \dfrac{1}{2\pi j} \oint_{C_u} \dfrac{H_1(z)H_1(z^{-1})z^{-1}\,dz}{D_2(z)D_2(z^{-1})}$

The use of the scaling formulas in (8.5-12) and (8.5-13) is best illustrated by the following example.

Example 8.5-3: The cascade implementation of

$$H(z) = H_1(z)H_2(z)$$

(8.5-14)

[†]The interested reader may refer to Appendix 8.4 for a derivation of this result.

where $H_2(z) = \dfrac{1 - 0.5z^{-1}}{1 - 0.8z^{-1}} = \dfrac{N_1(z)}{D_1(z)}$

$H_2(z) = \dfrac{1 + 0.5z^{-1}}{1 - 0.9z^{-1}} = \dfrac{N_2(z)}{D_2(z)}$

is shown in Fig. 8.5-5. Evaluate the scaling factors s_0 and s_1 to avoid overflow in adders 1 and 2, respectively.

Solution: From (8.5-12) and (8.5-14) we have

$$s_0^2 = \frac{1}{I_1} \qquad (8.5\text{-}15)$$

where $I_1 = \dfrac{1}{2\pi j} \oint_{C_u} \dfrac{z^{-1}\,dz}{(1 - 0.8z^{-1})(1 - 0.8z)}$

The value of I_1 is readily obtained by substituting $\alpha = 0.8$ in (8.2-14). Thus

$$I_1 = \frac{1}{1 - (0.8)^2} = 2.78$$

and hence (8.5-15) yields

$$s_0 = 0.6 \qquad (8.5\text{-}16)$$

Next, substitution for $H_1(z)$ and $D_2(z)$ in (8.5-13) leads to

$$s_1^2 = \frac{1}{s_0^2 I_2} \qquad (8.5\text{-}17)$$

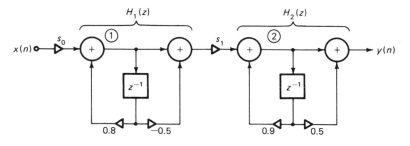

Fig. 8.5-5 Cascade implementation of $H(z)$ in (8.5-14).

where $\quad I_2 = \dfrac{1}{2\pi j} \displaystyle\oint_{C_u} \dfrac{z^{-1}(1 - 0.5z^{-1})(1 - 0.5z)\,dz}{(1 - 0.8z^{-1})(1 - 0.8z)(1 - 0.9z^{-1})(1 - 0.9z)}$

We note that there are two poles, at $z = 0.8$ and $z = 0.9$, respectively, that lie within the unit circle $|z| = 1$. Thus

$$I_2 = R_{z=0.8} + R_{z=0.9}$$

where $R_{z=0.8}$ and $R_{z=0.9}$ are the residues at the two poles. Evaluating $R_{z=0.8}$ and $R_{z=0.9}$, we obtain

$$R_{z=0.8} = -17.86$$

and

$$R_{z=0.9} = 41.35$$

which means that $I_2 = 23.49$.

With $I_2 = 23.49$ and $s_0 = 0.6$, (8.5-17) leads to

$$s_1 = 0.344 \tag{8.5-18}$$

The General Case [1]

The preceding results pertaining to two stages can be extended to a cascade of K stages as depicted in Fig. 8.5-6, where $K > 1$. Here the transfer function of the kth stage is given by

$$H_k(z) = \frac{a_{0k} + a_{1k}z^{-1} + a_{2k}z^{-2}}{1 + b_{1k}z^{-1} + b_{2k}z^{-2}} = \frac{N_k(z)}{D_k(z)} \tag{8.5-19}$$

for $1 \le k \le K$.

The scaling factor s_0 avoids overflow in the first (leftmost) adder of stage 1, while s_1 avoids overflow in the first adder of stage 2, and so on. The formulas for s_0 and s_1 are given in (8.5-4) and (8.5-13), respectively; that is,

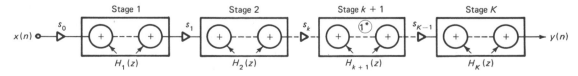

Fig. 8.5-6 Cascade implementation of a given $H(z)$ using K stages.

$$s_0^2 = \frac{1}{I_1}$$

where $\quad I_1 = \dfrac{1}{2\pi j} \oint_{C_u} \dfrac{z^{-1}\,dz}{D_1(z)D_1(z^{-1})}$

and $\qquad\qquad\qquad\qquad s_1^2 = \dfrac{1}{s_0^2 I_2}$

where $\quad I_2 = \dfrac{1}{2\pi j} \oint_{C_u} \dfrac{H_1(z)H_1(z^{-1})z^{-1}\,dz}{D_2(z)D_2(z^{-1})}$

In general, the scaling factor s_k avoids overflow in the first adder of stage $k + 1$. The general formula for s_k for a cascade of K stages is obtained as a simple extension of the case $K = 2$ discussed earlier. Thus we have

$$s_k^2 = \frac{1}{\left[\displaystyle\prod_{i=0}^{k-1} s_i^2\right] I_{k+1}}, \qquad 1 \le k \le K - 1 \tag{8.5-20}$$

where $\quad I_{k+1} = \dfrac{1}{2\pi j} \oint_{C_u} \dfrac{\left[\displaystyle\prod_{i=1}^{k} H_i(z)H_i(z^{-1})\right]z^{-1}}{D_{k+1}(z)D_{k+1}(z^{-1})}\,dz$

and s_0 is given by (8.5-4).

We note that s_k depends upon all the previous scaling factors s_i, $1 \le i \le k - 1$, and the transfer functions $H_i(z)$, $1 \le i \le k$, to account for processing through the past k stages. In addition, it depends on $1/D_{k+1}(z)$, which is the transfer function between the input to stage $k + 1$, and its first adder in Fig. 8.5-6.

8.6 LIMIT CYCLE OSCILLATIONS

There are two types of limit cycle oscillations. The first type is called *zero input limit cycle oscillations,* and refers to the phenomenon of getting a nonzero output in the absence of an input. This phenomenon occurs due to finite word sizes that are used to encode the data values, system coefficients, or both. To illustrate, we consider the first-order IIR system

$$y(n) = x(n) - 0.9y(n - 1), \qquad n \ge 0 \tag{8.6-1}$$

Table 8.6-1   Output with No Rounding

n	$x(n)$	$y(n-1)$	$y(n) = x(n) - 0.9y(n-1)$
0	10	0	10
1	0	10	-9
2	0	9	8.1
3	0	-8.1	-7.29
4	0	-7.29	6.561
5	0	6.561	-5.8949
6	0	-5.8949	4.80541
7	0	4.80541	-4.324869
8	0	.	.
.	.	.	.
∞	0	0	0

where
$$x(n) = \begin{cases} 10, & n = 0 \\ 0, & n > 0 \end{cases}$$

If infinite precision is available, then $y(n)$ tends to 0 with increasing values of n, as apparent from Table 8.6-1. Next, let us consider the case when only finite precision is available to encode data values. As such, we shall assume that each data value is *rounded to the nearest integer*. Again, let the input be the impulse sequence

$$x(n) = \begin{cases} 10, & n = 0 \\ 0, & n > 0 \end{cases} \tag{8.6-2}$$

Then the resulting output sequence is as summarized in Table 8.6-2, where the notation $[x]$ implies that the number x is rounded to the

Table 8.6-2   Output with Rounding

n	$x(n)$	$y^Q(n-1)$	$y^Q(n) = x(n) - 0.9y^Q(n-1)$
0	10	0	10
1	0	10	-9
2	0	-9	$[8.1] = 8$
3	0	8	$[-7.2] = -7$
4	0	-7	$[6.3] = 6$
5	0	6	$[-5.4] = -5$
6	0	-5	$[4.5] = 5$
7	0	5	$[-4.5] = -5$
8	0	-5	$[4.5] = 5$
.	.	.	.
.	.	.	.

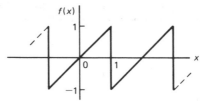

Fig. 8.6-1 Transfer characteristic of 2's complement adder.

nearest integer. From Table 8.6-2 it is clear that the output sequence is periodic beyond $n = 5$, although there is no input! This output is an example of the concept of zero input limit cycle oscillations. Clearly, it is a lack of sufficient accuracy that leads to this type of phenomenon. It is therefore important to have large enough word lengths so that such oscillations are kept small (e.g., see Problems 8-16 to 8-18).

Oscillations Due to Overflow

The second type of limit cycle oscillations is due to the overflow of adders that are used in the implementation of a DT system. The notion of this type of oscillations is best introduced by considering two positive data values x_1 and x_2, where $|x_i| < 1$, $i = 1, 2$. Since 2's complement arithmetic is assumed, it is clear that the sign bit associated with x_1 as well as x_2 is 0. Now suppose the sum of x_1 and x_2 exceeds 1 (i.e., an overflow occurs). Then it follows that a carry bit results to the left of the binary point. As such, the sign bit would become 1 and the resulting sum would be incorrectly interpreted as a *negative* number. It can be shown that this phenomenon could lead to large oscillations, which are undesirable [7].

We recall that the scaling procedure in Section 8.5 would assist in avoiding this adder overflow problem. An additional approach is also available by realizing that the transfer characteristic of a 2's complement adder is as illustrated in Fig. 8.6-1, where x is the input to the

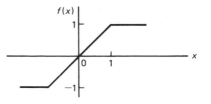

Fig. 8.6-2 Transfer characteristic of 2's complement adder that saturates.

adder, and $f(x)$ is the corresponding output. It can be shown that if an adder is modified to *saturate*, as indicated in Fig. 8.6-2, then the undesired overflow can be avoided. From a practical point of view, this means that one has to monitor overflows, and if overflow is detected, then the sum is set equal to the maximum allowable value.

8.7 TABLE LOOKUP IMPLEMENTATION METHOD

It is apparent that the implementations that we have considered thus far involve additions (subtractions) and multiplications. It is well known that multiplications are time-consuming operations. As such, special-purpose hardware multipliers may be required in real-time applications. In such cases it could be cost effective to consider the table lookup method [1], which enables one to implement first- and second-order sections of a cascade or parallel realization via additions (subtractions) and shifts only; that is, no multiplications are necessary. We devote this section to an introductory discussion of this useful implementation method, which was patented by Croisier et al. [8].

CASE 1:   FIRST-ORDER SYSTEM. If $x(n)$ is the input sequence to a first-order IIR system with transfer function

$$H(z) = \frac{a_0 + a_1 z^{-1}}{1 + b_1 z^{-1}} \tag{8.7-1}$$

then its output sequence $y(n)$ is given by

$$y(n) = a_0 x(n) + a_1 x(n - 1) - b_1 y(n - 1), \qquad n \geq 0 \tag{8.7-2}$$

For convenience, we introduce the notation

$$x(n) = x_n, \qquad y(n - 1) = y_{n-1}, \quad \text{etc.}$$

Then (8.7-2) becomes

$$y_n = a_0 x_n + a_1 x_{n-1} - b_1 y_{n-1}, \qquad n \geq 0 \tag{8.7-3}$$

It is assumed that the input is scaled so that $|x_n| < 1$, and word lengths of B bits are used to encode the data values and system coefficients. We shall also assume that a fixed-point implementation that employs 2's complement arithmetic is desired. Then from (8.1-5) it follows that x_n, x_{n-1}, y_n, and y_{n-1} can be written as

$$y_n = -y_n^0 + \sum_{j=1}^{B-1} y_n^j$$

$$y_{n-1} = -y_{n-1}^0 + \sum_{j=1}^{B-1} y_{n-1}^j 2^{-j} \qquad (8.7\text{-}4)$$

$$x_n = -x_n^0 + \sum_{j=1}^{B-1} x_n^j 2^{-j}$$

and
$$x_{n-1} = -x_{n-1}^0 + \sum_{j=1}^{B-1} x_{n-1}^j 2^{-j}$$

where the superscripts 0 and j denote the sign bit and jth data bit, respectively.

Substituting (8.7-4) in (8.7-3) and rearranging terms, we obtain

$$y_n = \sum_{j=1}^{B-1} F(x_n^j, x_{n-1}^j, y_{n-1}^j)2^{-j} - F(x_n^0, x_{n-1}^0, y_{n-1}^0) \qquad (8.7\text{-}5a)$$

where $F(x_n^j, x_{n-1}^j, y_{n-1}^j) = a_0 x_n^j + a_1 x_{n-1}^j - b_1 y_{n-1}^j \qquad (8.7\text{-}5b)$

FUNDAMENTAL REMARK. The right-hand side of (8.7-5b) has *exactly the same form* as the right-hand side of the input–output difference equation in (8.7-3). The only difference is that the quantities x_n, x_{n-1}, and y_{n-1} have been replaced by x_n^j, x_{n-1}^j, and y_{n-1}^j, respectively. This property of 2's complement notation is valid for higher-order systems also.

Next, examination of (8.7-5) reveals the following important properties pertaining to it:

1. $F(x_n^j, x_{n-1}^j, y_{n-1}^j)$ is a function of three *binary variables* x_n^j, x_{n-1}^j, y_{n-1}^j, and hence takes only 8 ($= 2^3$) values. Once these 8 values are computed, they can be stored in a device such as a read only memory (ROM) for future use.

2. Each output sample y_n can be computed by resorting to successive additions and shifting and a subtraction. Clearly, *no multiplications* are involved.

Table 8.7-1   Read Only Memory Addresses and Contents

Location	x_n^j	ROM Address x_{n-1}^j	y_{n-1}^j	ROM Contents $F(x_n^j, x_{n-1}^j, y_{n-1}^j)$
0	0	0	0	0
1	0	0	1	$-b_1$
2	0	1	0	a_1
3	0	1	1	$a_1 - b_1$
4	1	0	0	a_0
5	1	0	1	$a_0 - b_1$
6	1	1	0	$a_0 + a_1$
7	1	1	1	$a_0 + a_1 - b_1$

3. The bits x_n^j, x_{n-1}^j, y_{n-1}^j serve as the *address* to the contents of the ROM, which are the 8 values of the function $F(x_n^j, x_{n-1}^j, y_{n-1}^j)$, as summarized in Table 8.7-1.

From the preceding discussion it is now apparent that the output values y_n in (8.7-3) can be computed via (8.7-5) as depicted in Fig. 8.7-1. The register R1 shifts its contents j places to account for the term 2^{-j} in (8.7-5a), while the output is accumulated in register R2. The contents of R2 are then loaded in the output register at the end of $(M - 1)$ successive additions and shifts, followed by the subtraction of $F(x_n^0, x_{n-1}^0,$

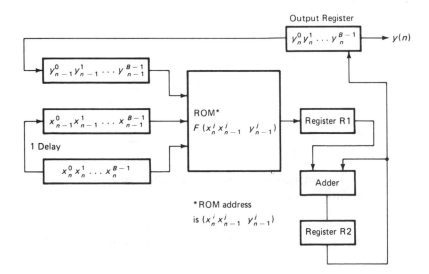

Fig. 8.7-1 Block diagram of table lookup implementation of a first-order IIR system.

y_{n-1}^0). Since the process of retrieving values of $F(x_n^j, x_{n-1}^j, y_{n-1}^j)$ with address $(x_n^j, x_{n-1}^j, y_{n-1}^j)$ from the ROM is equivalent to looking up a table of values, this approach is known as a table lookup implementation for first-order IIR filters.

CASE 2: SECOND-ORDER SYSTEMS. We now consider an IIR system whose transfer function is

$$H(z) = \frac{a_0 + a_1 z^{-1} + a_2 z^{-2}}{1 + b_1 z^{-1} + b_2 z^{-2}} \tag{8.7-6}$$

which results in the input–output equation

$$y_n = a_0 x_n + a_1 x_{n-1} + a_2 x_{n-2} - b_1 y_{n-1} - b_2 y_{n-2}, \qquad n \geq 0 \tag{8.7-7}$$

As in the first-order case, the data values in (8.7-7) can be represented in the following manner using 2's complement notation:

$$y_n = -y_n^0 + \sum_{j=1}^{B-1} y_n^j 2^{-j}$$

$$y_{n-1} = -y_{n-1}^0 + \sum_{j=1}^{B-1} y_{n-1}^j 2^{-j}$$

$$y_{n-2} = -y_{n-2}^0 + \sum_{j=1}^{B-1} y_{n-2}^j 2^{-j} \tag{8.7-8}$$

$$x_{n-1} = -x_{n-1}^0 + \sum_{j=1}^{B-1} y_{n-2}^j 2^{-j}$$

and

$$x_{n-2} = -x_{n-2}^0 + \sum_{j=1}^{B-1} x_{n-2}^j 2^{-j}$$

Substitution of (8.7-8) in (8.7-7) and subsequent rearrangement of terms lead to

$$y_n = \sum_{j=1}^{B-1} F(x_n^j, x_{n-1}^j, x_{n-2}^j, y_{n-1}^j, y_{n-2}^j) 2^{-j}$$
$$- F(x_n^0, x_{n-1}^0, x_{n-2}^0, y_{n-1}^0, y_{n-2}^0) \tag{8.7-9a}$$

where

$$F(x_n^j, x_{n-1}^j, x_{n-2}^j, y_{n-1}^j, y_{n-2}^j) = a_0 x_n^j + a_1 x_{n-1}^j + a_2 x_{n-2}^j \quad (8.7\text{-}9b)$$
$$- b_1 y_{n-1}^j - b_2 y_{n-2}^j$$

Once again we note that the right-hand side of (8.7-9b) is exactly the same as the right-hand side of the input–output difference equation in (8.7-7), and hence in agreement with the fundamental remark that followed (8.7-5b).

From (8.7-9b) it is clear that $F(x_n^j, x_{n-1}^j, \ldots y_{n-2}^j)$ is a Boolean function of the 5 binary variables $x_n^j, x_{n-1}^j, x_{x-2}^j, y_{n-1}^j$, and y_{n-2}^j. Thus it has 32 ($= 2^5$) values that can be stored in a ROM and addressed via the binary 5-tuple $(x_n^j, x_{n-1}^j, x_{n-2}^j, y_{n-1}^j, y_{n-2}^j)$. The corresponding block diagram of the implementation is obtained as a straightforward extension of that for the first-order case in Fig. 8.7-1. It is as shown in Fig. 8.7-2.

Scaling Considerations [9]

In the table lookup implementation just discussed, we could scale the contents of the ROM so that the output register shown in Figs. 8.7-1 and 8.7-2 does not overflow; that is, $|y_n| < 1$. To this end, we now develop a scaling procedure. For convenience we consider the first-order case,

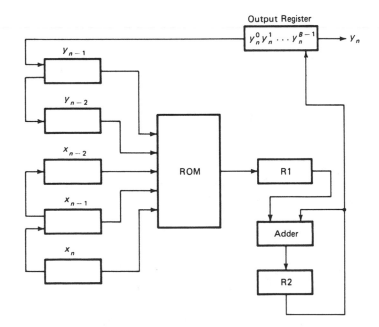

Fig. 8.7-2 Basic block diagram of table lookup implementation of a second-order IIR system.

and the results so obtained will also be applicable for the second-order case.

Let us commence by defining the parameters

$$\alpha = \max \{F(x_n^j, x_{n-1}^j, y_{n-1}^j)\} \tag{8.7-10}$$

and

$$\beta = \min \{F(x_n^j, x_{n-1}^j, y_{n-1}^j)\}$$

From the ROM contents in Table 8.7-1 it is clear that $\beta \le 0$.

Now, from (8.7-5) and (8.7-10), it follows that

$$y_n \le \alpha \sum_{j=1}^{B-1} 2^{-j} - \beta$$

which yields the strict inequality

$$y_n < \alpha \sum_{j=1}^{\infty} 2^{-j} - \beta \tag{8.7-11}$$

In (8.7-11) we note that

$$\sum_{j=1}^{\infty} 2^{-j} = \frac{1}{2} + \frac{1}{2^2} + \frac{1}{2^3} + \cdots = \frac{1}{2}\left(1 + \frac{1}{2} + \frac{1}{2^2} + \cdots\right)$$

$$= \frac{1}{2}\left(\frac{1}{1 - \frac{1}{2}}\right) = 1 \tag{8.7-12}$$

Thus (8.7-11) simplifies to yield

$$y_n < (\alpha - \beta) \tag{8.7-13}$$

Again, from (8.7-5) and (8.7-10) it is apparent that

$$y_n \ge \beta \sum_{j=1}^{B-1} 2^{-j} - \alpha$$

since $\beta \le 0$. That is,

$$y_n > \beta \sum_{j=1}^{\infty} 2^{-j} - \alpha \tag{8.7-14}$$

Substitution of (8.7-12) in (8.7-14) leads to

$$y_n > (\beta - \alpha) \tag{8.7-15}$$

Combining (8.7-13) and (8.7-15), we conclude that one way of ensuring that $|y_n| < 1$ is to scale each of the values $F(x_n^j, x_{n-1}^j, y_{n-1}^j)$ by a scaling factor, S, where

$$S > \alpha - \beta \tag{8.7-16}$$

and then storing the same in the ROM; note that S is positive, since $\beta \leq 0$ and $\alpha > \beta$. As such, the actual values stored in the ROM are given by

$$\hat{F}(x_n^j, x_{n-1}^j, y_n^j) = \frac{F(x_n^j, x_{n-1}^j, y_{n-1}^j)}{S} \tag{8.7-17}$$

In practice S should be chosen to be slightly larger than $\alpha - \beta$ so that the dynamic range of y_n is not sacrificed.

The preceding implementation and scaling procedure are now illustrated by the following example.

Example 8.7-1: The transfer function of a certain second-order IIR system is given by

$$H(z) = \frac{0.4z^{-1}}{1 - 1.85z^{-1} + 0.855z^{-2}} \tag{8.7-18}$$

If the system is implemented by the table lookup method, find the contents of the ROM. Account for scaling so that $|y_n| < 1$, where y_n is the output sequence.

Solution: With $H(z)$ in (8.7-18) substituted in the relation $Y(z) = H(z)X(z)$, we obtain the input–output relation

$$y_n = 0.4x_{n-1} + 1.85y_{n-1} - 0.855y_{n-2} \tag{8.7-19}$$

In accordance with the notation in (8.7-4), we have

$$y_n = -y_n^0 + \sum_{j=1}^{B-1} y_n^j 2^{-j}$$

$$y_{n-1} = -y_{n-1}^0 + \sum_{j=1}^{B-1} y_{n-1}^j 2^{-j}$$

$$(8.7\text{-}20)$$

$$y_{n-2} = -y_{n-2}^0 + \sum_{j=1}^{B-1} y_{n-1}^j 2^{-j}$$

and

$$x_{n-1} = -x_{n-1}^0 + \sum_{j=1}^{B-1} x_{n-1}^j 2^{-j}$$

Then from (8.7-9) it follows that substitution of (8.7-20) in (8.7-19) leads to

$$y_n = \sum_{j=1}^{B-1} F(x_{n-1}^j, y_{n-1}^j, y_{n-2}^j) 2^{-j} - F(x_{n-1}^0, y_{n-1}^0, y_{n-2}^0) \quad (8.7\text{-}21a)$$

where

$$F(x_{n-1}^j, y_{n-1}^j, y_{n-2}^j) = 0.4x_{n-1}^j + 1.85y_{n-1}^j - 0.855y_{n-2}^j \quad (8.7\text{-}21b)$$

We emphasize that the right-hand side of (8.7-21b) can be readily *written down* since it has exactly the same form as the right-hand side of the input–output equation in (8.7-19). The only difference is that x_{n-1}^j, y_{n-1}^j, and y_{n-2}^j take the place of x_{n-1}, y_{n-1}, and y_{n-2}, respectively.

It is apparent that $F(x_{n-1}^j, y_{n-1}^j, y_{n-2}^j)$ takes eight values as indicated in Table 8.7-2. These values are found by substituting the values of x_{n-1}^j, y_{n-1}^j, and y_{n-2}^j in the right-hand side of (8.7-21b).

Next we need to scale the values of $F(x_{n-1}^j, y_{n-1}^j, y_{n-2}^j)$ so that $|y_n|$ < 1. To this end we obtain α and β defined in (8.7-10) by examining the values of $F(x_{n-1}^j, y_{n-1}^j, y_{n-2}^j)$ listed in Table 8.7-2. It follows that $\alpha =$

Table 8.7-2   Eight Values Pertaining to Example 8.7-1

x_{n-1}^j	y_{n-1}^j	y_{n-2}^j	$F(x_{n-1}^j, y_{n-1}^j, y_{n-2}^j)$	ROM Contents: $\hat{F}(x_{n-1}^j, y_{n-1}^j, y_{n-2}^j)$
0	0	0	0	0.0000
0	0	1	−0.855	−0.2672
0	1	0	1.85	0.5781
0	1	1	0.995	0.3109
1	0	0	0.4	0.1250
1	0	1	−0.455	−0.1422
1	1	0	2.25	0.7031
1	1	1	1.395	0.4359

2.25 and $\beta = -0.855$, which yields $\alpha - \beta = 3.105$. Thus the scaling factor S in (8.7-16) must be chosen to be slightly larger than 3.105. For example, let $S = 3.2$. Then the contents of the ROM are given by (8.7-17) to be

$$\hat{F}(x_{n-1}^j, y_{n-1}^j, y_{n-2}^j) = \frac{F(x_{n-1}^j, y_{n-1}^j, y_{n-2}^j)}{3.2}$$

which yields the decimal values listed in Table 8.7-2. These values would be encoded in binary form and stored in the ROM. The pertinent word lengths may be determined via (8.1-16).

A computer program that yields the ROM contents for a given first- or second-order $H(z)$ is given in Appendix 8.2.

8.8 SUMMARY

This chapter was devoted to addressing some issues related to fixed-point implementations of IIR systems. To this end, a number of illustrative examples were presented in connection with input quantization, coefficient accuracy, product roundoff error, scaling, and limit cycles. Also, a useful implementation technique known as the table lookup method was presented and illustrated by examples.

We restricted our attention to fixed-point implementations in this chapter. However, more recently, the notion of block floating-point (BFP) implementations for DT filters has been gaining popularity with practicing engineers. This is because BFP implementations for DT filters are amenable to real-time applications using microprocessor-based systems. The interested reader may refer to Appendix 8.3 for a discussion of BFP notation.

In conclusion, we add that a heuristic optimization approach is available for pole zero pairing and scaling for minimum roundoff error for a given $H(z)$; for example, see [1], Chapter 6.

PROBLEMS

8–1 Find the fixed-point representation for the following decimal numbers using 7 data bits and a sign bit: 0.2; 0.435; -0.72; 0.81640625.

8–2 Truncate and round the fixed-point numbers obtained in Problem 8-1 to 6 data bits. Also compute the corresponding truncation and roundoff errors.

8–3 How many data bits would you use to encode each of the decimal numbers given in Problem 8-1 so that the resulting accuracy is at least that associated with the decimal representations?

8–4 A certain sequence $x(n)$ is encoded in binary form using 8 bits (i.e., $B = 8$). The resulting quantized sequence $x^Q(n)$ is then processed through a filter with transfer function

$$H(z) = \frac{1}{1 - 1.7z^{-1} + 0.72z^{-2}}$$

Find the steady-state noise power due to quantization that occurs at the output of $H(z)$.

8–5 Using truncation with 3 data bits and a sign bit to represent the b_j and a_i in (8.3-5), derive $H'(z)$ and $H''(z)$ in (8.3-6).

8–6 Show that I_2 and I_1 in (8.4-9) evaluate to the expressions given in (8.4-10a) and (8.4-10b), respectively.

8–7 Verify that I_1 and I_2 in (8.4-20) evaluate to 14.45 and 5.28, respectively.

8–8 Show that I_3 in (8.4-24) evaluates to 0.44.

8–9 Consider the transfer function

$$H(z) = \frac{1}{1 - 0.94z^{-1} + 0.64z^{-2}}$$

(a) Find the pole locations of $H(z)$ and plot its magnitude response $|\bar{H}(\nu)|$, $0 \le \nu \le 1$, using the computer program in Appendix 5.1.

(b) Find the fixed-point representations of the coefficients -0.94 and 0.64 using rounding to 3 data bits. This process will result in an approximation to $H(z)$, say $H'(z)$. Find the pole locations of $H'(z)$ and plot its magnitude response $|\bar{H}'(\nu)|$, $0 \le \nu \le 1$, and compare the same to the results obtained in part (a) with respect to $H(z)$.

8–10 Repeat the steps outlined in Problem 8-9 for the transfer function

$$H(z) = \frac{1}{(z - 0.9)(z^2 - 0.8z + 0.52)}$$

8–11 Show that the total steady-state noise power at the output of the direct form 1 realization in Fig. 5.6-1 is given by

$$\sigma_\epsilon^2 = \left[\frac{2^{-2B}}{3} \right] (2k + 1)I$$

where $I = \dfrac{1}{2\pi j} \oint_{C_u} \dfrac{z^{-1} \, dz}{D(z)D(z^{-1})}$

with $D(z)$ being the denominator polynomial of the given transfer function $H(z)$.

8–12 Consider the transfer function

$$H(z) = \frac{1}{(1 - 0.9z^{-1})(1 - 0.8z^{-1})}$$

whose cascade realization is shown in Fig. 8.4-3. Use (8.4-11) to evaluate the steady-state output noise power, $\sigma_{\epsilon_1}^2$, due to roundoff. Also compare $\sigma_{\epsilon_1}^2$ with the corresponding output noise powers, $\sigma_{\epsilon_2}^2$, for the parallel realizations shown in the following sketches.

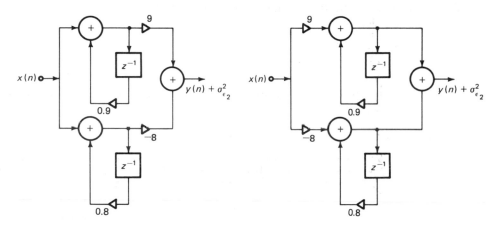

8–13 Given:

$$H(z) = \frac{1 + 0.875z^{-1}}{1 - 0.72z^{-1}}$$

which is implemented as shown.

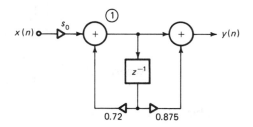

(a) Find s_0 to avoid overflow in adder 1.

(b) Encode the filter coefficients and the value of s_0 obtained in part (a) in binary (fixed-point form). Use a sufficient number of bits to ensure sufficient accuracy by using the relation in (8.1-16).

8–14   Show that the total steady-state noise power at the output of the direct form 2 realization in Fig. 5.6-2 is given by

$$\sigma_\epsilon^2 = \left[\frac{2^{-2B}}{3}\right] kI + (k + 1)\frac{2^{-2B}}{3}$$

where $\quad I = \dfrac{1}{2\pi j} \oint_{C_u} z^{-1}H(z)H(z^{-1})\, dz$

8–15   Consider the first-order transfer function

$$H(z) = \frac{1 + \alpha z^{-1}}{1 - \beta z^{-1}}, \qquad |\beta| < 1$$

whose direct form 1 and direct form 2 (or canonic) realizations are as follows.

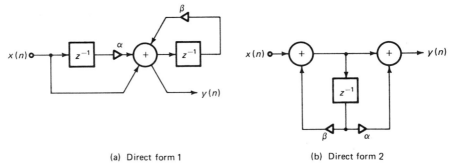

(a) Direct form 1   (b) Direct form 2

Let $\sigma_{\epsilon_1}^2$ and $\sigma_{\epsilon_2}^2$ denote the steady-state output noise powers due to round-off for realizations (a) and (b), respectively. Show that [6]

$$\sigma_{\epsilon_2}^2 = \sigma_{\epsilon_1}^2 + \left[\frac{2^{-2B}}{3}\right] P$$

where $\quad P = \dfrac{\alpha^2 + 2\alpha\beta - \beta^2}{1 - \beta^2}$

Hint: First show that

$$\sigma_{\epsilon_1}^2 = 2\left[\frac{2^{-2B}}{3}\right]\frac{1}{1 - \beta^2}$$

Comment: Note that direct form 1 and direct form 2 are preferred when $P > 0$ and $P < 0$, respectively. If $P = 0$, then either form produces the same roundoff noise.

8-16 Problems 8-16 through 8-18 concern the zero input limit cycle phenomenon, with respect to first-order IIR systems [1]. Consider the system

$$y(n) = x(n) + 0.5\,y(n - 1), \qquad n \geq 0 \qquad \text{(P8-16-1)}$$

Suppose $y(-1) = 0$, and

$$x(n) = \begin{cases} 0.5, & n = 0 \\ 0, & n > 0 \end{cases}$$

(a) Evaluate $y(0)$, $y(1)$, $y(2)$, $y(3)$, and $y(\infty)$, and summarize the results as in Table 8.6-1.

(b) Now suppose the data registers have 3 data bits plus a sign bit. Then, corresponding to (P8-16-1), we have

$$y^Q(n) = x(n) + 0.5y^Q(n - 1), \qquad n \geq 0$$

Evaluate $y^Q(n)$ for $n = 0, 1, 2, 3, 4$, and hence show that $y^Q(n) = 0.125$ for all $n \geq 2$. Summarize your results as in Table 8.6-2.

8-17 Repeat Problem 8-16 for the IIR system

$$y(n) = x(n) - 0.5y(n - 1), \qquad n \geq 0$$

In this case show that $y^Q(n)$ oscillates between ± 0.125 for $n \geq 2$.

8-18 Repeat Problem 8-17 with the same difference equation but use data registers with 4 data bits plus a sign bit. Show that the output $y^Q(n)$ now oscillates between ± 0.0625, for $n \geq 2$.

Note: One can generalize the results of Problems 8-16 through 8-18 in that $y^Q(n)$ oscillates between $\pm(0.5)^M$ when the data registers have M data bits plus a sign bit. Clearly, such oscillations can be made small by increasing M.

8-19 Show how the following second-order system can be implemented using the table lookup method:

$$H(z) = \frac{0.875z^{-1}}{1 - 0.75z^{-2}}$$

Show the ROM contents in binary form using 7 data bits and a sign bit.

8-20 Repeat Problem 8-19 for the transfer function

$$H(z) = \frac{0.4z^{-1}}{1 - 0.17z^{-1} + 0.72z^{-2}}$$

8–21 The cascade implementation of

$$H(z) = \frac{1}{1 - 1.7z^{-1} + 0.72z^{-2}}$$

is as shown, where s_0 and s_1 are scaling factors

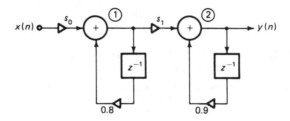

to avoid overflow in adders 1 and 2, respectively. Evaluate s_0 and s_1.

8–22 An implementation for

$$H(z) = \frac{1 + 0.5z^{-1}}{1 - 0.9z^{-1}}$$

is shown, where s_0 is included to avoid overflow in adder 1.

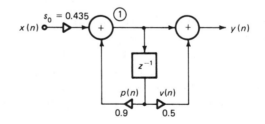

Suppose the input and output signal powers are denoted by σ_x^2 and σ_y^2, respectively.

(a) Find σ_y^2 in terms of σ_x^2 using the relation [1 to 4]

$$\sigma_y^2 = s_0^2 \sigma_x^2 I_1$$

where $\quad I_1 = \dfrac{1}{2\pi j} \oint_{C_u} z^{-1} H(z) H(z^{-1}) \, dz$

(b) Let $\sigma_{\epsilon_1}^2$ denote the output noise power due to input quantization (i.e., via a B-bit A/D converter). Evaluate $\sigma_{\epsilon_1}^2$ using the formula

$$\sigma_{\epsilon_1}^2 = \left\{ \frac{2^{-2B}}{3} \right\} s_0^2 I_1$$

where I_1 is as defined in part (a).

(c) Let $\sigma_{\epsilon_2}^2$ denote the output noise power due to rounding the products $s_0 x(n)$, $0.9 p(n)$, and $0.5 v(n)$ shown in the implementation, assuming B-bit registers. Find $\sigma_{\epsilon_2}^2$.

8–23 With respect to Problem 8-22, we can define the following signal-to-noise ratio (SNR), assuming that $\sigma_x^2 = 1$:

$$SNR = 10 \log_{10} \left[\frac{\sigma_y^2}{\sigma_{\epsilon_1}^2 + \sigma_{\epsilon_2}^2} \right] dB$$

Find the minimum value of B so that the SNR is at least 40 dB.

APPENDIX 8.1   FIXED-POINT AND FLOATING-POINT REPRESENTATION

Most numeric data within computers is stored using *fixed-point* or *floating-point representation*. The former is used to store integer and fractional numbers, and the latter to store any real number.

    Binary fixed-point schemes are those where the location of the binary point is fixed. A common scheme is the 2's complement representation, in which the leftmost (most significant) bit is called the *sign bit*. This is because the value of the sign bit indicates the sign of the number represented; for example, 0 implies a positive number, and 1 a negative number. The remaining bits are called data bits. For instance, the decimal numbers $(+123)_{10}$ and $(-123)_{10}$ have the following 16-bit fixed-point representations:

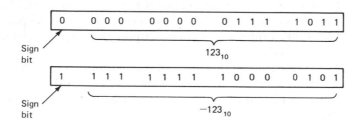

Thus we observe that the 2's complement representations of the largest and smallest integers are as follows:

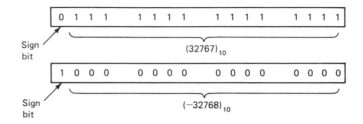

From the preceding illustrations it is apparent that the range of integers we can represent via the 16-bit representation is given by

$$-32768 \leq (N)_{10} \leq 32767$$

Thus it follows that the range of the numbers that can be stored via a fixed-point representation is quite restricted. Any attempt to store numbers that are outside this range results in the condition of *integer overflow.*

Next let us consider some aspects of floating-point representation, which are as follows:

| Sign | Fraction | X Radix $^{sign, exponent}$ |

To illustrate, let the radix be 10. Then, for example, the floating-point representation of the numbers .0023, -27416.843, and 202.611 are as follows.

$$(a) \ +.23 \times 10^{-2} = .0023$$

$$(b) \ -.27416843 \times 10^{+5} = -27416.843$$

$$(c) \ +.202611 \times 10^{+3} = 202.611$$

Thus we see that floating-point representation enables us to represent a much wider range of numbers, compared to fixed-point representation. The radix employed depends upon the specific computer (processor) that is used.

One type of floating-point representation is known as *binary floating-point* representation; it is shown in the following for the 32-bit case. It uses 6 bits and 24 bits to store the exponent and magnitude, respectively.

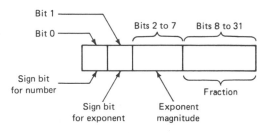

From this illustration if follows that numbers stored using this representation lie in the range

$$2^{-64} \leq (N)_{10} \leq 2^{64}$$

or
$$10^{-19} \leq (N)_{10} \leq 10^{19}$$

Clearly, this type of representation results in a much larger range than is the case with a fixed-point representation. However, it is important to note that the *accuracy* (or *precision*) with which we can store a floating-point number depends upon the number of bits that are allocated to the fractional part. Again, *floating-point overflow* occurs when one attempts to store a number larger than the maximum of the range, for example, $(N)_{10} > 10^{19}$ for a 32-bit representation. Conversely, numbers smaller than the minimum of the range result in a *floating-point underflow*, for example, $(N)_{10} < 10^{-19}$ for a 32-bit representation.

APPENDIX 8.2† COMPUTER PROGRAM FOR TABLE LOOKUP IMPLEMENTATION METHOD

The program† given in this appendix enables one to compute and scale the values of the function $F(x_n^j, x_{n-1}^j, x_{n-2}^j, y_{n-1}^j, y_{n-2}^j)$ corresponding to a given second-order IIR filter transfer function

$$H(z) = \frac{a_0 + a_1 z^{-1} + a_2 z^{-2}}{1 + b_1 z^{-1} + b_2 z^{-2}}$$

The output of this program is given in 2's complement notation in both its true and complemented form as a fixed-point 8-bit number. It is also converted to hexadecimal form for easy input to programmable read only memory (PROM) programmer stations. To illustrate, the output related to

---

†Program developed by David Hein and John Schmalzel.

$$H(z) = \frac{0.07 + 0.14z^{-1} + 0.07z^{-2}}{1 - 0.59z^{-1} + 0.3z^{-2}}$$

is listed.

```
C        THIS PROGRAM COMPUTES THE SCECOND ORDER FILTER COEFFICIENTS
C        USED IN DIGITAL FILTERS IMPLEMENTED USING THE TABLE LOOK UP  ,
C        METHOD ; DATA INPUT ARE THE COEFFICIENTS A(0),A(1),A(2),B(1),B(2)
C        FROM THE GENERALIZED TRANSFER FUNCTION;
C                              -1            -2
C            Y(Z)    A(0) + A(1)*Z   + A(2)*Z
C            ----- = ----------------------------
C                              -1            -2
C            X(Z)    1    + B(1)*Z   + B(2)*Z
C
C        WHOSE SOLUTION IS OF THE FORM:
C
C        Y(N) = A(0)*X(N)+A(1)*X(N-1)+A(2)*X(N-2)-B(1)*Y(N-1)-B(2)*Y(N-2)
C
C        THE PROGRAM COMPUTES THE FUNCTION F(----) AND SCALES IT ,
C        THE DECIMAL RESULTS ARE CONVERTED TO TWO'S COMPLEMENT;IT'S
C        COMPLEMENT IS ALSO AVAILABLE;THE MSB IS THE SIGN BIT ,
C
C        CONNECTIONS TO THE ROM ARE AS FOLLOWS:
C
C                        INPUT         ROM ADDRESS
C
C                        Y(N-2)            A4
C                        Y(N-1)            A3
C                        X(N-2)            A2
C                        X(N-1)            A1
C                        X(N)              A0
C
C
C
         INTEGER MSB,LSB,MSBC,LSBC
         INTEGER BIN(8),BINC(8)
         INTEGER X0,X1,X2,Y1,Y2
         REAL ROM(32),ROMS(32)
         REAL A0,A1,A2,B1,B2
C
    1    FORMAT(8F10.0)
C
    2    FORMAT('1 CONTENTS OF ROM SCALED BY ',F5.3)
C
    3    FORMAT(' ',I10,2F20.6,10X,I1,'.',3I1,1X,4I1,10X,2A1,
        * 10X,I1,'.',3I1,1X,4I1,10X,2A1)
C
    4    FORMAT('0',5X,'ADDRESS',13X,'ROM',15X,'SCALED',10X,'BINARY-TRUE',
        * 7X,'HEX-TRUE',3X,'BINARY-COMPLEMENT',1X,'HEX-COMPLEMENT')
C
    5    FORMAT('0',5X,'FILTER COEFFICIENTS: ',4X,'A(0)=',F5.2,4X,'A(1)=',
        * F5.2,4X,'A(2)=',F5.2,4X,'B(1)=',F5.2,4X,'B(2)=',F5.2)
C
C
```

```
500  READ(5,1,END=400)A0,A1,A2,B1,B2
     OFFSET=A0+A1+A2-B1-B2
     BETA=-1.0E10
     ALPHA=-BETA
     I=1
     DO 100 Y2=1,2
     DO 100 Y1=1,2
     DO 100 X2=1,2
     DO 100 X1=1,2
     DO 100 X0=1,2
     Y0=A0*X0+A1*X1+A2*X2-B1*Y1-B2*Y2-OFFSET
     IF(Y0.GT.BETA) BETA=Y0
     IF(Y0.LT.ALPHA)ALPHA=Y0
     ROM(I)=Y0
     I=I+1
100  CONTINUE
     SCALE=BETA-ALPHA
     DO 200 I=1,32
200  ROMS(I)=ROM(I)/SCALE
     WRITE(6,2)SCALE
     WRITE(6,5)A0,A1,A2,B1,B2
     WRITE(6,4)
     DO 300 J=1,32
     I=J-1
     CALL BINARY(ROMS(J),BIN,BINC,8)
     CALL HEX(BIN(1),MSB)
     CALL HEX(BIN(5),LSB)
     CALL HEX(BINC(1),MSBC)
     CALL HEX(BINC(5),LSBC)
     WRITE(6,3)I,ROM(J),ROMS(J),(BIN(K),K=1,8),MSB,LSB,
    * (BINC(K),K=1,8),MSBC,LSBC
300  CONTINUE
     GO TO 500
400  WRITE(6,2)SCALE
     STOP
     END

     SUBROUTINE BINARY(X,BIN,BINC,N)
     INTEGER BIN(1),N,BINC(1)
     REAL X,Y
     Y=X
     IF(Y.GT.1.0)Y=1.0
     IF(Y.LT.-1.0)Y=-1.0
     BIN(1)=0
     BINC(1)=1
     IF(Y.GE.0.0) GO TO 100
     BIN(1)=1
     BINC(1)=0
     Y=Y+1.0
100  DO 200 I=2,N
     Y=Y+Y
     BIN(I)=0.0
     BINC(I)=1
     IF(Y.LT.1.0) GO TO 200
     Y=Y-1.0
     BIN(I)=1
     BINC(I)=0
```

```
200  CONTINUE
     RETURN
     END
     SUBROUTINE HEX(BIN,DIGIT)
     INTEGER BIN(1),DIGITS(16),DIGIT
     DATA DIGITS/'0','1','2','3','4','5','6','7','8','9','A','B','C',
    *  'D','E','F'/
     J=1+BIN(4)+2*(BIN(3)+2*(BIN(2)+2*BIN(1)))
     IF(J.GT.16)J=16
     IF(J.LT.1)J=1
     DIGIT=DIGITS(J)
     RETURN
     END
```

CONTENTS OF ROM SCALED BY 1.170

FILTER COEFFICIENTS: A(0)= .07   A(1)= .14   A(2)= .07   B(1)= -.59   B(2)= .30

ADDRESS	ROM	SCALED	BINARY-TRUE	HEX-TRUE	BINARY-COMPLEMENT	HEX-COMPLEMENT
0	.000000	.000000	0.000 0000	00	1.111 1111	FF
1	.070000	.059829	0.000 0111	07	1.111 1000	F8
2	.139999	.119658	0.000 1111	0F	1.111 0000	F0
3	.210000	.179487	0.001 0110	16	1.110 1001	E9
4	.070000	.059829	0.000 0111	07	1.111 1000	F8
5	.139999	.119658	0.000 1111	0F	1.111 0000	F0
6	.210000	.179487	0.001 0110	16	1.110 1001	E9
7	.280000	.239316	0.001 1110	1E	1.110 0001	E1
8	.589999	.504273	0.100 0000	40	1.011 1111	BF
9	.659999	.564102	0.100 1000	48	1.011 0111	B7
10	.729998	.623931	0.100 1111	4F	1.011 0000	B0
11	.799999	.683761	0.101 0111	57	1.010 1000	A8
12	.659999	.564102	0.100 1000	48	1.011 0111	B7
13	.729998	.623931	0.100 1111	4F	1.011 0000	B0
14	.799999	.683761	0.101 0111	57	1.010 1000	A8
15	.869999	.743590	0.101 1111	5F	1.010 0000	A0
16	-.300000	-.256411	1.101 1111	DF	0.010 0000	20
17	-.230000	-.196582	1.110 0110	E6	0.001 1001	19
18	-.160001	-.136753	1.110 1110	EE	0.001 0001	11
19	-.090000	-.076923	1.111 0110	F6	0.000 1001	09
20	-.230000	-.196582	1.110 0110	E6	0.001 1001	19
21	-.160001	-.136753	1.110 1110	EE	0.001 0001	11
22	-.090000	-.076923	1.111 0110	F6	0.000 1001	09
23	-.020000	-.017094	1.111 1101	FD	0.000 0010	02
24	.289999	.247863	0.001 1111	1F	1.110 0000	E0
25	.359999	.307692	0.010 0111	27	1.101 1000	D8
26	.429999	.367521	0.010 1111	2F	1.101 0000	D0
27	.499999	.427350	0.011 0110	36	1.100 1001	C9
28	.359999	.307692	0.010 0111	27	1.101 1000	D8
29	.429999	.367521	0.010 1111	2F	1.101 0000	D0
30	.499999	.427350	0.011 0110	36	1.100 1001	C9
31	.569999	.487179	0.011 1110	3E	1.100 0001	C1

APPENDIX 8.3   BLOCK FLOATING-POINT NOTATION

This appendix presents a tutorial discussion of a block floating-point (BFP) notation and has been taken directly from a technical report.[†]

Introduction

Over the past 3 or 4 years Sandia's Information Systems Department 9230 has worked with digital signal processing in order to improve the operation of sensors used in perimeter security systems. One major task has been to implement real-time adaptive digital filters and associated detection processes in small, fieldable, microprocessor-based systems. Because the numerical range of the parameters and variables within these processors greatly exceeds the fixed-point capability of small (up to 16-bit) systems, a floating-point process is needed.

Floating-point processes differ from fixed-point processes in that floating-point processes are associated only with highly formalized and structured software or hardware schemes wherein numbers must be represented in normalized magnitude sign-exponent formats. However, early literature[1] indicates that in fixed-point implied integer or fractional processes, the decimal or binary point was at all times considered to be a fixed location in the processor word. Therefore, the process becomes floating-point if the implied binary point begins to move about in the processor.

In more recent literature,[2] processes termed block floating-point (BFP) are described wherein the floating-point formats are not explicit and the binary points and base exponents are implied. It is this class of processes that is addressed in this report.

The major advantages of BFP implementations are:

1. They can be designed to run at speeds to support real-time processes.

2. Numerical range and precision can be optimized in the process stages. (In formalized floating-point schemes the range/precision tradeoff is fixed.)

3. They can be readily used in small, microprocessor-based systems.

[†]"A Block Floating-Point Notation for Signal Processes," by James E. Simpson; Sandia National Laboratories technical report (SAND79-1823), March 1981, *courtesy of Sandia National Laboratories*. Albuquerque, N.M.

The major disadvantage of the BFP approach is the ever-present problem of keeping up with the binary point. BFP notation was developed in response to this challenge. A major use of the notation has been to plan, design, and analyze fast numerical processes, including real-time, single, and multiprecision signal-processing programs. BFP notation helps to:

1. Keep track of the binary point in arithmetic operations

2. Normalize number representations and aid in required scaling

3. Predict and possibly prevent overflow and underflow problems

4. Tailor floating-point processes for optimum balance of range, precision, and execution speed

5. Estimate or determine the effects of truncation, rounding, and related quantization noise in signal processes

6. Estimate and relate the expected results of running a process in differently structured machines or systems. This can be of great importance for converting initial results obtained on a large system that uses high-level language to a low-level implementation in a microprocessor-based system

7. Document programs such as listings and other program descriptions for increased ease of understanding and later modifications

A major attribute of the notation is its flexibility. It can be tailored to fit a range of individual user needs, from simple manual procedures to more sophisticated computer-aided schemes.

The Development of BFP Notation

Various methods have been used to indicate the magnitude or format of decimal number representations, one of which is I/F. In this representation I is the number of digits in the whole part of the number representation, and F is the number of digits in the fractional part. Thus 999.99 would be the maximum value of a 3/2 decimal number representation.

This same notation can be used in handling binary representation by defining the relation between unsigned numbers and their binary representations as

$$x \doteq X(I_x/F_x)$$

where x = unsigned number

X = bit pattern

(I_x/F_x) = pattern format

where $\doteq$ means "represented by." Then, for example, ten and one-half (10.5_{10}) would be represented by $10101_2(4/1)$ in the binary case.
     Use of this notation results in simple multiplication like

$$a * b \doteq A(I_a/F_a) * B(I_b/F_b) = C(I_c/F_c)$$

with $I_c = I_a + I_b$ and $F_c = F_a + F_{b'}$, assuming no restrictions on the format space because of processor word length.
     Now, to consider the simple addition of two numbers

$$a + b \doteq A(I_a/F_a) + B(I_b/F_b) = C(I_c/F_c)$$

In order to get a valid representation of the sum, the weights of the bit patterns must be properly aligned or, if left justification is assumed, I_a must equal I_b; sufficient word space must be provided to contain the sum representation. This indicates that $I_c = \text{MAX}\,(I_a, I_b) + 1$, and that A or B must be shifted if necessary so that $I_a = I_b = I_c - 1$.
     The preceding examples are simplistic and limited. However, the concept can be extended to accommodate twos-complement arithmetic operations. In this number representation system, at least one digit (bit) must be reserved for sign information. As these representations are shifted or otherwise manipulated within the processor, this sign information must be properly extended or preserved. Thus, twos-complement representations often have redundant sign bits that in effect limit the range of the represented number. In order to indicate this sign information and possible range limitation, a third format indicator was added to the previous format descriptor so that numbers are represented in twos-complement form as

$$x \doteq X(S_x/I_x/F_x),$$

$$S_x = \text{number of sign bits}$$

Also, it was advantageous to extend the range of this representation descriptor in order to (1) keep the I and F descriptors positive and within the double-precision range of the associated processor, and (2) preclude the occurrence of virtual binary points or embedded sign bits. The descriptor range was extended by adding an overall scaling descriptor to the representation to give

$$x \doteq X(S_x/I_x/F_x)Bn$$

where n is a positive or negative integer (including zero) indicating an overall scaling by 2^n. It is also possible to denote scaling by different bases, if desired, such as On for 8^n or Hn for 16^n.

The bit-pattern descriptor, X, can be a general variable or, in the case of process constants, a specific bit pattern (binary or otherwise). For example, in a 12-bit machine $0740_8(1/0/11)B-1$ or $00111100000(1/0/11)B-1$ would represent a constant of value 0.11719.

Another constraint that can be included in the format descriptor is that of polarity. If some process step yields numbers of only one sign, this information can be passed on in the format descriptor by placing the appropriate indicator directly ahead of the sign-bit descriptor. For instance,

$$x^2 \doteq X ** 2(+S_x/I_x/F_x)Bn$$

or
$$x^3 \doteq X ** 3(-S_x/I_x/F_x)Bn, \qquad X < 0, \text{ etc.}$$

Note also that mathematical operations can be included in the general bit-pattern descriptors, as further shown by

$$a * b = A(S_a/I_a/F_a) * B(S_b/I_b/F_b)$$

$$= A * B(S_r/I_r/F_r)$$

with $n = 0$ and the subscript r indicating result format descriptors. With this representation descriptor system established, its uses can be examined in a machine with word length N that uses twos-complement arithmetic. If the resulting $S + I + F > N$, after some operation, truncation is indicated that can be cause for concern. For instance, when sign-magnintude representations are truncated, the residue magnintude information represents less than or equal the original magnitude value. On the other hand, when twos-complement representations are truncated, the positive magnitude information is less than or equal the original value; when negative representations are truncated, the magnitude information is greater than or equal the original value. In averaging or filtering operations this can cause negative process components to be weighted greater than positive components, and can lead to drift problems or other process instabilities.

Note also that $S + I + F$ can be $< N$ when

1. Devices such as digitizers with fewer output bits than N are interfaced with an N-bit processor

2. Incoming or internal data are shifted left for scaling or alignment

3. The high-order word of double-precision results is left-shifted without bringing in low-order word information

For whatever reason, when $S + I + F$ is less than N, the processor word does not contain all the information or resolution that it is capable of representing. When the information entropy is critical, this BFP notation scheme provides a way to detect and possibly optimize the process.

The shifting of representations, mentioned before, is usually done to

1. Properly align binary points or bit weights before addition or subtraction

2. Scale the weight of one process component with respect to another before an arithmetic operation

3. Establish sufficient growth space for a mathematical result before a later operation

In general, shifting a representation to the right increases S; and when $I + F$ begins to decrease, truncation has begun. Valid shifting to the left is possible only if $S > 1$; and $I + F$ increases only if significant data are entered at the right concurrent with the shifting. Obviously the representation becomes invalid if $S = 0$ or $S = N$.

With the development sequence out of the way, detailed characteristics of BFP are now discussed.

Basic BFP Concepts

Before discussing mathematical processes, it is best to look at some of the basic concepts[†] of BFP notation such as valid, equivalent, normalized, and aligned formats.

Valid BFP Formats

Valid BFP formats exist when:

1. The number of sign bits, S, is in the range $1 \le S \le (N - 1)$

[†]Much of the BFP development related to concepts was done by R. J. Fogler, Division 9238, and is discussed in his thesis.[3]

2. The number of integer bits, I, is in the range $0 \le I \le (N - 1)$

3. The number of fraction bits, F, is in the range $0 \le F \le (N - 1)$

4. The exponential scale factor, B, O, or H, is an integer

5. The sum of $S + I + F \le N$

6. The sum of $I + F \ge 1$

Note that N is the number of bits in the processor word for the single-precision case, and an integer multiple of the processor word length for the extended-precision case.

Table 1 gives examples of valid-format descriptors and the components of the bit pattern, with specific bits shown as 0 or 1, variable bits as S, I, or F, and bit positions with no information as *. These are examples of single-precision representations with $N = 16$.

Table 1   Examples of Valid 16-Bit
              Representations

Notation Format	Pattern Format (X)
$X(1/0/15)$	SFFF FFFF FFFF FFFF
$X(+2/3/8)$	00II IFFF FFFF F***
$X(-6/3/0)$	1111 11II I*** ****

Note that a zero exponential scaling descriptor need not be shown, as Table 1 indicates. Also, this descriptor plays no part in determining valid BFP notation formats, so long as n is an integer.

Table 2 shows examples of invalid formats and the related condition that is violated, again for the single-precision case with $N = 16$.

Table 2   Examples of Invalid BFP Notation

Notation Format	Condition Violated
$X(5/-2/6)$	Range of I is $0 \le I \le (N - 1)$
$X(+1/5/12)$	$S + I + F$ is restricted to $\le N$ for single-precision case
$X(0/2/8)$	S cannot be zero in a twos-complement system
$X(3/0/0)$	No magnitude information is possible; $I + F$ must be > 0

The third example in Table 2 is sometimes referred to as arithmetic overflow, and the last example in this table is often called arithmetic underflow.

Equivalent BFP Notation

The exponential-scale factor adds not only to the flexibility of the notation but also to the possibility of equivalent formats that define the same numerical representation but can have different formats. As an example, $X(+1/0/15)$ and $X(+1/2/13)B-2$ define the same representation. They are therefore equivalent since both formats indicate the same number of significant bits with the same weight pattern (relative binary point), and both formats restrict the representation to positive values.

Thus, given two BFP format descriptions (less the pattern variable) $(aS_1/I_1/F_1)Bn_1$, and $(bS_2/I_2/F_2)Bn_2$, the formats are equivalent if, and only if,

$$a = b \,(+, \, -, \text{ or null})$$

$$S_1 = S_2$$

and
$$I_1 + F_1 + n_1 = I_2 + F_2 + n_2$$

Normalized BFP Formats

The BFP formats can be normalized in several different ways. Some prefer to work with a strictly fractional representation while adjusting with the exponential scale factor; others prefer integer representations. In this case it simplifies the documentation if the exponential scaling factor can be avoided. It therefore makes sense to drive this factor to zero when possible. Assume again that two equivalent notations are given as before. If $n_1 = 0$, then the format is considered normalized. If n_1 is positive, then the normalized form can be achieved by maintaining the same polarity and sign factors and letting, in sequence,

$$F_2 = \text{MAX}(0, F_1 - n_1)$$

$$I_2 = I_1 + F_1 - F_2$$

$$n_2 = n_1 + I_1 - I_2$$

If $n_1 < 0$, similar adjustments are, in sequence,

$$I_2 = \text{MAX}(0, I_1 + n_1)$$

$$F_2 = F_1 + I_1 - I_2$$

$$n_2 = n_1 + I_1 - I_2$$

Aligned BFP Formats

Since addition, subtraction, and comparison operations in twos-complement machines are normally accomplished with adders and complementers, these operations can be treated as the summation of aligned representations. Representations are considered properly aligned when relative bits of the operand bit patterns have the same binary weight. Given two numbers, x and y, to be added:

$$x \doteq X(S_x/I_x/F_x)Bn_x$$

$$y \doteq Y(S_y/I_y/F_y)Bn_y$$

their representation patterns X and Y are said to be aligned when $S_x + I_x + n_x = S_y + I_y + n_y$. If the representations are not aligned, then one or both must be shifted. If aligned, both representations may have to be shifted right to assure a valid result format, as shown later.

When necessary, formats are aligned by shifting operand representations left or right arithmetically wherein the sign bit is normally extended during right shifts, and zeros are shifted in during left shifts.

Note that when the representation pattern (X) is shifted for alignment, the format descriptor must also be adjusted to prevent scaling problems. For instance, if the representation for four (01000000(1/3/4)) is right-shifted two times without changing the format descriptor, it becomes 00010000(1/3/4), which represents one. The appropriate format descriptor would be (3/3/2).

BASIC MATHEMATICAL OPERATIONS
THROUGH THE USE OF BFP NOTATION

Summation

In order to prevent overflow in fast-summation processes, sufficient expansion space, or sign bits, must be established in the operands before doing the summation. To do this, consider the operation

$$\sum_{i=1}^{M} x_i = x_r$$

with

$$x_i \doteq XI(S_i/I_i/F_i)$$

$$x_r \overset{\bullet}{=} XR(S_r/I_r/F_r)$$

In the worst case the possible loss of sign bits can be expressed as

$$S_r - S_i - \text{INT}(\log_2 M + 0.5)$$

which indicates that, for general summation of an operand x_i, and in order to guarantee a valid result representation ($S_r \geq 1$),

$$S_i \geq 1 + \text{INT}(\log_2 M + 0.5)$$

before the summation (M times), after which

$$S_r = S_i - \text{INT}(\log_2 M + 0.5)$$
$$I_r = I_i + S_i - S_r$$
$$F_r = F_i$$
$$n_r = n_i$$

When only two operands are to be summed, besides being aligned, both operand representations must have at least two sign bits since $M = 2$, and $S_{\min} = 1 + \text{INT}(\log_2 2 + 0.5) = 2$.

   With this in mind, the general approach to adding two operands is

$$a + b \overset{\bullet}{=} A(S_a/I_a/F_a)Bn_a + B(S_b/I_b/F_b)Bn_b$$
$$= R(S_r/I_r/F_r)Bn_r$$

   First determine which representation format has the greatest $S_i + I_i + n_i$ ($i = a,b$). Increase the relevant S_i to 2 by a right shift if necessary. Refer to this adjusted format as $(S_x/I_x/F_x)Bn_x$ to determine the result format later.

   Next, adjust the representation of the other operand for proper alignment with $(S_x/I_x/F_x)Bn_x$ by shifting if required, and refer to this format as $(S_y/I_y/F_y)Bn_y$. The format of the operation result should then be

$$S_r = S_x - 1 \text{ (at least one)}$$
$$I_r = I_x + 1$$
$$F_r = \text{MIN}[\text{MAX}(F_x, F_y) + 1, n - S_r - I_r]$$
$$n_r = n_x$$

To illustrate how the summation rules apply, consider the addition of ($N = 8$):

$$a = 0.875 \doteq 0111****(1/1/2)\text{B-1}$$

and
$$b = 0.125 \doteq 001000**(2/0/4)\text{B-2}$$

Since

$$S_a + I_a + n_a = 1 > S_b + I_b + n_b = 0$$

the representation for a is shifted right once to become $00111***(2/1/2)\text{B-1}$ with

$$S_x = 2$$
$$I_x = 1$$
$$F_x = 2$$

and
$$n_x = -1$$

For proper alignment

$$(S_x + I_x + n_x = S_b + I_b + n_b)$$

the representation for b must be shifted right two times to become $00001000(4/0/4)\text{B-2}$ with

$$S_y = 4$$
$$I_y = 0$$
$$F_y = 4$$

and
$$n_y = -2$$

Adding the binary patterns and applying the format rules yields

$$S_r = S_x - 1 = 1$$
$$I_r = I_x + 1 = 2$$
$$F_r = \text{MIN}[\text{MAX}(2, 4) + 1, 8 - 1 - 2] = 5$$

and
$$n_r = n_x = -1$$

to give a result of $01000000(1/2/5)\text{B-1}$, which represents 1.000 or the correct answer. Note that since neither operand completely filled an 8-bit word with significant information, no precision was lost in the result because of the required right shifts. Note also that without the shifts overflow would occur, producing an invalid result.

An alternative to preventing possible overflow when summing is to clip the results if the process will allow this procedure. Clipping and special cases of summation are further discussed in Reference 3.

Multiplication

A BFP format definition for the multiplication process is complicated by the asymmetry of the twos-complement representation of numbers. There is one more negative than positive representation for a given word length. This representation, most commonly used to represent the smallest or most negative number within the basic word range, has always been a source of peculiar[4] problems such as lack of closure[5] for multiplication and absolute value results. For instance, an attempt to obtain the absolute, or positive, value of this problem representation (8000 hexadecimal for a 16-bit machine) by the usual complement-and-increment operations yields the same representation.

Most machines that now provide twos-complement multiplication place the representation of the product of two N-bit representations in a $2N$-bit space. These $2N$-bit representations usually contain two sign bits except when the problem representation is squared.

Because this one peculiar representation is troublesome and adds little to the machine range, it is often best to inhibit it either by trapping it in software or by preventing its generation in hardware as in some of the early microprocessor-support multipliers.[6,7] One common instance of the representation occurs when an N-bit result or product of two N-bit representations with opposite signs produces no magnitude information in the high-order product word.

When the case of squaring the peculiar representation is inhibited, the BFP format rules for multiplication become

$$a * b = A(S_a/I_a/F_a)Bn_a * B(S_b/I_b/F_b)Bn_b$$
$$= C(S_c/I_c/F_c)Bn_c$$

with

$$S_c = S_a + S_b$$
$$I_c = I_a + I_b$$
$$F_c = F_a + F_b$$

and
$$n_c = n_a + n_b$$

where C is a 2N-bit representation pattern.

If an N-bit representation of the product is desired, the format rules can be expressed with $F_c = 0$ as

$$S_c = S_a + S_b$$

$$I_c = \text{MIN}(I_a + I_b, N - S_c)$$

$$F_c = 0$$

and
$$n_c = I_a + I_b + n_a + n_b - I_c$$

Note that if $S_c \geq N$, underflow has occurred. The N-bit result should then be forced to zero to avoid the problem representation case and other undesirable scaling problems that can result from small negative results.

To demonstrate how the multiply format rules work, consider the following hexadecimal results with $N = 8$.

Example 1 (2N-Bit Result):

$$0.3046875 * (-0.890625) \doteq 4\text{E}(1/7/0)\text{B-8}^*$$
$$\text{C7}(2/6/0)\text{B-6} = \text{EEA2 } (S_c/I_c/F_c)Bn_c$$

$$S_c = 1 + 2 = 3$$

$$I_c = 7 + 6 = 13$$

$$F_c = 0 + 0 = 0$$

$$n_c = -8 - 6 = -14$$

to yield EEA2(3/13/0)B-14, which represents a result of -0.2713623, as expected.

Example 2 (Truncated and Rounded N-Bit Result):
Consider same inputs to give $\text{EE}(S_c/I_c/F_c)Bn_c$ with

$$S_c = 1 + 2 = 3$$

$$I_c = \text{MIN}(6 + 7, 8 - 3) = 5$$

$$F_c = 0 \text{ (assumed)}$$

$$n_c = 7 + 6 - 8 - 6 - 5 = -6$$

to yield EE(3/5/0)B-6. This represents a truncated result of -0.28125 (which is greater than the expected result, as discussed previously).

The result when rounded, based on the most significant bit of the low-order N-bits of the product pattern, is EF(3/5/0)B-6. This represents -0.265625, which is now less than and closer to the expected product.

Because using the N-bit result of the product of N-bit representations is a specific case of truncation, it is often desirable to round the product representations as illustrated above. Rounding operations are straightforward except for possible overflow conditions, elaborated on by Fogler.[3]

There is a widespread tendency to equate shifting within a processor with scaling of a representation by 2^m, where m is an integer with sign depending on the shift direction. Again, this is true only in a strictly fixed-point mode. In the BFP mode, on the other hand, binary scaling occurs only when representation patterns are shifted without an adjustment in the format descriptor, or vice versa. Representation patterns must often be shifted to prevent unwanted scaling, as previously discussed. Alternately, a representation can be scaled binarily by leaving the bit pattern unchanged and adjusting the format descriptor. For example, $a = 2.5 \doteq 0101(1/2/1)$ could be reformatted as 0101(1/3/0) to represent $2a = 5$ in a multiply or aligned summation.

Division

The last operation to be considered is division, probably the most difficult operation to control in BFP processes. The structure assumed for the twos-complement division process is fairly common[8] wherein a $2N$-bit dividend (NUMR) is divided by an N-bit divisor (DIVR) to yield an N-bit quotient (QUO) and an N-bit remainder (REM). The inverse operation would be represented by

$$|NUMR|(p_n S_n/I_n/F_n)Bn_n = DIVR(S_d/I_d/F_d)Bn_d{}^*$$

$$QUO(S_q/I_q/F_q)Bn_q + REM(p_r S_r/I_r/F_r)Bn_r$$

First, since $|DIVR^* QUO| \leq |NUMR|$, NUMR and REM have the same sign, or $p_r = p_n$. Also indicated in the relation is that REM and NUMR must have compatible (aligned) formats. This fact can be used to establish the resulting format descriptor for REM.

Note that the minimum number of sign bits occurs in QUO when the maximum number of sign bits are present in the representation for the inverse operation. This relation is expressed as

$$S_{n,\max} = S_d + S_q + 1$$

which indicates the fewest expected sign bits in the QUO format descriptor are

$$S_q = S_n - S_d - 1$$

Furthermore, in order to guarantee that the representation for QUO is always valid, $S_q \geq 1$, from which $S_n - S_d \geq 2$; otherwise |NUMR/DIVR| will not always fit in an N-bit word. The difficulty of structuring to meet this requirement consistently in a dynamic process is the reason division is hard to control in an unchecked BFP mode.

Since the minimum number of sign bits in a valid representation for DIVR is also one, then $S_n - 1 \geq 2$, or $S_n \geq 3$ in general. Establishing the rest of the format rules for QUO is simplified by assuming $F_q = 0$ and $I_q = N - S_q$, from which the format rules for the quotient representation can be summarized as

$$S_q = S_n - S_d - 1$$

$$I_q = N - S_q$$

$$F_q = 0$$

$$n_q = S_n + I_n + n_n - (S_d + I_d + n_d + N)$$

Now to address the format rules for REM. From the inverse relation, note that |DIVR * QUO| $\leq$ |NUMR| requires that NUMR and REM have the same sign, or $p_r = p_n$. Next, the fact that |REM| $\leq$ |DIVR| indicates that $S_r \geq S_d$, or at the worst $S_r = S_d$.

Also indicated is compatible (aligned) formats, for REM and NUMR from which the representation format rules for the remainder can be summarized. Again, for simplicity, F_r is assumed zero.

$$p_r = p_n$$

$$S_r = S_d$$

$$I_r = N - S_r$$

$$n_r = S_n + I_n + n_n - (S_r + I_e + N)$$

The N factor in the last relation is present because NUMR occupies a $2N$-bit space, whereas REM is in an N-bit space. Since $S_r + I_r = N$, the relation for n_r can be further compressed to

$$n_r = S_n + I_n + n_n - 2N$$

To illustrate how the division format rules apply, consider the following examples.

Example 1 (N = 8, Hexadecimal Representations):

Let $8191 \div 64 \stackrel{\bullet}{=} 1FFF(3/11/2)B2/40(1/3/4)B4$, yielding

$$7F(S_q/I_q/0)Bn_q$$

with remainder

$$3F(S_r/I_r/0)Bn_r$$

Using the QUO format rules yields

$$S_q = 3 - 1 - 1 = 1$$
$$I_r = 8 - 1 = 7$$
$$n_q = 3 + 11 + 2 - (1 + 3 + 4 + 8) = 0$$

Likewise, for REM:

$$S_r = S_d = 1$$
$$I_r = 8 - 1 = 7$$
$$n_r = 3 + 11 + 2 - 16 = 0$$

from which the quotient represented by $7F(1/7/0)$ is 127; the remainder represented by $3F(1/7/0)$ is 63; and $64^* 127 + 63 = 8191$, as expected.

Example 2 (Same Configuration):

Consider $0.24996 \div 4.1875 \stackrel{\bullet}{=} 1FFF(3/0/13)B-2/43(1/0/7)B3$, yielding

$$7A(S_q/I_q/0)Bn_q$$

with remainder

$$11(1/7/0)Bn_r$$

Again, using the division format rules:

$$S_q = 3 - 1 - 1 = 1$$

$$I_q = 8 - 1 = 7$$

$$n_q = 3 + 0 - 2 - (1 + 0 + 3 + 8) = -11$$

Also

$$S_r = S_d = 1$$
$$I_r = 8 - 1 = 7$$
$$n_r = 3 + 0 - 2 - 16 = -15$$

from which the quotient represented by 7A(1/7/0)B-11 is 0.05957 . . . , and the remainder represented by 11(1/7/0)B-15 is 0.0005188. . . . Again, results agree within the precision of the calculations.

In summary, if a division operation is well-defined and bounded so that the condition $S_n - S_d \geq 2$ can be assured, then a BFP process can be run unchecked. If not, and possible overflows are unacceptable, trapping and adjusting one or both operand representations is in order. In this case, the format rules can be used to plan and implement the recovery scheme.

APPENDIX A TO CONVERT FROM BFP NOTATION TO DECIMAL EQUIVALENTS

The following procedures are suggested in order to convert from BFP representations to decimal equivalents. First, force the BFP representation to a pure-integer-plus-scale form; i.e., $X(S_x/I_x/0)Bn_x$. Next, determine the decimal equivalent of X as an integer, X_D. Then determine x as $X_D * 2^m$, with $m = n_x + S_x + I_x - N$.

Example 1 (N = 8, Hexadecimal Format for X):
Given x as represented by

$$7F(1/0/7) = 7F(1/7/0)B\text{-}7$$

then

$$X_D = 127, \ m = -7 + 1 + 7 - 8 = -7$$

and

$$x = 127 * 2^{-7} = 0.9921875$$

Example 2 (Same Configuration):
Given x as represented by

$$C6(2/1/5) = C6(2/6/0)B\text{-}5$$

then

$$X_D = -58, m = -5 + 2 + 6 - 8 = -5$$

and

$$x = -58 * 2^{-5} = -1.8125$$

Example 3 (Same Configuration):
Given x as represented by

$$7C(1/0/5) = 7C(1/5/0)B\text{-}5$$

then

$$X_D - 124, m = -5 + 1 + 5 - 8 = -7$$

and

$$x = 124 * 2^{-7} = 0.96875$$

Note that in this example the representation for x has only 5 bits of precision instead of the usual 7 for an 8-bit representation space.

APPENDIX B TO DETERMINE A BFP REPRESENTATION FOR A DECIMAL CONSTANT

First, assume that the BFP representation format has been determined in order to provide for alignment in a later operation. That is, the constant x is represented by $X(S_x/I_x/F_x)Bn_x$ where the format descriptor is known.

Next, force the descriptor to a pure-integer-plus-scale form by forcing F_x to zero, and adjusting n_x if necessary. Now consider the relation for the magnitude of x as $|x| \doteq X(S_x/I_x/0)Bn_x$ from which $|x| = X_D$ $* 2^m$, where $m = n_x$, and X_D is the decimal equivalent of the bit pattern required for the BFP representation. Rearranging, $X_D = |x| * 2^{-m}$. In order to obtain an integer for the pattern, the relation $X_D = \text{INT}(|x| * 2^{-m} + 0.5)$ is suggested from which the representation pattern can be determined, if it is in the range of the format descriptor.

Example (N = 8):

Given a format descriptor of (1/7/0)B-7 and a constant of 0.0495, what are possible representation patterns and the resulting precision? Use of the above relation gives $X_D = \text{INT}(0.0495 * 2^7 + 0.5) = 6$; and x as represented by 06(1/7/0)B-7 = 0.046875 or 0.002625 less than the desired constant. Alternately, x as represented by 07(1/7/0)B-7 = 0.0546875, which is 0.005188 greater than the desired value and indicates the available precision with $N = 8$.

Note also that the range of constants that can be represented with this format descriptor constraint is from 2^{-7} to 127×2^{-7}, or from 0.0078125 to 0.9921875.

If the constant x happens to be negative, then use the twos-complement of X. For instance, in the above case -0.0495 would be represented by FA(1/7/0)B-7.

REFERENCES

1. D. N. LEESON, D. L. DIMITRY, and E. K. WALLSTEDT, *Basic Programming Concepts and the IBM 1620 Computer* (New York: Holt, Rinehart, and Winston, Inc., 1962), Chapter 11.

2. A. V. OPPENHEIM and R. W. SCHAFER, *Digital Signal Processing* (Englewood Cliffs, N.J.: Prentice-Hall, Inc., 1975), pp. 455–56.

3. R. J. FOGLER, *On a Block Floating-Point Implementation of an Intrusion-Detection Algorithm* (Master's Thesis, Kansas State University, October 1979).

4. I. FLORES, *The Logic of Computer Arithmetic* (Englewood Cliffs, N.J.: Prentice-Hall, Inc., 1963), p. 26.

5. Ibid., p. 244.

6. J. SIMPSON and R. WAYNE, *A High Speed Lookup Table Technique for Digital Multiplication*, SAND77-1866 (Albuquerque: Sandia Laboratories, March 1978).

7. J. SIMPSON, *An Array Multiplier for Twos-Complement Binary Numbers*, SAND78-0725 (Albuquerque: Sandia Laboratories, July 1978).

8. *MCS-86™ User's Manual* (Santa Clara, Calif.: Intel Corporation, July 1978).

APPENDIX 8.4 SOME DERIVATIONS

Derivation of (8.2-5)

We make the following assumptions regarding (8.2-4):

1. The error associated with each sample is uniformly distributed between $\pm q/2$. Then, if E denotes statistical expectation, it follows that

$$m_x = E\{e(n)\} = 0$$

and

$$\sigma_e^2 = E\{e^2(n)\} - m_x^2$$

$$= \frac{1}{q} \int_{-q/2}^{q/2} e^2 \, de$$

That is,

$$\sigma_e^2 = \frac{q^2}{12} = \frac{2^{-2B}}{3} \qquad \text{(A8.4-1)}$$

since $q = 2^{-B+1}$, which is (8.2-5).

2. The error at each sampling instant is statistically independent of the error at any other sampling instant.

   The preceding assumptions imply that the noise due to quantization is *white*, with zero mean and variance $\sigma_e^2 = 2^{-2B}/3$.

Derivation of (8.2-7)

We make the additional assumption that $x(n)$ and $e(n)$ in (8.2-4) are independent random variables. Then, since $H(z)$ in Fig. 8.2-3 is a linear time-invariant system, we can ignore $x(n)$ while computing the variance σ_ϵ^2 of the output noise $\epsilon(n)$ due to quantization; see (8.2-6).

The output noise sequence $\epsilon(n)$ in (8.2-6) is obtained by convolving $h(m)$ with $e(n)$, where $h(n) = Z^{-1}\{H(z)\}$ is the impulse response of the system in Fig. 8.2-3, and $e(n)$ is defined in (8.2-4). Thus we have

$$\epsilon(n) = \sum_{m=0}^{\infty} h(m)e(n - m) \qquad \text{(A8.4-2)}$$

where $h(m)$ is assumed to be zero for $m < 0$.

We define the autocorrelation function of the output sequence as

$$\phi_\epsilon(r) = E\{\epsilon(n)\epsilon(n - r)\} \qquad \text{(A8.4-3)}$$

where r is the size of the shift, and E denotes statistical expectation. Substitution of (A8.4-2) in (A8.4-3) leads to

$$\phi_\epsilon(r) = E\left\{\left[\sum_{m=0}^{\infty} h(m)e(n - m)\right]\left[\sum_{s=0}^{\infty} h(s)e(n - r - s)\right]\right\}$$

$$= \sum_{s=0}^{\infty} h(s) \sum_{m=0}^{\infty} h(m)E\{e(n - m)e(n - r - s)\}$$

That is,

$$\phi_\epsilon(r) = \sum_{s=0}^{\infty} h(s) \sum_{m=0}^{\infty} h(m)\phi_e(m - r - s) \qquad \text{(A8.4-4)}$$

where $\phi_e(\ )$ denotes the autocorrelation function of the input noise. However, since the input noise is assumed to be white with variance $\sigma_e^2 = 2^{-2B}/3$, we have

$$\phi_e(0) = \sigma_e^2 = 2^{-2B}/3$$

and $\phi_e(\ )$ is zero elsewhere. Thus (A8.4-4) simplifies to yield

$$\phi_\epsilon(r) = \sigma_e^2 \sum_{s=0}^{\infty} h(s)h(r + s) \qquad \text{(A8.4-5)}$$

We are interested in σ_ϵ^2, which is given by $\phi_\epsilon(r)$ at $r = 0$; that is,

$$\sigma_\epsilon^2 = \sigma_e^2 \sum_{s=0}^{\infty} h^2(s) \qquad \text{(A8.4-6)}$$

Next, using Parseval's theorem (Problem 3-15),

$$\sum_{s=0}^{\infty} h^2(s) = \frac{1}{2\pi j} \oint_{C_u} z^{-1}H(z)H(z^{-1}) \, dz \qquad \text{(A8.4-7)}$$

where it is assumed that all the poles of $H(z)$ lie within the unit circle; hence $\oint_{C_u}$ denotes integration around the unit circle in the counterclockwise direction.

Combining (A8.4-6) and (A8.4-7), we obtain

$$\sigma_\epsilon^2 = \sigma_e^2 \frac{1}{2\pi j} \oint_{C_u} z^{-1}H(z)H(z^{-1}) \, dz$$

which is (8.2-7).

Derivation of (8.5-4) and (8.5-5)

From Fig. 8.5-1 it is apparent that

$$V(z) = s_0 S(z) X(z)$$

where $S(z) = 1/D(z)$

Taking the IFT of $V(z)$ using (3.1-12b), we obtain

$$v(n) = \frac{s_0}{2\pi} \int_{-\pi}^{\pi} S(e^{j\theta}) X(e^{j\theta}) e^{jn\theta} \, d\theta \qquad \text{(A8.4-8)}$$

where $\theta = \omega T$, which implies that

$$v^2(n) = \frac{s_0^2}{(2\pi)^2} \left| \int_{-\pi}^{\pi} S(e^{j\theta}) X(e^{j\theta}) e^{jn\theta} \, d\theta \right|^2 \qquad \text{(A8.4-9)}$$

Let us now consider the *Schwarz inequality*, which states that[†]

$$\left| \int_a^b z(x) w(x) \, dx \right|^2 \leq \int_a^b |z(x)|^2 \, dx \int_a^b |w(x)|^2 \, dx \qquad \text{(A8.4-10)}$$

for any $z(x)$ and $w(x)$, where the equality sign holds if and only if $z(x)$ is proportional to $\tilde{w}(x)$, where $\sim$ denotes complex conjugate.

Application of (A8.4-10) to (A8.4-9) leads to

$$v^2(n) \leq s_0^2 \left[\frac{1}{2\pi} \int_{-\pi}^{\pi} |S(e^{j\theta})|^2 \, d\theta \right] \left[\frac{1}{2\pi} \int_{-\pi}^{\pi} |X(e^{j\theta})|^2 \, d\theta \right]$$

Next, applying Parseval's theorem (Problems 3-15 and 3-16) to the preceding inequality, it follows that

$$v^2(n) \leq s_0^2 \sum_{n=0}^{\infty} x^2(n) \cdot \frac{1}{2\pi j} \oint_{C_u} z^{-1} S(z) S(z^{-1}) \, dz$$

[†]For example, see A. Papoulis, *Signals, Systems, and Transforms*, McGraw-Hill Book Co., New York, 1977, p. 134.

Thus by choosing s_0 so that

$$\frac{s_0^2}{2\pi j} \oint_{C_u} z^{-1} S(z) S(z^{-1}) \, dz = 1 \qquad \text{(A8.4-11)}$$

we obtain

$$v^2(n) \leq \sum_{n=0}^{\infty} x^2(n)$$

which is (8.5-5).[†] We also see that (A8.4-11) is (8.5-4) with $S(z) = 1/D(z)$.

Derivation of (8.5-13)

Let the output of adder 1* in Fig. 8.5-4 be the sequence $v_2(n)$. Then we have

$$V_2(z) = s_0 s_1 S(z) X(z)$$

where $\quad S(z) = H_1(z) \cdot \dfrac{1}{D_2(z)}$

The IFT of $V_2(z)$ results in

$$v_2(n) = \frac{s_0 s_1}{2\pi} \int_{-\pi}^{\pi} S(e^{j\theta}) X(e^{j\theta}) e^{jn\theta} \, d\theta$$

where $\theta = \omega T$.
Proceeding as we did in connection with the derivation of (8.5-4) and (8.5-5), we obtain the following results:

$$v_2^2(n) \leq s_0^2 s_1^2 \left[\frac{1}{2\pi} \int_{-\pi}^{\pi} |S(e^{j\theta})|^2 \, d\theta \right] \left[\frac{1}{2\pi} \int_{-\pi}^{\pi} |X(e^{j\theta})|^2 \, d\theta \right]$$

$$v_2^2(n) \leq s_0^2 s_1^2 \sum_{n=0}^{\infty} x^2(n) \cdot \frac{1}{2\pi j} \oint_{C_u} z^{-1} S(z) S(z^{-1}) \, dz$$

Thus by choosing s_1 so that

[†]This is a special case of a more general formulation that is addressed in advanced texts; e.g., see pp. 318–320 of [2].

$$s_0^2 s_1^2 \cdot \frac{1}{2\pi j} \oint_{C_u} z^{-1} S(z) S(z^{-1}) \, dz = 1 \qquad (A8.4\text{-}12)$$

we obtain

$$v_2^2(n) \le \sum_{n=0}^{\infty} x^2(n)$$

which is the counterpart of (8.5-5). Equation (A8.4-12) is clearly equivalent to (8.5-13).

REFERENCES

1. A. PELED and B. LIU, *Digital Signal Processing*, Wiley, New York, 1976, Chaps. 4–6.

2. L. R. RABINER and B. GOLD, *Theory and Application of Digital Signal Processing*, Prentice Hall, Englewood Cliffs, N.J., 1975, Chap. 5.

3. A. V. OPENHEIM and R. SCHAFER, *Digital Signal Processing*, Prentice-Hall, Englewood Cliffs, N.J., 1975, Chap. 9.

4. S. A. WHITE, "Digital Filter Mechanization and Application," classroom notes of a National Engineering Consortium, Inc., Professional Growth in Engineering Seminar, 1974, pp. 3, 5-1-3, 5-1-13.

5. J. F. KAISER, *Digital Filters in System Analysis by Digital Computer*, F. F. Kuo and J. Kaiser, eds., Wiley, New York, 1966.

6. V. B. LAWRENCE, *Introduction to Digital Filtering*, R. E. Bogner and A. G. Constantinides, eds., Wiley, New York, 1975, Chap. 10.

7. P. M. EBERT, J. E. MAZO, and M. G. TAYLOR, "Overflow Oscillations in Digital Filters," *Bell Sys. Tech. J.*, Vol. 48, 1969, pp. 2999–3020.

8. A. CROISIER et. al, "Digital Filter for PCM Encoded Signal," U.S. Patent 3777130, Dec. 3, 1973.

9. J. SCHMALZEL, D. HEIN, and N. AHMED, "Some Pedagogical Considerations of Digital Filter Hardware Implementation," *IEEE Circuits and Systems Magazine*, Vol. 2, No. 1, 1980, pp. 4–13.

10. A. ANTONIOU, *Digital Filters: Analysis and Design*, McGraw-Hill, New York, 1979, Chap. 11.

11. S. A. TRETTER, *Introduction to Discrete-time Signal Processing*, Wiley, New York, 1976, Chap. 9.

12. L. B. JACKSON, "On the Interaction of Round-off Noise and Dynamic Range in Digital Filters," *Bell Sys. Tech. J.*, Vol. 49, 1970, pp. 159–184.

13. ———, "Round-off Noise Analysis for Fixed-point Digital Filters Realized in Cascade or Parallel Form," *IEEE Trans. Audio Electroacoust.*, Vol. AU-18, 1970, pp. 107–122.

Answers
to Selected
Problems

Chapter 1

1–1(a). $x(t) = \dfrac{1}{2} \displaystyle\sum_{n=-\infty}^{\infty} c_n\, e^{jn\pi t}$

where $c_n \begin{cases} 0, & n = 0 \\ 0, & n \text{ even} \\ 4/(jn\pi), & n \text{ odd} \end{cases}$

(b). $x(t) = \dfrac{1}{2} \displaystyle\sum_{n=-\infty}^{\infty} c_n\, e^{jn\pi t}$

where $c_n = \begin{cases} 0, & n = 0 \\ 0, & n \text{ even} \\ 8/(n^2\pi^2), & n \text{ odd} \end{cases}$

1–5. $Z(\omega) = \dfrac{1 - \cos[(\omega + \omega_0)\epsilon]}{(\omega + \omega_0)^2 \epsilon} + \dfrac{1 - \cos[(\omega - \omega_0)\epsilon]}{(\omega - \omega_0)^2 \epsilon}$

1–16(b). $y(t) = e^{-t} - 0.5e^{-2t},\ t \geq 0$
(c). Natural response $= e^{-t},\ t \geq 0$
Forced response $= -0.5\, e^{-2t},\ t \geq 0$

1–17(b). $H(s) = 0.5/(s + 1)$
$H(\omega) = 0.5/(1 + j\omega)$
$y_{ss}(t) = 0.05 \sin(100t - 30°)$

1–18(a).
$$y(t) = \begin{cases} 0.5(e^{-t} + t - 1), 0 \le t \le 1 \\ 0.5e^{-t}, t \ge 1 \\ 0, t < 0 \end{cases}$$

1–20(b). System is marginally stable for $k = 0$ and stable for $k > 0$.

1–23(b). $y_{ss}(t) = 100 \sin (100t + 29.97°)$

Chapter 2

2–1(a). $f(n) = (16)2^n - 13 - 3n$

(b). $f(n) = 0.5n^2 + n$

(c). $f(n) = 6(0.8)^n + n(0.8)^n$

(d). $f(n) = 2 + n - (0.5)^n$

2–2(a). $y(0) = 1.5, y(1) = 1.4167, y(2) = 1.4142, y(3) = 1.4142, y(4) = 1.4142$

(b). $\lim_{n \to \infty} y(n) = \sqrt{2}$

2–3. $y_p(n) = \frac{7}{17} \sin(n\pi/2) - \frac{6}{17} \cos (n\pi/2)$

2–4. $y(n) - \cos (n\pi/2)$ for $k = 1$
and $y(n) = (0.5)^n \cos (n\pi/2)$ for $k = \frac{1}{4}$

Chapter 3

3–1. $G(z) = \dfrac{2z}{z - e^{-6}} - \dfrac{2z}{z - e^{-3}} + \dfrac{24ze^{-6}}{(z - e^{-6})^2}, |z| > e^{-6}$

3–5. $f(0) - 0.165, f(1) = 0.549, f(2) = 0.796, f(3) = 0.917$

3–6(a). $f(n) = \dfrac{e^{-an} - e^{-bn}}{e^{-a} - e^{-b}}$

(b). $f(n) = 5 - 4(0.8)^n$

3–7(a). $f(n) = n(0.5)^n(7n - 5)$

(b). $f(n) = (-1)^n \left(\dfrac{10}{9} - \dfrac{2n}{3} \right) + \dfrac{8}{9} (0.5)^n$

3–9(a). $f(n) = \dfrac{84}{5} (2)^n + \dfrac{1}{5} \left(\dfrac{1}{3} \right)^n - 2n - 10$

(b). $f(n) = 5 + (0.9)^{n-1} (n + 1)$

Chapter 4

4–4. $|\bar{X}(f)| = 1.181$, where $f = 80$ Hz

4–7. $X(5) = -j, X(6) = 3 - 2j, X(7) = 2 - j$

4–9. $x(0) = 1$, $x(1) = 2$, $x(2) = -1$, $x(3) = 3$

4–10.   0, 8, 4, 12, 2, 10, 6, 14, 1, 9, 5, 13, 3, 11, 7, 15

4–11. $X(0) = 13$, $X(1) = -1.293 + j0.707$, $X(2) = 2 - j$, $X(3) = -2.707 + j0.707$, $X(4) = -1$, $X(5) = -2.707 - j0.707$, $X(6) = 2 + j$, $X(7) = -1.293 - j0.707$

Chapter 5

5–3(b). $|\tilde{H}(\nu)|^2 = \dfrac{0.09}{1.49 - 1.4 \cos \pi\nu}$, $0 \le \nu \le 1$

(c). $B = 0.115 f_N$.

5–4.   System is marginally stable.

5–5.   System is marginally stable.

5–7.   Steady-state value of $y_s(n)$ is $1/(1 - K^2)$.

5–8(a). $H(z) = \dfrac{\beta}{1 - \alpha}\left[1 - \alpha\left(\dfrac{z - 1}{z - \alpha}\right)\right]$

5–9. $|\tilde{H}(\nu)|^2 = 1.5 \dfrac{(1 + \cos \pi\nu)}{(2 + \cos \pi\nu)}$, $0 \le \nu \le 1$

and the corresponding bandwidth is 100 Hz.

5–11(a). $|\tilde{H}(\nu)|^2 = \dfrac{(1 - K)^2}{1 + K^2 - 2K \cos \pi\nu}$, $0 \le \nu \le 1$

(b). $K = 0.327$

Chapter 6

6–2(a). $H(z) = \dfrac{1}{2}\left[\dfrac{z}{z - e^{-T}} + \dfrac{z}{z - e^{-3T}}\right]$

6–3(b). $|H(\omega)|^2 = \dfrac{(1 - \omega^2)^2}{(1 - \omega^2)^2 + \omega^2}$

6–7. $H(z) = \dfrac{(z - 1)^2}{1.3661z^2 - 1.9001z + 0.7339}$

6–8. $H(z) = \dfrac{0.402(z^2 - 1)}{1.638z^2 - 1.528z + 0.834}$

6–11(d). $H(z) = \dfrac{1.236z^2 - 1.528z + 1.236}{1.638z^2 - 1.528z + 0.834}$

Chapter 7

7–7. $h(0) = -\frac{1}{8}, h(1) = \frac{1}{4}, h(2) = \frac{3}{4}$

7–8(a). $a_0 = 0, a_1 = 0.0354, a_2 = 0, a_3 = -0.0455, a_4 = 0, a_5 = 0.0637, a_6 = 0, a_7 = -0.1061, a_8 = 0, a_9 = 0.3183, a_{10} = 0.5, a_i = a_{20-i}, 11 \leq i \leq 20$

7–8(b). $a_0' = 0, a_1' = 0.0036, a_2' = 0, a_3' = -0.0123, a_4' = 0, a_5' = 0.0344,$
$a_6' = 0, a_7' = -0.0860, a_8' = 0, a_9' = 0.3111,$
$a_{10}' = 0.5, a_i' = a_{20-i}', = 11 \leq i \leq 20$

7–9(a). $a_0 = 0, a_1 = 0.0494, a_2 = 0.0757, a_3 = 0.0573, a_4 = 0, a_5 = -0.0637,$
$a_6 = -0.0935, a_7 = -0.0681, a_8 = 0, a_9 = 0.0704, a_{10} = 0.1, a_i = a_{20-i}, 11 \leq i \leq 20$

7–9(b). $a_0' = 0, a_1' = 0.0051, a_2' = 0.0127, a_3' = 0.0154, a_4' = 0, a_5' = -0.0344,$
$a_6' = -0.0638, a_7' = -0.0552, a_8' = 0, a_9' = 0.0688,$
$a_{10}' = 0.1, a_i' = a_{20-i}', 11 \leq i \leq 20$

7–10(a). $a_0 = 0, a_1 = 0.0022, a_2 = 0.0072, a_3 = 0.0094, a_4 = 0.0049, a_5 = 0, a_6 = 0.0109, a_7 = 0.0509, a_8 = 0.1146, a_9 = 0.1750, a_{10} = 0.2, a_i = a_{20-i}, 11 \leq i \leq 20$

7–14(a). $a_0 = 0, a_1 = -0.0494, a_2 = -0.0757, a_3 = -0.0573, a_4 = 0, a_5 = 0.0637, a_6 = 0.0935, a_7 = 0.0681, a_8 = 0, a_9 = -0.0704, a_{10} = 0.9, a_i = a_{20-i}, 11 \leq i \leq 20$

Chapter 8

8–1. 0.0011001; 0.0110111; 1.0100100; 0.1101000

8–2. Truncation: 0.001100; 0.011011; 1.010010; 0.110100
Error:　　　0.0125; 0.01325; 0.00125; 0.0039
Roundoff:　0.001101; 0.011100; 1.010010; 0.110100
Error:　　　0.003125; 0.0025; 0.00125; 0.0039

8–3. 4, 10, 7, and 27 bits, respectively

8–4. $\sigma_\epsilon^2 = \dfrac{2^{-16}}{3} (89.808) = 0.000457$

8–12. $\sigma_{\epsilon_1}^2 = 92.59 \left[\dfrac{2^{-2B}}{3} \right]$

$\sigma_{\epsilon_2}^2 = 606 \left[\dfrac{2^{-2B}}{3} \right]$.

$\sigma^2_{\epsilon_2}$ = output noise power for parallel realization .

$$= 16.08 \left[\frac{2^{-2B}}{3} \right]$$

8–13(a). $s_0 = 0.694$

8–21. $s_0 = 0.6; \; s_1 = 0.176$

8–22(a). $\sigma^2_y = 2.25\sigma^2_x$

(b). $\sigma^2_{\epsilon_1} = 2.318 \left[\frac{2^{-2B}}{3} \right]$

(c). $\sigma^2_{\epsilon_2} = 24.92 \left[\frac{2^{-2B}}{3} \right]$

Index